W9-DFR-260

Applied Probability
Control
Economics
Information and Communication
Modeling and Identification
Numerical Techniques
Optimization

Applications of Mathematics

18

Edited by A. V. Balakrishnan

Applications of Mathematics

1 Fleming/Rishel, **Deterministic and Stochastic Optimal Control** (1975)
2 Marchuk, **Methods of Numerical Mathematics,** Second Ed. (1982)
3 Balakrishnan, **Applied Functional Analysis,** Second Ed. (1981)
4 Borovkov, **Stochastic Processes in Queueing Theory** (1976)
5 Lipster/Shiryayev, **Statistics of Random Processes I: General Theory** (1977)
6 Lipster/Shiryayev, **Statistics of Random Processes II: Applications** (1978)
7 Vorob'ev, **Game Theory: Lectures for Economists and Systems Scientists** (1977)
8 Shiryayev, **Optimal Stopping Rules** (1978)
9 Ibragimov/Rozanov, **Gaussian Random Processes** (1978)
10 Wonham, **Linear Multivariable Control: A Geometric Approach** (1979)
11 Hida, **Brownian Motion** (1980)
12 Hestenes, **Conjugate Direction Methods in Optimization** (1980)
13 Kallianpur, **Stochastic Filtering Theory** (1980)
14 Krylov, **Controlled Diffusion Processes** (1980)
15 Prabhu, **Stochastic Storage Processes: Queues, Insurance Risk, and Dams** (1980)
16 Ibragimov/Has'minskii, **Statistical Estimation: Asymptotic Theory** (1981)
17 Cesari, **Optimization: Theory and Applications** (1982)
18 Elliott, **Stochastic Calculus and Applications** (1982)
19 Marchuk/Shaydourov, **Difference Methods and Their Extrapolations** (in prep.)

Robert J. Elliott

Stochastic Calculus and Applications

Springer-Verlag
New York Heidelberg Berlin

Robert J. Elliott
Department of Pure Mathematics
The University of Hull
22 Newland Park
Cottingham Road
Hull HU5 2DW
England

Managing Editor
A. V. Balakrishnan
Systems Science Department
University of California
Los Angeles, CA 90024
U.S.A.

AMS Subject Classifications (1980): 49A60, 60G35, 93EXX, 94A99, 60H05, 60H10, 93E20, 93E11, 60G07, 60G40, 60G42, 60G44, 60G55, 60J10, 60J65, 60J75, 60J60

Library of Congress Cataloging in Publication Data

Elliott, Robert James.
Stochastic calculus and applications.

1. Stochastic analysis. I. Title.
QA274.2.E44 1982 519.2′32 82-16816
ISBN 0-387-90763-7 (New York)

Typeset by J. W. Arrowsmith Ltd., Bristol, England.
Printed and bound by R. R. Donnelley & Sons, Harrisonburg, VA.
Printed in the United States of America.

9 8 7 6 5 4 3 2 1

ISBN 0-387-90763-7 Springer-Verlag New York Heidelberg Berlin
ISBN 3-540-90763-7 Springer-Verlag Berlin Heidelberg New York

To Ann

Preface

The object of this book is to take a reader, who has been exposed to only the usual courses in probability, measure theory and stochastic processes, through the modern French general theory of random processes, to the point where it is being applied by systems theorists and electronic engineers. It is surprising and unfortunate that, although this general theory is found so useful by theoretical engineers, it is not (with a few significant exceptions) widely taught or appreciated in the English-speaking world. Such natural and basic concepts as the stochastic integral with respect to semimartingales, the general differentiation rule and the dual predictable projection should be familiar to a larger audience, so that still more applications and results might be found.

This book is, therefore, at a first-year graduate level. The first part is, of course, largely drawn from the original works of the French school, particularly those of Dellacherie, Jacod and Meyer, but the development is hopefully almost self-contained. Most proofs are carefully given in full (an exception, for example, being the proof of the section theorem). However, the aim is to reach the results of the stochastic calculus in as direct a manner as possible, so embellishments and extensions of the theory are not usually given. Also the original approach and definitions of the French authors are followed when these appear more intuitive than the even more abstract (though beautiful) recent treatments in, for example, the second editions of Dellacherie and Meyer's *Probabilités et Potentiel.* (So a predictable stopping time is a stopping time which is announced by a sequence of earlier stopping times, rather than a stopping time T for which $[\![T, \infty[\![$ belongs to the σ-field generated by processes adapted to the filtration $\{\mathscr{F}_{t-}\}$.) In its treatment of strong Markov solutions of stochastic differential equations and Girsanov's theorem, this book combines the approaches of Kallianpur, Liptser and Shiryayev, and Neveu.

The use of martingale methods in stochastic control was first developed by Benes, Davis, Duncan, Haussmann and Varaiya, *inter alia*. The chapters of this book, dealing with the stochastic control of continuous and jump processes, are based on the formulation of this approach due to Davis and the author. The chapter on filtering uses the canonical decomposition of a special semimartingale and an idea of Wong to obtain the general nonlinear filtering equation and Zakai's equation for the unnormalized distribution. This technique appears to be new. The book is more elementary than those of Dellacherie and Meyer, and unlike the treatments of Kallianpur and Liptser and Shiryayev, it presents the general theory of processes and stochastic calculus in full, including discontinuous processes. The martingale approach to optimal control has not yet been described in any text. Such a self-contained treatment of stochastic calculus and its applications does not, so far, exist, and hopefully this book fills a gap in the literature.

Acknowledgments

This book has grown out of graduate courses I gave at the University of Alberta and the University of Kentucky during the academic year 1977/78. I wish to thank Professor Ghurye and Professor A. Al-Hussaini of the University of Alberta and Professor R. Rishel and Professor R. Wets of the University of Kentucky for arranging my visits and, in addition, the audiences of my lectures for mathematical stimulation and encouragement. Dr. E. Kopp and Dr. W. Kendall of the University of Hull have read sections of the manuscript and suggested many improvements. I am particularly indebted to Dr. M. H. A. Davis of Imperial College, London, for invaluable discussions and advice over the years. Gill Turpin of the Department of Pure Mathematics of the University of Hull produced a beautiful typed version (which I, nevertheless, chopped and changed.) Finally, I wish to thank my family for their constant support.

Contents

CHAPTER 1

Conditional Expectation. Uniform Integrability

In this chapter the basic concepts and results concerning conditional expectation and uniform integrability are presented. A proof of the monotone class theorem is also given.

Conditional Expectation

The notion of conditional expectation is quite fundamental for what follows. It is easy to understand in elementary situations, such as probabilities associated with the throwing of a fair die. For example, the probability of the top face showing four is then $1/6$; the conditional probability of four showing, given that an even number has turned up, is $1/6(1/2)^{-1} = 1/3$. On a general probability space $(\Omega, \mathscr{F}, P)$ the conditional probability of event $B \in \mathscr{F}$ occurring, given that event $A \in \mathscr{F}$ has occurred, is $P(B|A) = P(A \cap B)/P(A)$, as long as $P(A) \neq 0$.

Again, on a probability space $(\Omega, \mathscr{F}, P)$ suppose a measurable function f with values in $N = \{1, 2, \ldots\}$ is defined. Write

$$A_n = \{\omega : f(\omega) = n\} \in \mathscr{F},$$

and suppose X is another (real) random variable defined on Ω. A natural definition for the conditional expectation of X, given the function f, is the random variable defined as

$$Y(\omega) = Y_n = \frac{\int_{A_n} X \, dP}{P(A_n)},$$

if $\omega \in A_n$ and $P(A_n) \neq 0$. If $P(A_n) = 0$ Y_n can be given an arbitrary value (though, following a convention $0/0 = 0$, it is usually defined to be 0). Roughly, X has been averaged over the sets on which f is constant.

Consider now a more general situation.

Theorem 1.1. *Suppose $(\Omega, \mathscr{F}, P)$ is a probability space and f a measurable function defined on $(\Omega, \mathscr{F})$ with values in the measurable space $(E, \mathscr{E})$.*

Write Q for the probability measure induced on $(E, \mathscr{E})$ by f (i.e. $Q(A) = P(f^{-1}(A)), A \in \mathscr{E}$), and let X be a P-integrable random variable on $(\Omega, \mathscr{F})$. Then there exists a Q-integrable random variable Y on $(E, \mathscr{E})$ such that for every $A \in \mathscr{E}$

$$\int_A Y(x)\, dQ(x) = \int_{f^{-1}(A)} X(\omega)\, dP(\omega). \tag{1.1}$$

Furthermore, if Y_1 also satisfies (1.1) then $Y_1 = Y$ a.s.

PROOF. *Uniqueness.* Suppose for all $A \in \mathscr{E}$, Y_1 and Y satisfy (1.1). Then $\int_A (Y_1 - Y)\, dQ = 0$ for all $A \in \mathscr{E}$ so $Y_1 = Y$ a.s.

Existence. First suppose that $X \in L^2(\Omega, \mathscr{F}, P)$, that is X is $\mathscr{F}$ measurable, and $EX^2 < \infty$. Then to each $Z \in L^2(E, \mathscr{F}, Q)$ we can associate the number

$$\phi(Z) = \int_\Omega (Z \circ f) X\, dP.$$

Clearly ϕ is a linear functional on $L^2(E, \mathscr{E}, Q)$ and $\|\phi\| \le \|X\|_{L^2(P)}$. Therefore, because the dual of $L^2(E, \mathscr{E}, Q)$ is again $L^2(E, \mathscr{E}, Q)$, there is a function $Y \in L^2(E, \mathscr{E}, Q)$ such that

$$\begin{aligned} \phi(Z) &= \int_\Omega (Z \circ f) X\, dP \\ &= \int_E ZY\, dQ \quad \text{for all } Z \in L^2(E, \mathscr{E}, Q). \end{aligned}$$

Y is $\mathscr{E}$ measurable, and, taking Z to be indicator functions of sets $A \in \mathscr{E}$, we see Y satisfies (1.1).

Note that if $X \in L^2(\Omega, \mathscr{F}, P)$ and X is nonnegative, then $\int_A Y\, dQ \ge 0$ for all $A \in \mathscr{E}$, so $Y \ge 0$ a.s.

Turning to the general case, suppose X is just integrable. Then X^+ and X^- are both integrable, and $X_n^+ = X^+ \wedge n \in L^2(\Omega, \mathscr{F}, P)$ for each $n \in N$. By the proof above, the random variables Y_n^+ exist which satisfy (1.1) with X_n replacing X. The random variables Y_n^+ are nonnegative, increasing and their integrals are bounded above by $E[X^+]$. The random variable $Y^+ = \lim_{n\to\infty} Y_n^+$, therefore, exists and, by Lebesgue's theorem, $E[Y^+] = \lim_{n\to\infty} E[Y_n^+] = E[X^+]$. Similarly, $Y^- = \lim_{n\to\infty} Y_n^-$ is integrable, where Y_n^- is obtained in the same manner from $X_n^- = X^- \wedge n$.

Consequently the random variable $Y = Y^+ - Y^-$ is $\mathscr{E}$ measurable and satisfies (1.1). □

Remark 1.2. Y is sometimes called the conditional expectation of X given f. As it is defined only almost surely, any function satisfying (1.1) should, strictly speaking, be called a version of the conditional expectation of X. This is an important point when we discuss continuous time martingales.

Conditioning with respect to a Sub-σ-Field

Suppose $(\Omega, \mathscr{F}, P)$ is a probability space, that E is Ω and that $\mathscr{E}$ is a sub-σ-field of $\mathscr{F}$. Let f be the identity map on Ω, then the measure Q induced on $(E, \mathscr{E})$ by f is just the restriction of P to $\mathscr{E}$.

Definition 1.3. Suppose X is a real valued integrable random variable defined on $(\Omega, \mathscr{F}, P)$. Then the conditional expectation of X given $\mathscr{E}$ is any $\mathscr{E}$-measurable integrable random variable Y such that, for all $A \in \mathscr{E}$,

$$\int_A X(\omega)\, dP(\omega) = \int_A Y(\omega)\, dP(\omega).$$

Notation 1.4. Y is denoted by $E[X|\mathscr{E}]$.

Remarks 1.5. The existence of Y follows from Theorem 1.1. Again Y is defined only almost surely, so Y should be called a version of the conditional expectation. Roughly, Y is an averaging of X over the coarser sets of $\mathscr{E}$.

If $\mathscr{E}$ is the σ-field generated by a family f_i, $i \in I$, of random variables, that is, $\mathscr{E} = \sigma\{f_i : i \in I\}$, Y will be called the conditional expectation of X given the f_i.

If $X = \chi(A)$, Y is called the conditional probability of A given $\mathscr{E}$. Note that Y is not a number but a random variable.

Properties of Conditional Expectations

Let $(\Omega, \mathscr{F}, P)$ be a probability space on which the random variables below are defined, and let $\mathscr{E}$ be any sub-σ-field of $\mathscr{F}$.

Lemma 1.6. *Suppose X and Y are integrable random variables, and $\alpha, \beta, \gamma \in R$ are constants. Then*

$$E[\alpha X + \beta Y + \gamma|\mathscr{E}] = \alpha E[X|\mathscr{E}] + \beta E[Y|\mathscr{E}] + \gamma \quad \text{a.s.}$$

PROOF. (a.s. here means that any version of the left-hand side equals a version of the right-hand side.) For any $A \in \mathscr{E}$

$$\begin{aligned}\int_A E[\alpha X + \beta Y + \gamma|\mathscr{E}]\, dP &= \int_A (\alpha X + \beta Y + \gamma)\, dP \\ &= \alpha \int_A X\, dP + \beta \int_A Y\, dP + \gamma \int_A dP \\ &= \alpha \int_A E[X|\mathscr{E}]\, dP + \beta \int_A E[Y|\mathscr{E}]\, dP + \gamma \int_A dP \\ &= \int_A (\alpha E[X|\mathscr{E}] + \beta E[Y|\mathscr{E}] + \gamma)\, dP.\end{aligned}$$

The integrand in the final integral is $\mathscr{E}$ measurable and has the same integral over all sets $A \in \mathscr{E}$ as the $\mathscr{E}$-measurable integrand in the first integral. Therefore, they are equal almost surely. □

Lemma 1.7. *Suppose X and Y are integrable random variables and $X \leq Y$ a.s. Then*

$$E[X|\mathscr{E}] \leq E[Y|\mathscr{E}] \quad a.s.$$

PROOF. The proof is obtained immediately by integrating over arbitrary sets $A \in \mathscr{E}$. □

Lemma 1.8. *Suppose X_n, $n \in N$, is an increasing sequence of integrable random variables which converge to an integrable random variable X. Then*

$$E[X|\mathscr{E}] = \lim_n E[X_n|\mathscr{E}] \quad a.s.$$

PROOF. By Lebesgue's monotone convergence theorem, for any $A \in \mathscr{E}$:

$$\int_A \lim_n E[X_n|\mathscr{E}]\, dP = \lim_n \int_A E[X_n|\mathscr{E}]\, dP$$
$$= \lim_n \int_A X_n\, dP = \int_A X\, dP = \int_A E[X|\mathscr{E}]\, dP.$$

□

Lemma 1.9 (Jensen's Inequality). *Suppose ϕ is a convex map of R into R and suppose X is an integrable random variable such that $\phi \circ X$ is integrable. Then*

$$\phi \circ E[X|\mathscr{E}] \leq E[\phi \circ X|\mathscr{E}] \quad \text{a.s.}$$

PROOF. ϕ is the upper envelope of a countable family of affine functions

$$\Lambda_n(x) = \alpha_n x + \beta_n, \qquad x \in R, \quad n \in N.$$

The random variables $\Lambda_n \circ X$ are integrable and

$$\Lambda_n \circ E[X|\mathscr{E}] = E[\Lambda_n \circ X|\mathscr{E}] \leq E[\phi \circ X|\mathscr{E}] \quad \text{a.s.}$$

Taking the supremum on the left the result follows. □

Lemma 1.10. *Suppose X is an integrable random variable and $\mathscr{E}$ is a sub-σ-field of $\mathscr{F}$. Then $X = E[X|\mathscr{E}]$ a.s. if and only if X is $\mathscr{E}$ measurable.*

PROOF. By definition, $E[X|\mathscr{E}]$ is $\mathscr{E}$ measurable. Conversely, if X is $\mathscr{E}$ measurable then by uniqueness (almost surely) $X = E[X|\mathscr{E}]$. □

Lemma 1.11. *Suppose $\mathscr{D}$ and $\mathscr{E}$ are sub-σ-fields of $\mathscr{F}$ such that $\mathscr{D} \subset \mathscr{E} \subset \mathscr{F}$. Then, for any integrable random variable X*

$$E[E[X|\mathscr{E}]\mathscr{D}] = E[X|\mathscr{D}] \quad a.s.$$

PROOF. Again, the two sides are $\mathscr{D}$ measurable and have the same integral over any $A \in \mathscr{D}$. $\square$

Corollary 1.12. *Suppose $\mathscr{D}$ is the trivial σ-field (Ω, ϕ), so $E[X|\mathscr{D}] = E[X]$. Then $E[E[X|\mathscr{E}]\mathscr{D}] = E[E[X|\mathscr{E}]] = E[X]$.*

Lemma 1.13. *Suppose X is an integrable random variable, and Y an $\mathscr{E}$-measurable random variable such that the product XY is integrable. Then $E[XY|\mathscr{E}] = YE[X|\mathscr{E}]$ a.s.*

PROOF. Suppose first that Y is a simple function, that is, Y takes only countably many values $\{a_i, i \in N\}$. Write $A_i = Y^{-1}(a_i) \in \mathscr{E}$. Then for any $A \in \mathscr{E}$

$$\begin{aligned}\int_A E[XY|\mathscr{E}]\,dP &= \int_A XY\,dP \\ &= \sum_i \int_{A\cap A_i} a_i X\,dP = \sum_i \int_{A\cap A_i} a_i E[X|\mathscr{E}]\,dP \\ &= \int_A YE[X|\mathscr{E}]\,dP.\end{aligned}$$

Because an integrable random variable is the limit of a monotone increasing sequence of simple functions the general result follows from Lemma 1.8. $\square$

Remarks 1.14. Applying Jensen's inequality to the convex function

$$\phi(x) = |x|^p, \qquad 1 \le p < \infty,$$

we have

$$\|E[X|\mathscr{E}]\|_p \le \|X\|_p \quad \text{for } X \in L^p(\Omega, \mathscr{F}, P).$$

Also

$$\text{ess. sup.}|E[X|\mathscr{E}]| \le \text{ess. sup.}|X|,$$

so

$$\|E[X|\mathscr{E}]\|_\infty \le \|X\|_\infty.$$

The map $X \to E[X|\mathscr{E}]$ is, therefore, a continuous map of norm ≤ 1 from $L^p(\Omega, \mathscr{F}, P)$ to itself for $1 \le p \le \infty$. It is, therefore, continuous when L^p is given the weak topology and, in particular, it is continuous in the topologies $\sigma(L^1, L^\infty)$ and $\sigma(L^2, L^2)$.

Uniform Integrability

The important concept of uniform integrability is often not covered in measure theory courses. Again $(\Omega, \mathscr{F}, P)$ is a probability space, and $L^1(\Omega, \mathscr{F}, P)$ is the space of (equivalence classes) of real random variables X such that $\|X\|_1 = E[|X|] < \infty$.

Definition 1.15. Suppose $K \subset L^1(\Omega, \mathcal{F}, P)$. Then K is said to be a *uniformly integrable* subset of $L^1(\Omega, \mathcal{F}, P)$ if

$$\int_{\{|X|\geq c\}} |X(\omega)|\, dP(\omega)$$

converges to 0 uniformly in $X \in K$ as $c \to +\infty$.

Notation 1.16. If X is a random variable and $c > 0$ define

$$X^c(\omega) = \begin{cases} X(\omega), & \text{if } |X(\omega)| \leq c, \\ 0, & \text{if } |X(\omega)| > c, \end{cases}$$

and

$$X_c(\omega) = X(\omega) - X^c(\omega).$$

Remarks 1.17. $K \subset L^1(\Omega, \mathcal{F}, P)$ is then uniformly integrable if for any $\varepsilon > 0$ there is a $c > 0$ such that $\|X_c\|_1 < \varepsilon$ for all $X \in K$.

Note that every set of random variables whose elements are bounded above in modulus by an integrable random variable is uniformly integrable.

Theorem 1.18. *Suppose K is a subset of $L^1(\Omega, \mathcal{F}, P)$. Then K is uniformly integrable if and only if*

(i) *there is a number $A < \infty$ such that for all $X \in K$ $E[|X|] < A$*, and
(ii) *for any $\varepsilon > 0$ there is a $\delta > 0$ such that for all $A \in \mathcal{F}$ with $P(A) \leq \delta$ the integral $\int_A |X(\omega)|\, dP(\omega)$ is less than ε for all $X \in K$.*

PROOF. *Necessity*. Note that for any integrable X and any set $A \in \mathcal{F}$

$$\int_A |X(\omega)|\, dP(\omega) \leq cP(A) + E[|X_c|].$$

Consider $\varepsilon > 0$. If K is uniformly integrable we can find a $c > 0$ such that $E[|X_c|] < \varepsilon/2$ for all $X \in K$. Then $E[|X|] \leq c + \varepsilon/2$ for all $X \in K$, establishing (i).

For the same c, if $P(A) \leq \delta = \varepsilon/2c$ we have $\int_A |X(\omega)|\, dP(\omega) < \varepsilon$, proving (ii).

Sufficiency. Suppose conditions (i) and (ii) are satisfied, and consider $\varepsilon > 0$. There is then a δ such that (ii) is satisfied. Take

$$c = \delta^{-1} . \sup_{X \in K} E[|X|].$$

Suppose $A_X = \{|X| \geq c\}$, so that $P(A_X) = P\{\omega : |X(\omega)| \geq c\} \leq c^{-1} . E[|X|] \leq \delta$. Then

$$\int_{\{|X|\geq c\}} |X(\omega)|\, dP(\omega) = \int_{A_X} |X(\omega)|\, dP(\omega) < \varepsilon,$$

for all $X \in K$, so K is uniformly integrable. □

Corollary 1.19. *K is a subset of $L^1(\Omega, \mathcal{F}, P)$. Suppose there is a positive and increasing function $\phi(t)$ defined on $[0, \infty[$ such that $\lim_{t\to\infty} t^{-1}\phi(t) = +\infty$ and $\sup_{X\in K} E[\phi \circ |X|] < \infty$. Then K is uniformly integrable.*

Proof. Write $A = \sup_{X\in K} E[\phi \circ |X|]$ and consider $\varepsilon > 0$. Put $a = \varepsilon^{-1}A$ and choose c large enough such that $t^{-1}\phi(t) \geq a$ if $t \geq c$. Then on the set $\{|X| \geq c\}$ we have

$$|X| \leq a^{-1} . (\phi \circ |X|),$$

so

$$\int_{\{|X|\geq c\}} |X(\omega)|\, dP(\omega) \leq a^{-1} \int_{\{|X|\geq c\}} \phi \circ |X|\, dP$$
$$\leq a^{-1} E[\phi \circ |X|] \leq \varepsilon.$$

Therefore, K is uniformly integrable. □

Remarks 1.20. The following theorem states when almost sure convergence is equivalent to convergence in $L^1(\Omega, \mathcal{F}, P)$.

Theorem 1.21. *Suppose X_n is a sequence of integrable random variables which converge almost surely to the random variable X. Then X is integrable, and the X_n converge to X in the norm of $L^1(\Omega, \mathcal{F}, P)$, if and only if the X_n are uniformly integrable.*

If for each n, $X_n \geq 0$ a.s. it is necessary and sufficient that $\lim_n E[X_n] = E[X] < \infty$.

Proof. *Necessity.* Suppose the X_n converge to X in the norm of $L^1(\Omega, \mathcal{F}, P)$, so that X itself is certainly in $L^1(\Omega, \mathcal{F}, P)$, then for any set $A \in \mathcal{F}$

$$\int_A |X_n(\omega)|\, dP(\omega) \leq \int_A |X(\omega)|\, dP(\omega) + \|X_n - X\|_1 \tag{1.2}$$

Taking $A = \Omega$ we see that the expectations $E[|X_n|]$ are uniformly bounded.

For any $\varepsilon > 0$ there is an integer N such that if $n > N$ we have $\|X_n - X\|_1 \leq \varepsilon/2$.

Consider the finite set of random variables $\mathcal{S} = \{X_1, \ldots, X_N, X\}$. There is a $\delta > 0$ such that if $A \in \mathcal{F}$, $P(A) \leq \delta$ and $Y \in \mathcal{S}$ then $\int_A |Y|\, dP \leq \varepsilon/2$.

Consequently, from (1.2), if $A \in \mathcal{F}$ and $P(A) \leq \delta$ then $\int_A |X_n|\, dP \leq \varepsilon$ for all X_n.

Therefore, the set $\{X_n\}$ satisfies conditions (i) and (ii) of Theorem 1.18 and is uniformly integrable.

Sufficiency. Conversely, suppose the $\{X_n\}$ are uniformly integrable. Then the set of expectations $E[|X_n|]$ is bounded and so, by Fatou's lemma,

$$E[|X|] = E[\lim_n |X_n|] \leq \liminf_n E[|X_n|]$$
$$< \infty$$

Now

$$\|X_n - X\|_1 \le \|X_n^c - X^c\|_1 + \|X_{nc}\|_1 + \|X_c\|_1.$$

Consider any $\varepsilon > 0$. Because the $\{X_n\}$ are uniformly integrable there exists a number $c > 0$ such that $\|X_c\|_1 < \varepsilon/3$ and $\|X_{nc}\|_1 > \varepsilon/3$ for all n. Now X_n^c converges to X^c almost surely and $|X_n^c - X^c| \le 2c$ so by the Lebesgue dominated convergence theorem $\lim_n \|X_n^c - X^c\|_1 = 0$. There is, therefore, an integer N such that $\|X_n^c - X^c\|_1 \le \varepsilon/3$ if $n > N$. Consequently, if $n > N$ $\|X_n - X\|_1 < \varepsilon$, and the X_n convergence to X in $L^1(\Omega, \mathcal{F}, P)$. Because

$$|\|X_n\|_1 - \|X\|_1| \le \|X_n - X\|_1,$$

$E[|X_n|]$ converges to $E[|X|]$.

Suppose that for each n, $X_n \ge 0$ and that $\lim_n E[X_n] = E[X] < \infty$. Now

$$X_n + X = (X \vee X_n) + (X \wedge X_n) \quad \text{and} \quad |X_n - X| = (X \vee X_n) - (X \wedge X_n).$$

From Lebesgue's convergence theorem

$$\lim E[X \vee X_n] = E[X].$$

Also by hypothesis

$$\lim_n E[X + X_n] = 2E[X].$$

Consequently,

$$\lim_n E[X \wedge X_n] = E[X] \quad \text{and} \quad \lim_n E[|X_n - X|] = \lim_n \|X_n - X\|_1 = 0. \quad \square$$

The Monotone Class Theorem

We now prove a fundamental result known as the *monotone class theorem.*

Theorem 1.22. *Let E be a set, and $\mathcal{N}$ a family of subsets of E which is closed under finite unions, finite intersections and complements, and such that $\phi \in \mathcal{N}$. Suppose $\mathcal{M}$ is a family of subsets of E which contains $\mathcal{N}$, and which is closed under countable unions (resp. countable intersections) of monotonic increasing (resp. decreasing) sequences of its members, that is, $\mathcal{M}$ is a monotone class. Then $\mathcal{M}$ contains the σ-field $\mathcal{R}$ generated by $\mathcal{N}$.*

PROOF. Following the terminology of Dellacherie and Meyer [23], a family of subsets closed under finite intersections and finite unions will be called a hord. Using Zorn's lemma, there is a maximal hord $\mathcal{H}$ which contains $\mathcal{N}$ and which is contained in $\mathcal{M}$. Because $\mathcal{H}$ is a hord, to show $\mathcal{H}$ is closed under countable intersections it is enough to consider a decreasing sequence $\{A_n\}$ of sets $A_n \in \mathcal{H}$. Write $A = \bigcap_n A_n$. Consider the family $\mathcal{K}$ of all sets of the form $K = (H \cap A) \cup H'$, where $H \in \mathcal{H} \cup \{E\}$ and $H' \in \mathcal{H}$. Then $\mathcal{K}$ is a hord contained in $\mathcal{M}$ and taking $H = \phi$ we see $\mathcal{H} \subset \mathcal{K}$. Taking $H = \Omega$ and $H' = \phi$ we see $A \in \mathcal{K}$. Because $\mathcal{H}$ is maximal, $\mathcal{H} = \mathcal{K}$, so $A \in \mathcal{H}$. Therefore, $\mathcal{H}$ is closed under countable intersections, so $\mathcal{M}$ contains all sets generated by countable intersections of elements $\mathcal{N}$. The proof for countable unions is similar.

Write $\mathcal{S}$ for the family of sets generated by all countable unions and countable intersections of elements of $\mathcal{N}$, so that $\mathcal{S} \subset \mathcal{M}$, and $\mathcal{R}$ for the family of $A \in \mathcal{S}$ such that $A^c \in \mathcal{S}$. Clearly $\mathcal{N} \subset \mathcal{R} \subset \mathcal{M}$, and $\mathcal{R}$ is a σ-field, so the result is proved. □

CHAPTER 2

Filtrations, Stopping Times and Stochastic Processes

Suppose $(\Omega, \mathscr{F})$ is a measurable space. We wish to model mathematically the idea of the development in time of information about some random phenomenon. This is done by considering an increasing family of sub-σ-fields of $\mathscr{F}$.

Definition 2.1. Suppose the time index set τ is either $[0, \infty]$, (i.e. continuous time), or $N = \{0, 1, 2, \ldots\}$, (discrete time). A *filtration* $\{\mathscr{F}_t\}$ of $(\Omega, \mathscr{F})$ is a family of sub-σ-fields $\mathscr{F}_t$, $t \in \tau$, of $\mathscr{F}$ such that if $s \leq t$ then $\mathscr{F}_s \subset \mathscr{F}_t$. We define $\mathscr{F}_\infty = \bigvee_t \mathscr{F}_t$.

Remark 2.2. The family of σ-fields $\{\mathscr{F}_t\}$ can be considered as describing the history of some phenomenon and $\mathscr{F}_t$ is sometimes called the σ-field of events prior to time t.

Definition 2.3. Write $\mathscr{F}_{t+} = \bigcap_{s>t} \mathscr{F}_s$ and, for $t > 0$, $\mathscr{F}_{t-} = \bigvee_{s<t} \mathscr{F}_s$. The filtration $\{\mathscr{F}_t\}$ is right continuous if $\mathscr{F}_t = \mathscr{F}_{t+}$. (Note that the filtration $\mathscr{F}_{t+}$ is always right continuous.)

When $\tau = N$, $\mathscr{F}_{n+} = \mathscr{F}_{n+1}$ (so a right continuous filtration is constant), and $\mathscr{F}_{n-} = \mathscr{F}_{n-1}$.

Definition 2.4. Suppose $(\Omega, \mathscr{F})$ is a measurable space with a filtration $\{\mathscr{F}_t\}$, $t \in \tau$. A random variable $T: \Omega \to \tau$ is said to be a *stopping time* if for every $t \in \tau$: $\{T \leq t\} = \{\omega : T(\omega) \leq t\} \in \mathscr{F}_t$.

Remarks 2.5. A stopping time is a random time T such that the event "T has occurred up to time t" depends only on the history up to time t, and not on any information about the future.

Examples 2.6.

(1) A constant random variable $T = t \in \tau$ is a stopping time.

(2) If T is a stopping time and $s \in \tau$ then $T + s$ is a stopping time.

Lemma 2.7. *Suppose S and T are stopping times. Then $S \wedge T$ and $S \vee T$ are stopping times. If $\{T_n\}$, $n \in N$, is a sequence of stopping times, then $\bigwedge_n T_n$ and $\bigvee_n T_n$ are stopping times.*

PROOF

$$\{S \wedge T \leq t\} = \{S \leq t\} \cup \{T \leq t\} \in \mathscr{F}_t.$$

$$\{S \vee T \leq t\} = \{S \leq t\} \cap \{T \leq t\} \in \mathscr{F}_t. \qquad \square$$

Definition 2.8. Suppose T is a stopping time with respect to the filtration $\{\mathscr{F}_t\}$. Then the σ-field $\mathscr{F}_T$ of events occurring up to time T is the σ-field of events $A \in \mathscr{F}$ such that

$$A \cap \{T \leq t\} \in \mathscr{F}_t \quad \text{for every } t.$$

Remark 2.9. Clearly T is $\mathscr{F}_T$ measurable. Note that if $T = t$ then $\mathscr{F}_T = \mathscr{F}_t$.

Theorem 2.10. *Suppose S and T are stopping times.*

(i) *If $S \leq T$ a.s. then $\mathscr{F}_S \subset \mathscr{F}_T$.*
(ii) *If $A \in \mathscr{F}_S$ then $A \cap \{S \leq T\} \in \mathscr{F}_T$.*

PROOF. (i) Suppose $B \in \mathscr{F}_S$ and $t \in \tau$. Then

$$B \cap \{T \leq t\} = B \cap \{S \leq t\} \cap \{T \leq t\} \in \mathscr{F}_t.$$

(ii) Suppose $A \in \mathscr{F}_S$. Then

$$\begin{aligned} &A \cap \{S \leq T\} \cap \{T \leq t\} \\ &\quad = (A \cap \{S \leq t\}) \cap \{T \leq t\} \cap \{S \wedge t \leq T \wedge t\}. \end{aligned}$$

Each of these three sets is in $\mathscr{F}_t$: the first because $A \in \mathscr{F}_S$, the second because T is a stopping time, and the third because $S \wedge t$ and $T \wedge t$ are $\mathscr{F}_t$ measuring random variables. $\square$

Lemma 2.11. *If S and T are stopping times then*

$$\mathscr{F}_{S \wedge T} = \mathscr{F}_S \cap \mathscr{F}_T.$$

PROOF.

$$S \wedge T \leq S \quad \text{and} \quad S \wedge T \leq T.$$

Therefore, by Theorem 2.10(i)

$$\mathscr{F}_{S \wedge T} \subset \mathscr{F}_S \cap \mathscr{F}_T.$$

Conversely, suppose $A \in \mathscr{F}_S \cap \mathscr{F}_T$. Then

$$\begin{aligned} A \cap \{S \wedge T \leq t\} &= A \cap (\{S \leq t\} \cup \{T \leq t\}) \\ &= (A \cap \{S \leq t\}) \cup (A \cap \{T \leq t\}) \in \mathscr{F}_t, \quad \text{so } A \in \mathscr{F}_{S \wedge T}. \end{aligned}$$ □

Theorem 2.12. *Suppose S and T are stopping times. Then the events $\{S < T\}$, $\{S = T\}$, $\{S > T\}$ belong to both $\mathscr{F}_S$ and $\mathscr{F}_T$.*

PROOF. From part (ii) of Theorem 2.10 above $\{S \leq T\} \in \mathscr{F}_T$, so $\{S > T\} \in \mathscr{F}_T$. Write $R = S \wedge T$, so that R is a stopping time and $\mathscr{F}_R$ measurable. From 2.10(i) $\mathscr{F}_R \subset \mathscr{F}_T$ so

$$\{R = T\} = \{S = T\} \in \mathscr{F}_T \quad \text{and} \quad \{R < T\} = \{S < T\} \in \mathscr{F}_T.$$

Interchanging the roles of S and T we see these events also belong to $\mathscr{F}_S$. □

Lemma 2.13. *Suppose $\tau = N$ and $\{\mathscr{F}_n\}$, $n \in \tau$, is a discrete filtration of the measure space $(\Omega, \mathscr{F})$. If $\{X_n\}$ is a sequence of random variables, $n \in \tau$, such that each X_n is $\mathscr{F}_n$ measurable the random variable $X_T = X_{T(\omega)}(\omega)$ is $\mathscr{F}_T$ measurable.*

PROOF. We must show that $\{X_T < \alpha\} \in \mathscr{F}_T$ for every real number α. That is, we must show that for each $n \in \tau$: $\{X_T < \alpha\} \cap \{T \leq n\} \in \mathscr{F}_n$. However,

$$\{X_T < \alpha\} \cap \{T \leq n\} = \bigcup_{k=1}^{n} \{X_T < \alpha\} \cap \{T = k\}$$

and

$$\{X_T < \alpha\} \cap \{T = k\} \in \mathscr{F}_k \subset \mathscr{F}_n.$$ □

Remarks 2.14. We now wish to describe a mathematical model for a process whose evolution in time is random. Such an object is called a stochastic process.

Definition 2.15. Suppose the time index set τ is either $[0, \infty[$, $[0, \infty]$, N or $N \cup \{\infty\}$, and $(\Omega, \mathscr{F})$ is a measurable space. If $(E, \mathscr{E})$ is a second measurable space, a *stochastic process* defined on $(\Omega, \mathscr{F})$ with values in $(E, \mathscr{E})$ is a family $\{X_t\}$ of E valued random variables indexed by $t \in \tau$. $(\Omega, \mathscr{F})$ is called the *base space* and $(E, \mathscr{E})$ is the *state space*. For $t \in \tau$, X_t is the *state* at time t. For a fixed $\omega \in \Omega$ $\{X_t(\omega); t \in \tau\}$ is called the *sample path* associated with ω.

Convention 2.16. Suppose $(\Omega, \mathscr{F}, P)$ is a probability space and $\{X_t\}$ is a stochastic process. If "Π" is some property of a sample path (for example, "Π" might be right continuity, of having left-hand limits for every $t \in \tau$, or being of bounded variation), then we say the process has property "Π" if, almost surely, every sample path $X_t(\omega)$ has property Π. That is, the event $\{\omega : t \to X_t(\omega) \text{ has } \Pi\}$ has probability one. Note the subtlety of this concept:

it is possible for a continuous time process to be continuous at every time t with probability one, but for its paths to be discontinuous with probability one. See Example 2.19 below.

When the state space $(E, \mathscr{E})$ is a topological space we shall be particularly concerned with processes which are both continuous on the right and have limits on the left; such a process will be said to be CORLOL. Similarly, the term COLLOR would be used to describe a process which is continuous on the left and has limits on the right. The French terms are CADLAG and CAGLAD, respectively.

Remarks 2.17. It is natural to ask when two stochastic processes are, in fact, modelling the same phenomenon. A possible definition is the following.

Definition 2.18. Suppose $\{X_t\}$, $\{Y_t\} t \in \tau$, are two processes defined on the same probability space $(\Omega, \mathscr{F}, P)$ with values in $(E, \mathscr{E})$. Then $\{Y_t\}$ is said to be a *modification* of $\{X_t\}$ if for every $t \in \tau$, $X_t = Y_t$ a.s.

Example 2.19. Suppose that $\Omega = [0, 1]$, $\mathscr{F}$ is the Borel σ-field on Ω and P is Lebesgue measure. With $\tau = [0, \infty[$ define the process $\{X_t\}$ by putting $X_t(\omega) = 0$ for all ω and all t. Define $\{Y_t\}$ by putting $Y_t(\omega) = 0$ if $t - [t] \neq \omega$, and $Y_t(\omega) = 1$ if $t - [t] = \omega$. ($[t]$ denotes the greatest integer less than or equal to t.) Clearly $\{Y_t\}$ is a modification of $\{X_t\}$, but whilst all trajectories of $\{X_t\}$ are continuous, all trajectories of $\{Y_t\}$ are discontinuous.

A stronger definition that two processes are the same is as follows.

Definition 2.20. Again, $\{X_t\}$ and $\{Y_t\}$, $t \in \tau$, are two processes defined on the probability space $(\Omega, \mathscr{F}, P)$ with values in $(E, \mathscr{E})$. $\{X_t\}$ and $\{Y_t\}$ are said to be *indistinguishable* if for almost every $\omega \in \Omega$

$$X_t(\omega) = Y_t(\omega) \quad \text{for all } t \in \tau.$$

The point is that in Definition 2.18 the set of zero measure on which X_t and Y_t may differ depends on $t \in \tau$, whilst in Definition 2.20 there is just one set of zero measure outside which $X_t(\omega) = Y_t(\omega)$ for all t.

When $\tau = N$ or $N \cup \{\infty\}$ the two definitions are equivalent, but the above example exhibits they are different in continuous time.

Lemma 2.21. *Suppose that* $\{X_t\}$ *is a modification of* $\{Y_t\}$, $t \in \tau = [0, \infty[$ *or* $[0, \infty]$, *and that both processes are right continuous. Then* $\{X_t\}$ *and* $\{Y_t\}$ *are indistinguishable.*

Proof. For each rational number $r \in Q$ the set $\{X_r \neq Y_r\}$ has zero measure. Consequently, the set $D = \bigcup_{r \in Q} \{\omega ; X_r(\omega) \neq Y_r(\omega)\}$ has measure zero. By right continuity $\{X_t \neq Y_t\} \subset D$ for any $t \in \tau$, so $\{X_t\}$ and $\{Y_t\}$ are indistinguishable. $\square$

ition 2.22. Suppose A is a subset of $\tau \times \Omega$ and that $I_A(t, \omega) = I_A$ is the function of A. Then A is said to be *evanescent* if I_A is indistinguish- the zero process.

In fact A is evanescent if and only if the projection of A on Ω is a set of measure zero.

Definition 2.23. Suppose $\{\mathcal{F}_t\}$, $t \in \tau$, is a filtration of the measurable space $(\Omega, \mathcal{F})$, and that $\{X_t\}$ is a process defined on $(\Omega, \mathcal{F})$ with values in $(E, \mathcal{E})$. Then $\{X_t\}$ is said to be *adapted* to $\{\mathcal{F}_t\}$ if X_t is $\mathcal{F}_t$ measurable for each $t \in \tau$.

Remarks 2.24. If $(\Omega, \mathcal{F}, P)$ is a probability space and $\{X_t\}$ is adapted to the filtration $\{\mathcal{F}_t\}$, then a modification $\{Y_t\}$ of $\{X_t\}$ is also adapted if each $\mathcal{F}_t$ contains all null sets of $\mathcal{F}$. This remark illustrates the importance of the following definition.

Definition 2.25. Suppose $(\Omega, \mathcal{F}, P)$ is a probability space with a filtration $\{\mathcal{F}_t\}$, $t \in \tau$. $\mathcal{F}^P$ will denote the completion of $\mathcal{F}$, and $\mathcal{F}_t^P$, $t \in \tau$, the σ-field generated by $\mathcal{F}_t$ and the P-null sets of $\mathcal{F}^P$. Then $\{\mathcal{F}_t^P\}$ is a filtration on $(\Omega, \mathcal{F}^P, P)$ and is called the *completion* of the filtration $\{\mathcal{F}_t\}$. A filtration is said to be *complete* if $\mathcal{F}$ is complete and each $\mathcal{F}_t$ contains all P-null sets of $\mathcal{F}$.

Example 2.26. Suppose $\{X_t\}$, $t \in \tau$, is a stochastic process defined on the probability space $(\Omega, \mathcal{F}, P)$. Then $\{X_t\}$ is certainly adapted to the filtration $\{\mathcal{F}_t\}$ where $\mathcal{F}_t = \sigma\{X_s : s \leq t\}$ (the σ-field on Ω generated by all the random variables X_s, $s \leq t$). This filtration may not be, in general, complete, but it can be completed by adding all P-null sets of $\mathcal{F}$ as in the above definition.

Notation 2.27. $\mathcal{B} = \mathcal{B}([0, \infty[)$, $\mathcal{B}([0, t])$, $t \in \tau = [0, \infty[$, will denote the Borel σ-fields.

Definition 2.28. Suppose $\tau = [0, \infty[$ and $\{X_t\}$ is a stochastic process defined on the measurable space $(\Omega, \mathcal{F})$, with values in $(E, \mathcal{E})$. Then $\{X_t\}$ is said to be a *measurable* process if the map $(t, \omega) \mapsto X_t(\omega)$ is a measurable map when $\tau \times \Omega$ is given the product σ-field $\mathcal{B} \otimes \mathcal{F}$.

Remarks 2.29. If $\{\mathcal{F}_t\}$, $t \in \tau = [0, \infty[$, is a filtration on $(\Omega, \mathcal{F})$, to say $\{X_t\}$ is adapted says something about the measurability in ω of $X_t(\omega)$. To say $\{X_t\}$ is a measurable process is a very weak statement about measurability in t and ω. We now give a definition which relates measurability in t and ω with the filtration.

Definition 2.30. Suppose $\{\mathcal{F}_t\}$, $t \in \tau = [0, \infty[$ is a continuous time filtration on $(\Omega, \mathcal{F})$ and that $\{X_t\}$ is a stochastic process defined on $(\Omega, \mathcal{F})$. Then $\{X_t\}$

is said to be *progressively measurable* or *progressive* if for every $t \in \tau$ the map $(s, \omega) \to X_s(\omega)$ of $[0, t] \times \Omega$ into $(E, \mathscr{E})$ is measurable, when $[0, t] \times \Omega$ is given the product of σ-field $\mathscr{B}([0, t]) \times \mathscr{F}_t$.

Remarks 2.31. A progressive process is adapted. However, again a "diagonal" process gives a counter example. Suppose $\Omega = [0, \infty[$ with the Borel σ-field, and that a probability measure equal to e^{-x} times Lebesgue measure is given on Ω. For $t \in [0, \infty[$ define $X_t(\omega) = 0$ if $t \neq \omega$, and $X_t(t) = 1$. Then if for every t, $\mathscr{F}_t$ is the σ-field generated by the points of $[0, \infty[$, $\{X_t\}$ is adapted but not progressive. The following theorem gives a positive result in this direction.

Theorem 2.32. *Suppose $\{\mathscr{F}_t\}$, $t \in [0, \infty[$, is a filtration on $(\Omega, \mathscr{F})$, and $\{X_t\}$ is an adapted right continuous process with values in the metric space E (which has the Borel σ-field $\mathscr{E}$). Then $\{X_t\}$ is progressively measurable. The same result is true if $\{X_t\}$ is adapted and left continuous.*

PROOF. For any $t \in [0, \infty[$ consider a partition of $[0, t]$ into 2^n equal intervals. For $s \in [(k-1)2^{-n}, k2^{-n}t[$, $1 \leq k \leq 2^n$, write

$$X_s^n(\omega) = X_{k2^{-n}t}(\omega),$$

and put

$$X_t^n(\omega) = X_t(\omega).$$

Clearly X^n is a map of $[0, t] \times \Omega$ into E which is is measurable when $[0, t] \times \Omega$ is given the σ-field $\mathscr{B}([0, t]) \times \mathscr{F}_t$. Letting n tend to infinity we see the map $(s, \omega) \mapsto X_s(\omega)$ is measurable for this σ-field, so X is progressive. □

Theorem 2.33. *Suppose $\{X_t\}$, $t \in [0, \infty[$, is a progressive process on the space $(\Omega, \mathscr{F})$ with the filtration $\{\mathscr{F}_t\}$. If S is an $\{\mathscr{F}_t\}$ stopping time the random variable $X_S = X_{S(\omega)}(\omega)$ is $\mathscr{F}_S$ measurable, and the process stopped at S, $X_t^S = X_{t \wedge S}$ is progressive.*

PROOF. To establish the first result we show show that, if $(E, \mathscr{E})$ is the state space of $\{X_t\}$, then for every $B \in \mathscr{E}$ the set $\{X_S \in B\} \cap \{S \leq t\}$ is in $\mathscr{F}_t$. However, $\{X_S \in B\} \cap \{S \leq t\} = \{X_{t \wedge S} \in B\} \cap \{S \leq t\}$ so it is enough to prove the second part of the theorem. Now $t \wedge S$ is a stopping time less than the constant stopping time t, so $t \wedge S$ is $\mathscr{F}_t$ measurable. Therefore, the map $(s, \omega) \to (s \wedge S(\omega), \omega)$ is measurable as a map from $([0, t] \times \Omega, \mathscr{B}([0, t]) \times \mathscr{F}_t)$ to itself, and so the map $(s, \omega) \to X_{s \wedge S(\omega)}(\omega)$ is measurable, as the composition of two measurable maps. Consequently $\{X_t^S\}$ is progressive. □

CHAPTER 3

Martingales: Discrete Time Results

In this chapter and the next we consider one of the most important kinds of stochastic process, the martingale. Their significance was first emphasized in the classical book of Doob [27]. Results for discrete time martingales are established in this chapter and extended to continuous time martingales in the next.

Remarks 3.1. The term "martingale" has an interesting history. A martingale is, in fact, part of a horse's harness which prevents the horse from raising its head too high. Through horse racing the word became a gambling term, and the mathematical definition below can be thought of as representing a fair game of chance, where the conditional expectation of the reward at a later time equals one's present reward.

A probability space $(\Omega, \mathcal{F}, P)$ with a filtration $\{\mathcal{F}_t\}$, $t \in \tau$, is given.

Definition 3.2. A real valued stochastic process $\{X_t\}$, $t \in \tau$, that is a sequence $\{X_t\}$, $t \in \tau$, of real random variables is said to be a *supermartingale* (resp. a *submartingale*) on $(\Omega, \mathcal{F}, P)$ with respect to the filtration $\{\mathcal{F}_t\}$ if:

(i) each X_t is $\mathcal{F}_t$ measurable, i.e. $\{X_t\}$ is adapted to $\{\mathcal{F}_t\}$,
(ii) $E[|X_t|] < \infty$, $t \in \tau$,
(iii) $E[X_t|\mathcal{F}_s] \leq X_s$ a.s. if $t \geqq s$ (resp. $E[X_t|\mathcal{F}_s] \geq X_s$ a.s. if $t \geq s$).

If the sequence $\{X_t\}$ is both a supermartingale and a submartingale then it is said to be a *martingale*.

Note that $\{X_t\}$ is a submartingale if and only if $\{-X_t\}$ is a supermartingale.

Example 3.3 (W. Kendall). Suppose one's alarm clock is set to ring at 6 a.m. for an early start. However, one awakes in the night and is unable to get back to sleep. One does not know the time and one is too lazy to lift one's head to see the clock face. Let X_t be the conditional probability t minutes after waking that the alarm will, in fact, ring 60 minutes after waking, (conditional on one's experience $\mathcal{F}_t$ from waking to time t, that is, on whether the alarm has rung or not). Then $E[X_{t+h}|\mathcal{F}_t]=X_t$, that is, $\{X_t\}$ is a martingale.

Examples 3.4.

(1) Suppose $\tau=N$, $Y\in L^1(\Omega,\mathcal{F},P)$ and $X_n=E[Y|\mathcal{F}_n]$. Then $\{X_n\}$ is a martingale.

(2) Suppose $\{Y_i\}$, $i\in N$, is a sequence of integrable independent random variables defined on $(\Omega,\mathcal{F})$, such that $EY_i=0$ for each i. Define $X_n=Y_0+\cdots+Y_n$, and $\mathcal{F}_n$ equal to the σ-field generated by $Y_0,\ldots,Y_n$. Then $\{X_n\}$ is an $\{\mathcal{F}_n\}$ martingale.

Lemma 3.5. *Suppose $\{X_t\}$ is an $\{\mathcal{F}_t\}$ martingale (resp. submartingale) and ϕ is a convex (resp. convex increasing) function defined on R such that the random variables $\phi\circ X_t$ are integrable. Then $\{\phi\circ X_t\}$ is an $\{\mathcal{F}_n\}$ submartingale.*

PROOF. Write $Y_t=\phi\circ X_t$, $t\in\tau$, and suppose $t\geq s$. Then by Jensen's inequality, Lemma 1.9,

$$\begin{aligned}E[Y_t|\mathcal{F}_s]&=E[\phi\circ X_t|\mathcal{F}_s]\\&\geq\phi\circ E[X_t|\mathcal{F}_s]\\&\geq\phi\circ X_s=Y_s\quad\text{a.s.}\end{aligned}$$

□

Remarks 3.6. Examples of functions satisfying the above conditions are:

$$\phi(x)=|x|^p,\qquad p\geq 1,$$

$$\phi(x)=x\vee 0=x^+,$$

$$\phi(x)=(x-\alpha)^+\quad\text{for }\alpha\in R.$$

Convention 3.7. From now on in this chapter τ will denote either N or $N\cup\{\infty\}$.

Theorem 3.8 (Optional Stopping). *Suppose that $\{X_n\}$, $n\in N$, is an $\{\mathcal{F}_n\}$ supermartingale.*

If S and T are bounded $\{\mathcal{F}_n\}$ stopping times and $S\leq T$ a.s. then $E[X_T|\mathcal{F}_S]\leq X_S$ a.s.

PROOF. Suppose $S \vee T \leq M$ a.s. We must show that for every $A \in \mathcal{F}_S$

$$\int_A X_S \, dP \geq \int_A X_T \, dP.$$

Suppose first that $S \leq T \leq S+1$ and write

$$\begin{aligned} B_n &= A \cap \{S = n\} \cap \{T > S\} \\ &= A \cap \{S = n\} \cap \{T > n\}. \end{aligned}$$

Now $A \cap \{S = n\} \in \mathcal{F}_n$ from this definition of $\mathcal{F}_S$, and $\{T > n\}$ is the complement of $\{T \leq n\} \in \mathcal{F}_n$. Consequently, each $B_n \in \mathcal{F}_n$ and $A = \bigcup_{n=1}^{M} B_n$. Therefore

$$\begin{aligned} \int_A (X_S - X_T) \, dP &= \sum_{n=0}^{M} \int_{B_n} (X_S - X_T) \, dP \\ &= \sum_{n=0}^{M} \int_{B_n} (X_n - X_{n+1}) \, dP \geq 0, \end{aligned}$$

because $E[X_{n+1} | \mathcal{F}_n] \leq X_n$ a.s. The result is, therefore, proved when $T - S$ is at most equal to 1.

In the general case write

$$R_n = T \wedge (S + n), \qquad n = 0, 1, 2, \ldots, M,$$

so that from Example 2.6(2) and Lemma 2.7 the R_n are $\{\mathcal{F}_n\}$ stopping times and, because $S \leq R_n$ for each n, $\mathcal{F}_S \subset \mathcal{F}_{R_n}$. Consequently, $A \in \mathcal{F}_{R_n}$ for each n and $R_{n+1} - R_n$ is at most equal to 1. Now $R_0 = S$ and $R_M = T$, so from the case discussed above:

$$\int_A X_S \, dP = \int_A X_{R_0} \, dP \geq \int_A X_{R_1} \, dP \geq \cdots \geq \int_A X_{R_M} \, dP = \int_A X_T \, dP. \qquad \square$$

Notation 3.9. If $Y = \{Y_n\}$, $n \in N$, is a sequence of random variables and T is an $\mathcal{F}_n$ stopping time, $Y^S = \{Y_{S \wedge n}\}$ denotes the sequence "stopped" at time S.

Corollary 3.10. *Suppose $X = \{X_n\}$ is an $\{\mathcal{F}_n\}$ supermartingale and S is a bounded stopping time, then X^S is an $\{\mathcal{F}_n\}$ supermartingale.*

Remarks 3.11. We now establish Doob's upcrossing and downcrossing inequalities, which are used to establish convergence results. Suppose $\{Z_n\}$, $n \in \tau = N$ or $N \cup \{\infty\}$ is a submartingale and that $S(\leq M$ a.s.) is a bounded stopping time. Then if $X_n = Z_n^S = Z_{S \wedge n}$, $\{X_n\}$ is a submartingale which is constant after time S. Suppose α, β are real numbers with $\alpha < \beta$. $M(\omega, X; [\alpha, \beta])$ (resp. $D(\omega, X; [\alpha, \beta])$ will denote the number of upcrossings (resp. downcrossings) of the interval $[\alpha, \beta]$ by the function $n \mapsto X_n(\omega)$.

From Lemma 3.5, if $Y_n = (X_n - \alpha)^+$ then $\{Y_n\}$ is an $\mathscr{F}_n$ submartingale and:

$$\begin{aligned} M(\omega, Y; [0, \beta - \alpha]) &= M(\omega, X; [\alpha, \beta]) \\ &= M, \quad \text{say}, \\ D(\omega, Y; [0, \beta - \alpha]) &= D(\omega, X; [\alpha, \beta]) \\ &= D, \quad \text{say}. \end{aligned}$$

Theorem 3.12. *Using the above notation*

$$E[M|\mathscr{F}_0] \leq (\beta - \alpha)^{-1}(E[Y_S|\mathscr{F}_0] - Y_0),$$

$$E[D|\mathscr{F}_0] \leq (\beta - \alpha)^{-1}E[(X_S - \beta)^+\mathscr{F}_0].$$

PROOF. We shall prove these inequalities hold when M and D are replaced by the larger numbers M^1 and D^1, the number of upcrossings (resp. downcrossings) of the open interval $]0, \beta - \alpha[$ by Y_n.

Define a sequence of stopping times $\{T_n\}$ as follows:

$T_0(\omega) = 0 \quad \text{for all } \omega \in \Omega,$

$T_1(\omega) = \min\{n : n > T_0(\omega) \text{ and } Y_n(\omega) = 0\}$

(or S, if this set is empty).

$T_2(\omega) = \min\{n : n > T_1(\omega) \text{ and } Y_n(\omega) \geq \beta - \alpha\}$

(or S, if this set is empty),

and so on, so that

$$T_{2k+1}(\omega) = S \wedge \min\{n : n > T_{2k}(\omega) \text{ and } Y_n(\omega) = 0\},$$

$$T_{2k+2}(\omega) = S \wedge \min\{n : n > T_{2k+1}(\omega) \text{ and } Y_n(\omega) \geq \beta - \alpha\}.$$

Eventually we reach $T_{2p}(\omega) = S$ a.s. because we are considering a discrete time process. Then

$$\begin{aligned} Y_S(\omega) - Y_0(\omega) = &[Y_{T_1}(\omega) - Y_{T_0}(\omega)] + [Y_{T_2}(\omega) - Y_{T_1}(\omega)] \\ &+ \cdots + [Y_{T_{2p}}(\omega) - Y_{T_{2p-1}}(\omega)]. \end{aligned}$$

Consider the terms $Y_{T_{2k}}(\omega) - Y_{T_{2k-1}}(\omega)$ in this sum. These either correspond to a jump from 0 to a value greater than $\beta - \alpha$ or to a partial jump followed by zero terms. In any case, their sum is at least $(\beta - \alpha)M^1$, and so

$$E[Y_s|\mathscr{F}_0] - Y_0 \geq (\beta - \alpha)E[M^1|\mathscr{F}_0] + \sum_{k<p} E[Y_{T_{2k+1}} - Y_{T_{2k}}|\mathscr{F}_0].$$

Because $\{Y_n\}$ is a submartingale, from Theorem 3.8 each of the expectations in the final sum above is positive. Therefore,

$$\begin{aligned} (\beta - \alpha)E[M|\mathscr{F}_0] &\leq (\beta - \alpha)E[M^1|\mathscr{F}_0] \\ &\leq E[Y_S|\mathscr{F}_0] - Y_0. \end{aligned}$$

To prove the second inequality, define a sequence of stopping times $\{S_n\}$ by:

$$S_0(\omega)=0 \text{ for all } \omega\in\Omega,$$

$$S_1(\omega)=\min\{n: n>S_0(\omega) \text{ and } Y_n(\omega)\geq\beta-\alpha\}$$

(or S, if this set is empty).

$$S_2(\omega)=\min\{n: n>S_1(\omega) \text{ and } Y_n(\omega)=0\}$$

(or S, if this set is empty).

Continue in this manner, so that eventually $S_{2p}(\omega)=S(\omega)$. By Theorem 3.8

$$E[(Y_{S_2}-Y_{S_1})+\cdots+(Y_{S_{2p}}-Y_{S_{2p-1}})|\mathcal{F}_0]\geqq 0.$$

However, each term in this sum except possibly the final one corresponds to a descent to 0 from a value greater than or equal to $(\beta-\alpha)$; the final term has a value at most equal to

$$(Y_S-(\beta-\alpha))^+=(X_S-\beta)^+.$$

Consequently

$$E[(X_s-\beta)^+|\mathcal{F}_0]-(\beta-\alpha)E[D^1|\mathcal{F}_0]\geq 0,$$

and so

$$E[D|\mathcal{F}_0]\leq(\beta-\alpha)^{-1}E[(X_S-\beta)^+|\mathcal{F}_0].$$

□

Corollary 3.13. *Suppose $\{X_n\}$ is a supermartingale stopped at the bounded stopping time S. Applying the above inequalities to the submartingale $\{-X_n\}$ over the interval $[-\beta,-\alpha]$ and taking expectations we have:*

$$\begin{aligned}E[M(\omega,-X;[-\beta,-\alpha])]&=E[D(\omega,X;[\alpha,\beta])]\\&\leq(\beta-\alpha)^{-1}E[(-X_S+\beta)^+-(-X_0+\beta)^+]\\&\leq(\beta-\alpha)^{-1}E[X_0\wedge\beta-X_S\wedge\beta],\end{aligned}$$

and

$$\begin{aligned}E[D(\omega,-X;[-\beta,-\alpha])]&=E[M(\omega,X;[\alpha,\beta])]\\&\leq(\beta-\alpha)^{-1}E[(-X_S+\alpha)^+]\\&=(\beta-\alpha)^{-1}E[(X_S-\alpha)^-]\\&\leq(\beta-\alpha)^{-1}(E[X_S^-]+|\alpha|)\\&\leq(\beta-\alpha)^{-1}(E[|X_S|]+|\alpha|).\end{aligned}$$

Using these inequalities we now prove the following convergence theorem.

Theorem 3.14. *Suppose $\{X_n\}$, $n\in N$, is an $\{\mathcal{F}_n\}$ supermartingale such that*

$$\sup_n E[X_n^-]<\infty.$$

Then the sequence X_n converges almost surely to an integrable random variable X_∞.

PROOF. For any finite integer k and any pair of real numbers $\alpha, \beta, \alpha<\beta$, consider the number $M(\omega, k; [\alpha, \beta])$, that is the number of upcrossings of $[\alpha, \beta]$ by the finite supermartingale $X_1, \ldots, X_k$. By the above Corollary 3.13

$$E[M(\omega, k; [\alpha, \beta])] \leq (\beta-\alpha)^{-1}(E(X_k^-)+|\alpha|).$$

Now the sequence $M(\omega, k; [\alpha, \beta])$ is monotonic increasing to $M(\omega, X; [\alpha, \beta])$ so

$$E[M(\omega, X; [\alpha, \beta])] = \lim_k E[M(\omega, k; [\alpha, \beta])].$$

Therefore,

$$E[M(\omega, X; [\alpha, \beta] \leq (\beta-\alpha)^{-1}\left(\sup_k E[X_k^-]+|\alpha|\right) < \infty,$$

and so $M(\omega, X; [\alpha, \beta]) < \infty$ for almost every $\omega \in \Omega$.

Consequently, the set

$$\begin{aligned} H_{\alpha,\beta} &= \{\omega : M(\omega, X; [\alpha, \beta] = \infty\} \\ &= \{\omega : \limsup X_n(\omega) \geq \beta \text{ and } \liminf X_n(\omega) \leq \alpha\} \end{aligned}$$

has measure zero.

Suppose H is the union of all sets $H_{\alpha,\beta}$, where α, β are rational numbers, $\alpha<\beta$. Then H has measure zero. Therefore, if $\omega \notin H$, $\liminf X_n(\omega) = \limsup X_n(\omega)$. That is, $\lim_{n\to\infty} X_n(\omega) = X_\infty(\omega)$ exists if $\omega \notin H$. □

Corollary 3.15. *If the above hypotheses are satisfied $E[|X_\infty|] < \infty$.*

PROOF. Now

$$\begin{aligned} \|X_n\|_1 &= E[X_n^+ + X_n^-] \\ &= E[X_n] + 2E[X_n^-] \\ &\leq E[X_0] + 2E[X_n^-]. \end{aligned}$$

So, by Fatou's lemma

$$\begin{aligned} E[|X_\infty|] = E\left[\lim_n |X_n|\right] &\leq \liminf E[|X_n|] \\ &\leq \sup \|X_n\|_1 < \infty. \end{aligned}$$

□

Remark 3.16.

$$E[X_n^-] \leq \|X_n\|_1 \leq E[X_0] + 2E[X_n^-],$$

so the hypothesis of the theorem is equivalent to saying $\sup_n \|X_n\|_1 < \infty$.

Corollary 3.17. *Suppose $\{X_n\}$, $n \in N$, is a positive supermartingale, i.e. $X_n \geq 0$ a.s., then certainly $\sup_n E[X_n^-] = 0 < \infty$ so the sequence $\{X_n\}$ converges almost surely to an integrable random variable X_∞.*

Corollary 3.18. *Suppose $\{X_n\}$, $n \in N$, is a uniformly integrable supermartingale, that is the random variables $\{X_n\}$ are uniformly integrable, so, in particular, $\sup_n \|X_n\|_1 < \infty$. Then if $\mathscr{F}_\infty = \bigvee_n \mathscr{F}_n$ the process $\{X_n\}$, $n \in N \cup \{\infty\}$, is a supermartingale and $\lim_n \|X_n - X_\infty\|_1 = 0$.*

PROOF. Because $\sup_n \|X_n\|_1 < \infty$ the random variables X_n converge almost surely to the integrable random variable X_∞. Because the $\{X_n\}$ are uniformly integrable, the convergence also takes place in L^1. If $\mathscr{F}_\infty = \bigvee_n \mathscr{F}_n$ then X_∞ is certainly $\mathscr{F}_\infty$ measurable. Suppose $m < n$ and $A \in \mathscr{F}_m$. Then $\int_A X_m \, dP \geq \int_A X_n \, dP$, so by Fatou's lemma, letting $n \to \infty$, we see that $E[X_\infty | \mathscr{F}_m] \leq X_m$ a.s. □

Corollary 3.19. *Suppose $\{X_n\}$, $n \in N$, is a uniformly integrable martingale. Then $\lim_n X_n = X_\infty$ a.s. and in L^1, and $X_n = E[X_\infty | \mathscr{F}_n]$ a.s. for each $n \in N$.*

PROOF. Apply the above corollary to the supermartingales $\{X_n\}$ and $\{-X_n\}$. □

(See Example 3.4(1)). A converse to this corollary is the following result.

Theorem 3.20. *Suppose $\{\mathscr{F}_n\}$, $n \in N$ is a filtration of $(\Omega, \mathscr{F})$ and that $\mathscr{F}_\infty = \bigvee_n \mathscr{F}_n$. If $Y \in L^1(\Omega, \mathscr{F}_\infty, P)$ then $\{X_n\}$ is a uniformly integrable martingale when $X_n = E[Y | \mathscr{F}_n]$ a.s. Furthermore $\lim_{n\to\infty} X_n = Y$ a.s.*

PROOF. By Jensen's inequality $|X_n| \leq E[|Y| \, | \mathscr{F}_n]$ a.s. so

$$E[|X_n|] \leq E[|Y|], \quad \text{and for } m \geq 0$$

$$E[X_{n+m} | \mathscr{F}_n] = E[E[Y | \mathscr{F}_{n+m}] \mathscr{F}_n] = X_n \text{ a.s.}$$

For $\lambda \geq 0$:

$$I(n, \lambda) = \int_{\{|X_n| > \lambda\}} |X_n| \, dP \leq \int_{\{|X_n| > \lambda\}} |Y| \, dP$$

and

$$P\{|X_n| > \lambda\} \leq \frac{E[|X_n|]}{\lambda} \leq \frac{E[|Y|]}{\lambda},$$

so $I(n, \lambda)$ tends to zero as $\lambda \to \infty$ uniformly in n. Therefore, from Definition 1.15 the set $\{X_n\}$ is uniformly integrable.

From Corollary 3.19 we know that $\lim_n X_n = X_\infty$ exists and that $X_n = E[X_\infty | \mathscr{F}_n]$.

Write $\mathscr{G}$ for the family of events $A \in \mathscr{F}_\infty$ such that

$$\int_A Y \, dP = \int_A X_\infty \, dP.$$

Now for each $n \in N$, $\mathcal{F}_n \subset \mathcal{G}$ and $\mathcal{G}$ is closed under countable unions and intersections. Therefore, by the monotone class Theorem 1.22, $\mathcal{G} = \mathcal{F}_\infty$. □

Lemma 3.21. *Suppose $\{X_n\}$, $n \in N$, is a supermartingale. For every $\alpha \geq 0$*

$$\alpha P\Big(\Big\{\sup_n X_n \geq \alpha\Big\}\Big) \leq E[X_0] + \sup_n E[X_n^-] \leq 2 \sup_n \|X_n\|_1.$$

PROOF. Put $T(\omega) = \inf\{n : X_n(\omega) \geq \alpha\}$ and define a sequence of stopping times $T_k = T \wedge k$, $k \in N$. By Theorem 3.8

$$E[X_{T_k}] \leq E[X_0].$$

Either

$$X_{T_k}(\omega) \geq \alpha \quad \text{or} \quad X_{T_k}(\omega) = X_k(\omega).$$

Therefore

$$\alpha P(\{\sup_{n \leq k} X_n \geq \alpha\}) + \int_{\{\sup_{n \leq k} X_n < \alpha\}} X_k \, dP \leq E(X_0),$$

and

$$\alpha P\Big(\Big\{\sup_{n \leq k} X_n \geq \alpha\Big\}\Big) \leq E(X_0) + E[X_k^-], \tag{3.1}$$

so letting $k \to \infty$:

$$\alpha P\Big(\Big\{\sup_n X_n \geq \alpha\Big\}\Big) \leq E[X_0] + \sup_n \int X_n^- \, dP$$
$$\leq 2 \sup_n \|X_n\|_1.$$ □

Lemma 3.22. *Suppose $\{X_n\}$, $n \in N$, is a supermartingale. For every $\alpha \geq 0$ $\alpha P(\{\inf_n X_n \leq -\alpha\}) \leq \sup_n E[X_n^-]$.*

PROOF. Put $S(\omega) = \inf\{n : X_n(\omega) \leq -\alpha\}$ and define a sequence of stopping times

$$S_k = S \wedge k, \qquad k \in N.$$

By Theorem 3.8 for every $k \in N$: $E[X_{S_k}] \geq E[X_k]$. Therefore,

$$E[X_k] \leq -\alpha P\Big(\Big\{\inf_{n \leq k} X_n \leq -\alpha\Big\}\Big) + \int_{\{\inf_{n \leq k} X_n > -\alpha\}} X_k \, dP,$$

so

$$\alpha P\Big(\Big\{\inf_{n \leq k} X_n \leq -\alpha\Big\}\Big) \leq E[-X_k] + \int_{\{\inf_{n \leq k} X_n > -\alpha\}} X_k \, dP$$
$$= \int_{\{\inf_{n \leq k} X_n \leq -\alpha\}} -X_k \, dP \leq E[X_k^-]. \tag{3.2}$$

Letting $k \to \infty$, the result follows. □

A corollary of the above two lemmas is the following result:

Corollary 3.23. *Suppose* $\{X_n\}$, $n \in N$, *is a supermartingale. For every* $\alpha \geq 0$, $\alpha P(\{\sup_n |X_n| \geq \alpha\}) \leq 3 \sup_n \|X_n\|_1$.

Furthermore, Doob's maximal theorem is a corollary of Lemma 3.22 above:

Corollary 3.24 (Doob's Maximal Theorem). *Suppose* $\{X_n\}$, $n \in N$, *is a martingale. For every* $\alpha \leq 0$, $\alpha P(\{\sup_n |X_n| \geq \alpha\}) \leq \sup_n \|X_n\|_1$.

PROOF. From Lemma 3.5 if $Y_n = -|X_n|$, $\{Y_n\}$ is a (negative) supermartingale and

$$\|Y_n\|_1 = \|X_n\|_1 = E[Y_n^-].$$

Also,

$$\left\{\inf_n Y_n \leq -\alpha\right\} = \left\{\sup_n |X_n| \geq \alpha\right\},$$

so the result follows from Lemma 3.22. □

To establish the L^p inequality of Doob we first prove the following result:

Lemma 3.25. *Suppose* X *and* Y *are two positive random variables defined on the probability space* $(\Omega, \mathscr{F}, P)$ *such that* $X \in L^p$ *for some* p, $1 < p < \infty$, *and for every* $\alpha > 0$, $\alpha P(\{Y \geq \alpha\}) \leq \int_{\{Y \geq \alpha\}} X\, dP$. *Then* $\|Y\|_p \leq q\|X\|_p$, *where* $p^{-1} + q^{-1} = 1$.

PROOF. Let $F(\lambda) = P(\{Y > \lambda\})$ be (1 minus) the distribution function of Y. Then, by integration by parts,

$$E[Y^p] = -\int_0^\infty \lambda^p\, dF(\lambda)$$

$$= \int_0^\infty F(\lambda)\, d(\lambda^p) - \lim_{h \to \infty} [\lambda^p F(\lambda)]_0^h$$

$$\leq \int_0^\infty F(\lambda)\, d(\lambda^p) \leq \int_0^\infty \lambda^{-1} \left(\int_{\{Y \geq \lambda\}} X\, dP \right) d(\lambda^p),$$

by hypothesis,

$$= E\left[X \int_0^Y \lambda^{-1}\, d(\lambda^p) \right] \quad \text{by Fubini's theorem}$$

$$= \left(\frac{p}{p-1} \right) E[XY^{p-1}]$$

$$\leq q\|X\|_p \|Y^{p-1}\|_q \quad \text{by Hölder's inequality.}$$

We have, therefore, proved that

$$E[Y^p] \le q\|X\|_p \, . \, (E[Y^{pq-q}])^{1/q}.$$

If we know that $\|Y\|_p$ is finite, because $pq-q=p$, the inequality follows immediately. Otherwise, the random variable $Y_n = Y \wedge n$ satisfies the hypothesis and is in L^p for every n. Therefore,

$$\|Y_n\|_p \le q\|X\|_p$$

and the result follows by letting n tend to infinity. □

Theorem 3.26. *Suppose $\{X_n\}$ is a positive submartingale, and put $X^*(\omega) = \sup_n X_n(\omega)$. For $1 < p \le \infty$, $X^* \in L^p$ if and only if $\sup_n \|X_n\|_p < \infty$. Furthermore, for $1 < p < \infty$ and $q^{-1} = 1 - p^{-1}$ we have $\|X^*\|_p \le q \sup_n \|X_n\|_p$.*

PROOF. When $p = \infty$ the first part of the theorem is immediate. Clearly, for $1 < p \le \infty$ if $X^* \in L^p$ then $\sup_n \|X_n\|_p \le \|X^*\|_p < \infty$.

From Corollary 3.17 the random variables $\{X_n\}$ are uniformly integrable, so $\lim_{n\to\infty} X_n(\omega) = X_\infty(\omega)$ exists, and by Fatou's lemma

$$E\left[\lim_n X_n^p\right] \le \liminf_n E[X_n^p]$$

$$\le \sup_n E[X_n^p] < \infty,$$

so

$$X_\infty \in L^p \quad \text{and} \quad \|X_\infty\|_p \le \sup_n \|X_n\|_p.$$

Now $\{-X_n\}$ is a supermartingale, and writing $X_k^* = \sup_{n\le k} X_n$, we have from equation (3.2) that for any $\alpha > 0$

$$\alpha P\left(\left\{\inf_{n\le k}(-X_n) \le -\alpha\right\}\right) = \alpha P(\{X_k^* \ge \alpha\}) = \int_{\{X_k^* \ge \alpha\}} X_k \, dP \le \int_{\{X^* \ge \alpha\}} X_k \, dP.$$

Letting $k \to \infty$ we have that for any $\alpha > 0$

$$\alpha P(\{X^* \ge \alpha\}) \le \int_{\{X^* \ge \alpha\}} X_\infty \, dP.$$

Consequently we can apply Lemma 3.25 with $Y = X^*$ and $X = X_\infty$ to deduce that

$$\|X^*\|_p \le q\|X_\infty\|_p. \qquad \square$$

To establish certain right continuity properties, and other results, it is necessary to consider sequences of σ-fields $\{\mathscr{F}_n\}$ which get smaller as n becomes larger.

Definition 3.27. Suppose $(\Omega, \mathscr{F}, P)$ is a probability space and $\{\mathscr{F}_n\}$, $n \in N$, a decreasing sequence of sub-σ-fields of $\mathscr{F}$. That is, if $n \ge m$ then $\mathscr{F}_n \subset \mathscr{F}_m$.

A sequence $\{X_n\}$, $n \in N$, of random variables is said to be a *reverse supermartingale* if

(i) each X_n is $\mathscr{F}_n$ measurable,
(ii) $E[|X_n|] < \infty$, $n \in N$,
(iii) $E[X_m | \mathscr{F}_n] \leq X_n$ a.s. if $n \geq m$.

Reverse submartingales and reverse martingales are defined analogously.

Theorem 3.28. *Suppose $\{X_n\}$ is a reverse supermartingale with respect to the decreasing sequence of σ-fields $\{\mathscr{F}_n\}$, $n \in N$. If $\lim_n E[X_n]$ is finite the set $\{X_n\}$ is uniformly integrable.*

PROOF. Because $E[X_n]$ is an increasing function of n, $\alpha = \lim_n E[X_n]$ exists. Suppose $\alpha < \infty$ and consider any $\varepsilon > 0$. There is an integer k such that $\alpha - E[X_k] \leq \varepsilon/2$ and, in fact,

$$0 \leq E[X_n] - E[X_k] \leq \varepsilon/2, \tag{3.3}$$

if $n \geq k$. Consider any $\lambda > 0$ and suppose $n \geq k$. Then

$$\begin{aligned} I(n,\lambda) &= \int_{\{|X_n| > \lambda\}} |X_n| \, dP \\ &= \int_{\{X_n < -\lambda\}} (-X_n) \, dP + \int_{\{X_n > \lambda\}} X_n \, dP \\ &= -\int_{\{X_n < -\lambda\}} X_n \, dP + E[X_n] - \int_{\{X_n \leq \lambda\}} X_n \, dP \\ &\leq -\int_{\{X_n < -\lambda\}} X_k \, dP + E[X_k] - \int_{\{X_n \leq \lambda\}} X_k \, dP + \varepsilon/2, \end{aligned}$$

because $\{X_n\}$ is a reverse supermartingale, and using (3.3) above. Therefore,

$$I(n,\lambda) \leq \int_{\{|X_n| > \lambda\}} |X_k| \, dP + \varepsilon/2.$$

Now the process $\{X_n^-\}$ is a reverse submartingale, so

$$\begin{aligned} E[|X_n|] &= E[X_n] + 2E[X_n^-] \\ &\leq \alpha + 2E[X_0^-] = \beta, \quad \text{say}. \end{aligned}$$

However,

$$\begin{aligned} P(\{|X_n| > \lambda\}) &\leq E[|X_n|]/\lambda \\ &\leq \beta/\lambda, \end{aligned}$$

so there is a $\lambda_0 > 0$ such that

$$\int_{\{|X_n| > \lambda\}} |X_k| \, dP \leq \varepsilon/2$$

if $\lambda \geq \lambda_0$ and $n \geq k$. That is,

$$I(n, \lambda) \leq \varepsilon \quad \text{if } \lambda \geq \lambda_0 \quad \text{and} \quad n \geq k.$$

For $n < k$ there is a λ_1 such that $I(n, \lambda) \leq \varepsilon$ if $\lambda \geq \lambda_1$, so if $\lambda \geq \lambda_0 \vee \lambda_1$. Definition 1.15 is satisfied, and $\{X_n\}$ is uniform integrable. □

Corollary 3.29. *Suppose $\{X_n\}$, $n \in N$, is a reverse supermartingale such that $\lim_n E[X_n] < \infty$. Then the analogue of Theorem 3.12 can be established, to show that both the number of upcrossings, and the number of downcrossings, of the sequence $\{X_n(\omega)\}$ is almost surely finite. Consequently $\lim_n X_n(\omega) = X_\infty(\omega)$ exists almost surely, and, because from Theorem 3.28 the sequence $\{X_n\}$ of random variables is uniformly integrable, the convergence also takes place in $L^1(\Omega, \mathscr{F}, P)$.*

Definition 3.30. Suppose $\{X_n\}$, $n \in N$, is a positive supermartingale. Then $\{X_n\}$ is said to be a potential if $\lim_n E[X_n] = 0$. An example of a potential is provided by a man condemned to play poker in an unfair game until he loses all his money.

From Corollary 3.17 we know that $\lim X_n(\omega) = X_\infty(\omega)$ exists almost surely, and by Fatou's lemma $E[X_\infty] = 0$. Consequently $X_\infty(\omega) = 0$ a.s., and the convergence also takes place in $L^1(\Omega, \mathscr{F}, P)$.

The following Riesz decomposition for supermartingales can now be established.

Theorem 3.31 (Riesz Decomposition). *Suppose $\{X_n\}$, $n \in N$, is a supermartingale. Then the following two conditions are equivalent:*

(i) *there is a submartingale $\{Y'_n\}$ such that $Y'_n \leq X_n$ a.s. for all $n \in N$,*
(ii) *there is a martingale $\{Y_n\}$ and a potential $\{Z_n\}$ such that for each $n \in N$: $X_n = Y_n + Z_n$.*

These two processes $\{Y_n\}$ and $\{Z_n\}$ are then unique, and if $\{Y'_n\}$ is any submartingale such that $Y'_n \leq X_n$ a.s. for all $n \in N$, then $Y'_n \leq Y_n$ a.s.

Proof. If (ii) is satisfied, then $Y_n \leq X_n$ so (i) is true.

Conversely, suppose (i) holds. For $p \in N$ write

$$X_{n,p} = E[X_{n+p} | \mathscr{F}_n] \leq X_n \quad \text{a.s.,}$$

so

$$\begin{aligned} X_{n,p+1} &= E[E[X_{n+p+1} | \mathscr{F}_{n+p}] \mathscr{F}_n] \\ &\leq E[X_{n+p} | \mathscr{F}_n.] = X_{n,p}. \quad \text{a.s.} \end{aligned}$$

Therefore, $X_{n,p}$ is, almost surely, decreasing in p. However,

$$X_{n,p} \geq E[Y'_{n+p} | \mathscr{F}_n] \geq Y'_n \quad \text{a.s.}$$

Define Y_n to be the $\mathscr{F}_n$ measurable random variable which is, almost surely, equal to $\lim_p X_{n,p}$, so

$$X_n \geq \lim_p X_{n,p} = Y_n \geq Y'_n \quad \text{a.s.}$$

and for $m \geq 0$:

$$E[Y_{n+m}|\mathscr{F}_n] = \lim_p E[X_{n+m,p}|\mathscr{F}_n]$$
$$= \lim_p E[X_{n+m+p}|\mathscr{F}_n] = Y_n \quad \text{a.s.}$$

Therefore, $\{Y_n\}$ is a martingale. Write

$$Z_n = X_n - Y_n.$$

Clearly $Z_n(\omega) \geq 0$ a.s., and so $\{Z_n\}$ is a positive supermartingale. From the definition of $\{Y_n\}$:

$$\lim_p E[Z_{n+p}|\mathscr{F}_n] = \lim E[X_{n+p}|\mathscr{F}_n] - Y_n = 0 \quad \text{a.s.}$$

for each $n \in N$. By Lebesgue's convergence theorem, because $E[Z_{n+p}] \leq E[Z_n]$,

$$\lim_p E[Z_{n+p}] = 0,$$

so $\{Z_n\}$ is a potential.

Finally, suppose $X_n = Y_n^* + Z_n^*$ is a second decomposition with $\{Y_n^*\}$ a martingale and $\{Z_n^*\}$ a potential. Then

$$E[X_{n+p}|\mathscr{F}_n] = E[Y_{n+p}^*|\mathscr{F}_n] + E[Z_{n+p}^*|\mathscr{F}_n]$$
$$= Y_n^* + E[Z_{n+p}^*|\mathscr{F}_n].$$

Letting p tend to infinity, $\lim_p E[X_{n+p}|\mathscr{F}_n] = Y_n$ a.s. and $\lim E[Z_{n+p}^*|\mathscr{F}_n] = 0$ a.s. so the result follows. □

We can now extend Theorem 3.8 to possibly infinite stopping times.

Theorem 3.32. *Suppose $\{X_n\}$ is an $\{\mathscr{F}_n\}$ supermartingale for $n \in N \cup \{\infty\}$, where $\mathscr{F}_\infty = \bigvee_n \mathscr{F}_n$.*

If S and T are two stopping times such that $S \leq T$ a.s., then X_S and X_T are integrable random variables and $E[X_T|\mathscr{F}_S] \leq X_S$ a.s. If $\{X_n\}$ is a martingale then $X_S = E[X_T|\mathscr{F}_S] = E[X_\infty|\mathscr{F}_S]$.

PROOF. Put $Y_n = E[X_\infty|\mathscr{F}_n] \leq X_n$ a.s. and $Z_n = X_n - Y_n$, so that $\{Y_n\}$ is a uniformly integrable martingale and $\{Z_n\}$ is a potential.

The result will be proved for $\{Y_n\}$ and $\{Z_n\}$. Consider first the uniformly integrable martingale $\{Y_n\}$ and define the stopping time S_n by

$$S_n(\omega) = \begin{cases} S(\omega) & \text{on } \{S \leq n\}, \\ \infty, & \text{on } \{S > n\}. \end{cases}$$

Now the finite sequence $\{0, 1, 2, \ldots, n, \infty\}$ is order isomorphic to the finite sequence $\{0, 1, 2, \ldots, n, n+1\}$, so we can apply Theorem 3.8 and deduce that

$$Y_{S_n} = E[Y_\infty | \mathcal{F}_{S_n}] \quad \text{a.s.}$$

On the set $\{S \leq n\} \cup \{S = \infty\}$ $Y_{S_n} = Y_S$, so as n tends to infinity $\lim_n Y_{S_n} = Y_S$. Furthermore, $\mathcal{F}_S = \bigcap_n \mathcal{F}_{S_n}$, so the random variables $E[Y_\infty | \mathcal{F}_{S_n}]$ are a reverse martingale. Consequently from Corollary 3.29

$$\lim_n E[Y_\infty | \mathcal{F}_{S_n}] = E[Y_\infty | \mathcal{F}_S] \quad \text{a.s.,}$$

so

$$Y_S = E[Y_\infty | \mathcal{F}_S] \quad \text{a.s.}$$

and Y_S is certainly integrable. If $S \leq T$ we know that $\mathcal{F}_S \subset \mathcal{F}_T$ so

$$Y_S = E[E[Y_\infty | \mathcal{F}_T] \mathcal{F}_S] = E[Y_T | \mathcal{F}_S] \quad \text{a.s.}$$

Consider now the potential $\{Z_n\}$.

Define the stopping times S_n as above, and T_n similarly. Because $S_{n+1} \leq S_n$ a.s. and $\mathcal{F}_S = \bigcap_n \mathcal{F}_{S_n}$ we have from Theorem 3.8 that

$$Z_{S_{n+1}} \leq E[Z_{S_n} | \mathcal{F}_{S_{n+1}}] \quad \text{a.s.}$$

Therefore,

$$E[Z_{S_{n+1}} | \mathcal{F}_S] \leq E[Z_{S_n} | \mathcal{F}_S] \quad \text{and} \quad \lim_n E[Z_{S_n} | \mathcal{F}_S] = Z_S \text{ a.s.}$$

Consequently, by Lebesgue's monotone convergence theorem

$$0 \leq E[Z_S] = \lim E[Z_{S_n}] \leq E[Z_0].$$

Similarly

$$0 \leq E[Z_T] = \lim E[Z_{T_n}] \leq E[Z_0].$$

Therefore, Z_S and Z_T are integrable and the convergence takes place in the topology of $L_1(\Omega, \mathcal{F}, P)$.

Now by Theorem 3.8

$$Z_{S_n} \geq E[Z_{T_n} | \mathcal{F}_{S_n}] \quad \text{a.s.} \tag{3.4}$$

Now $Z_{T_{n+1}} - Z_{T_n} = 0$ except on $\{T = n+1\}$, and there it is equal to $Z_{n+1} - Z_\infty = Z_{n+1} \geq 0$ a.s. because Z is a potential. Consequently

$$E[Z_{T_{n+1}} - Z_{T_n} | \mathcal{F}_{S_{n+1}}] \geq 0 \quad \text{a.s.} \tag{3.5}$$

Write

$$W_n = E[Z_{T_n} | \mathcal{F}_{S_n}].$$

Then

$$\begin{aligned} W_{n+1} &= E[Z_{T_{n+1}} | \mathcal{F}_{S_{n+1}}] \\ &\geq E[Z_{T_n} | \mathcal{F}_{S_{n+1}}] = E[E[Z_{T_n} | \mathcal{F}_{S_n}] \mathcal{F}_{S_{n+1}}] \\ &= E[W_n | \mathcal{F}_{S_{n+1}}], \end{aligned}$$

by (3.5) because $\{\mathcal{F}_{S_n}\}$ is a decreasing family of σ-fields as $n \to \infty$. Therefore, $\{W_n, \mathcal{F}_{S_n}\}$ is a reverse supermartingale and

$$\lim_n E[W_n] = \lim_n E[Z_{T_n}] \le E[Z_0].$$

From Corollary 3.29 $\{W_n\}$ is uniformly integrable and converges to a random variable W, both almost surely and in $L^1(\Omega, \mathcal{F}, P)$.

For any $A \in \mathcal{F}_{S_n}$

$$\int_A W_n \, dP = \int_A Z_{T_n} \, dP$$

and

$$\lim_n \int_A W_n \, dP = \int_A W \, dP,$$

$$\lim_n \int_A Z_{T_n} \, dP = \int_A Z_T \, dP.$$

However, $\mathcal{F}_S = \bigcap_n \mathcal{F}_{S_n}$, so for $A \in \mathcal{F}_S$

$$\int_A W \, dP = \int_A Z_T \, dP.$$

That is,

$$W = \lim_n E[Z_{T_n} | \mathcal{F}_{S_n}] = E[Z_T | \mathcal{F}_S].$$

From (3.4), therefore,

$$Z_S \ge E[Z_T | \mathcal{F}_S] \quad \text{a.s.}$$

□

Corollary 3.33. *Suppose that $\{\mathcal{F}_n\}$, $n \in N$, is a filtration, that $\mathcal{F}_\infty = \mathcal{F}$ and that S and T are two stopping times. (We do not suppose $S \le T$.) Then the projection operators $E[\cdot | \mathcal{F}_S]$ and $E[\cdot | \mathcal{F}_T]$ commute, and their product is $E[\cdot | \mathcal{F}_{S \wedge T}]$.*

Proof. Suppose Y is any integrable $\mathcal{F}_\infty$ measurable random variable. Consider the martingale $X_n = E[Y | \mathcal{F}_n]$, for $n \in N \cup \{\infty\}$, so $X_\infty = Y$. Then by Theorem 3.32

$$\begin{aligned} X_n^T = X_{T \wedge n} &= E[X_\infty | \mathcal{F}_{T \wedge n}] \\ &= E[E[X_\infty | \mathcal{F}_T] \mathcal{F}_{T \wedge n}] = E[X_T | \mathcal{F}_{T \wedge n}] \\ &= X_T I_{\{T < n\}} + E[X_T | \mathcal{F}_n] I_{\{T \ge n\}} \\ &= E[X_T | \mathcal{F}_n]. \end{aligned}$$

Therefore, the martingale $\{X_n^T\}$ has values $E[X_T|\mathscr{F}_n]$. Again applying Theorem 3.32,

$$\begin{aligned} X_{T\wedge S} &= X_S^T = E[X_T|\mathscr{F}_S] \\ &= E[E[X_\infty|\mathscr{F}_T]\mathscr{F}_S]. \end{aligned}$$

However, the left-hand side is symmetric in S and T and has the value $E[X_\infty|\mathscr{F}_{S\wedge T}]$. Because $X_\infty = Y$ is arbitrary the result is proved. □

Remark 3.34. Once the optional stopping theorem is proved for continuous time filtrations (see Theorem 4.12 and Corollary 4.13), it is immediate that the result of Corollary 3.33 is valid for stopping times with respect to a continuous time filtration. The above proof applies almost word for word.

CHAPTER 4

Martingales: Continuous Time Results

In this section $\tau = [0, \infty[$ or $[0, \infty]$ and the results of the previous chapter will be extended to processes whose time parameter set is such a τ. A probability space $(\Omega, \mathscr{F}, P)$ with a filtration $\{\mathscr{F}_t\}$, $t \in \tau$, is given.

Notation 4.1. Suppose $\{X_t\}$, $t \in \tau$, is a real valued stochastic process, $\alpha < \beta$, and I is a subset of $[0, \infty[$. Similarly to the notation of the Remarks 3.11 $M(\omega, X, I, [\alpha, \beta])$ will denote the number of upcrossings of $[\alpha, \beta]$ by the function $X_t(\omega)$ as t runs through I in an increasing manner. $D(\omega, X, I; [\alpha, \beta])$ will denote the number of downcrossings. We first extend to continuous parameter martingales the inequalities of the last chapter. Q will denote a countable, dense subset of τ, e.g. the rational numbers.

Theorem 4.2. *Let $\{X_t\}$, $t \in \tau$, be a right continuous supermartingale in* (1) *to* (4) *below, and a right continuous submartingale in* (5), *and $J = [u, v]$ an interval contained in τ. Then for any $\lambda > 0$ and any numbers α, β with $\alpha < \beta$ the following inequalities hold*:

(1) $\lambda P(\{\sup_{t \in J} X_t \geq \lambda\}) \leq E[X_u] + E[X_v^-]$.
(2) $\lambda P(\{\inf_{t \in J} X_t \leq -\lambda\}) \leq E[|X_v|]$.
(3) $E[M(\omega, X, J; [\alpha, \beta])] \leq (\beta - \alpha)^{-1} E[(X_v - \alpha)^-]$.
(4) $E[D(\omega, X, J; [\alpha, \beta])] \leq (\beta - \alpha)^{-1} E[(X_u \wedge \beta) - (X_v \wedge \beta)]$.
(5) *If* $1 < p < \infty$, $p^{-1} + q^{-1} = 1$ *and* $X_v \in L^p$ *then*

$$\|\sup_{t \in J} X_t\|_p \leq q \sup_{t \in J} \|X_t\|_p = q \|X_v\|_p.$$

(*N.B. in* (5) $\{X_t\}$ *is a submartingale*).

Proof. Suppose F is a finite subset of $Q_J = (Q \cap J) \cup \{u\} \cup \{v\}$. Then the inequalities (1) and (2), with F replacing J, follow from equations (3.1) of Lemma 3.21 and (3.2) of Lemma 3.22. Inequalities (3), (4) and (5) with F replacing J, are just the results of Corollary 3.13 and Theorem 3.26. By considering an increasing family of finite subsets of Q_J, whose union is the whole of Q_J, we see the inequalities hold when J is replaced by Q_J. By right continuity, the left-hand sides of the inequalities are unchanged when Q_J is replaced by J, and so the results follow. □

We next investigate the behaviour of the trajectories of a supermartingale.

Theorem 4.3. *Suppose $\{X_t\}$, $t \in \tau$, is a right continuous supermartingale. Then (almost surely) $\{X_t\}$ has left-hand limits, that is $\{X_t\}$ is corlol, and (almost) every sample path is bounded on every compact interval.*

Proof. The boundedness follows immediately from inequalities (1) and (2) of Theorem 4.2. With $J = [0, n]$ and α, β rational numbers, $\alpha < \beta$, we have from inequality (3) of Theorem 4.2 that $E[M(\omega, X, [0, n]; [\alpha, \beta])] < \infty$ a.s. Therefore, the set

$$H_{n,\alpha,\beta} = \{\omega : M(\omega, X, [0, n]; [\alpha, \beta]) = \infty\}$$

has measure zero. Suppose H_n is the union of all sets $H_{n,\alpha,\beta}$, where α, β are rational numbers $\alpha < \beta$. Then H_n also has measure zero. But H_n corresponds to those sample paths $X_t(\omega)$ for which there is a $t \in [0, n]$ and

$$\liminf_{\substack{s \to t \\ s < t}} X_s(\omega) < \limsup_{\substack{s \to t \\ s < t}} X_s(\omega).$$

Therefore, if $\omega \notin H_n$

$$\liminf_{\substack{s \to t \\ s < t}} X_s(\omega) = \lim_{\substack{s \to t \\ s < t}} X_s(\omega) = X_{t-}(\omega).$$ □

A related result is the following.

Lemma 4.4. *Suppose $\{X_t\}$, $t \in \tau$, is an arbitrary supermartingale and Q is a countable, dense subset of τ. (For example, Q might be the rationals in τ.) Then the restriction to Q of the map $s \mapsto X_s(\omega)$ has a left and right limit at every point $t \in \tau$ for almost every $\omega \in \Omega$.*

Proof. Adapt the notation of Theorem 4.3 above and write

$$H_{n,\alpha,\beta} = \{\omega : M(\omega, X, S \cap [0, n]; [\alpha, \beta]) = \infty\} \quad \text{so that } H_n = \bigcup_{\alpha,\beta} H_{n,\alpha,\beta}$$

has measure zero. Therefore, for $\omega \notin H_n$ and $t \in [0, n[$

$$\limsup_{\substack{s \to t \\ s \in S \\ s < t}} X_s(\omega) = \liminf_{\substack{s \to t \\ s \in S \\ s < t}} X_s(\omega) = X_{t-}(\omega)$$

and

$$\limsup_{\substack{s \to t \\ s \in S \\ s > t}} X_s(\omega) = \liminf_{\substack{s \to t \\ s \in S \\ s > t}} X_s(\omega) = X_{t+}(\omega). \qquad \square$$

Notation 4.5. For $\omega \in H = \bigcup_n H_n$ define $X_{t+}(\omega) = X_{t-}(\omega) = 0$. If the filtration $\{\mathscr{F}_t\}$ is complete, that is each $\mathscr{F}_t$ contains all the P-null sets of $\mathscr{F}$, then X_{t+} is $\mathscr{F}_{t+}$ measurable, and X_{t-} is $\mathscr{F}_{t-}$ measurable.

Theorem 4.6. *Suppose $\{X_t\}$, $t \in \tau$, is a supermartingale and Q is a countable dense subset of τ. Using the above notation we have*

(1) $X_t \geq E[X_{t+} | \mathscr{F}_t]$ *a.s.*,
$X_{t-} \geq E[X_t | \mathscr{F}_{t-}]$ *a.s.*
(2) *The process $\{X_{t+}\}$, $t \in \tau$, is a supermartingale with respect to the filtration $\{\mathscr{F}_{t+}\}$.*
(3) *Suppose the filtration is right continuous, that is $\mathscr{F}_{t+} = \mathscr{F}_t$ for all $t \in \tau$. Then the supermartingale $\{X_t\}$ has a right continuous modification if and only if the function $E[X_t]$ is right continuous in t.*

Proof. (1) Consider a sequence $\{t_n\} \subset Q$ such that $t_n > t$ and t_n converges to t. Then $\{X_{t_n}\}$ is a uniformly integrable reverse supermartingale on the σ-fields $\{\mathscr{F}_{t_n}\}$, so for any $A \in \mathscr{F}_t$

$$\int_A X_t \, dP \geq \int_A X_{t_n} \, dP.$$

Letting t_n converge to t we have

$$\int_A X_t \, dP \geq \int_A X_{t+} \, dP,$$

so

$$X_t \geq E[X_{t+} | \mathscr{F}_t] \quad \text{a.s.}$$

Next consider a sequence $\{s_n\} \subset Q$ such that $s_n < t$ and s_n converges to t. Write $Y_n = X_{s_n} - E[X_t | \mathscr{F}_{s_n}]$. Then $\{Y_n\}$ is a positive supermartingale such that $\lim_{n \to \infty} Y_n = Y_\infty = X_{t-} - E[X_t | \mathscr{F}_{t-}]$. $Y_\infty \geq 0$ and so the result follows.

(2) Suppose $s < t$ and $\{s_n\}$, $\{t_n\}$ are sequences in Q such that $s < s_n < t$, $t < t_n$, $\lim_{n \to \infty} s_n = s$ and $\lim_{n \to \infty} t_n = t$. Then $\{X_{s_n}\}$ (resp. $\{X_{t_n}\}$) is a uniformly integrable $\{\mathscr{F}_{s_n}\}$ (resp. $\{\mathscr{F}_{t_n}\}$) reverse supermartingale, so for any $A \in \mathscr{F}_{s+} = \bigcap_{u > s} \mathscr{F}_u$ we have

$$\int_A X_{s_n} \, dP \geq \int_A X_{t_n} \, dP.$$

Letting n tend to infinity we see that

$$\int_A X_{s+} \, dP \geq \int_A X_{t+} \, dP,$$

that is, $X_{s+} \geq E[X_{t+} | \mathscr{F}_{s+}]$ a.s.

(3) We now suppose that $\mathscr{F}_t = \mathscr{F}_{t+}$, $t \in \tau$.

Then by part (1), $X_t \geq X_{t+}$ a.s., so these random variables are equal if and only if they have the same expectation. Suppose $\{t_n\} \subset \tau$ is a sequence such that $t_n > t$ and $\lim_{n\to\infty} t_n = t$. Then, as in part (1), the random variables $\{X_{t_n}\}$ are uniformly integrable so

$$E[X_{t+}] = \lim_{n\to\infty} [X_{t_n}].$$

Therefore, $X_t = X_{t+}$ a.s. if and only if this limit equals $E[X_t]$. Because the expectation function is monotonic this is the same as saying it is right continuous at every point $t \in \tau$. Consequently, the process $\{X_{t+}\}$ is a right continuous modification of $\{X_t\}$.

If $\{Y_t\}$ is a right continuous modification of $\{X_t\}$ then $E[Y_t] = E[X_t]$ for all $t \in \tau$, and so from the above argument $E[X_t]$ is right continuous. □

We now establish convergence results when $\tau = [0, \infty[$. The proofs are similar to those of Theorem 3.14 and Corollaries 3.17–3.20.

Theorem 4.7. *Suppose $\{X_t\}$, $t \in \tau = [0, \infty[$, is a right continuous supermartingale.*

If $\sup_{t\in\tau} E[X_t^-] < \infty$ then X_t converges almost surely to an integrable random variable as t tends to ∞.

PROOF. From (3) of Theorem 4.2

$$E[M(\omega, X, [0, n]; [\alpha, \beta])] \leq (\beta - \alpha)^{-1}(E[X_n^-] + |\alpha|),$$

so letting n tend to ∞

$$E[M(\omega, X, \tau; [\alpha, \beta])] \leq (\beta - \alpha)^{-1} \sup_n (E[X_n^-] + |\alpha|) < \infty.$$

If $H_{\alpha,\beta} = \{\omega : M(\omega, X, \tau; [\alpha, \beta]) = \infty\}$ then $H_{\alpha,\beta}$ has measure zero. Writing H for the union of all sets $H_{\alpha,\beta}$, where α and β are rational numbers $\alpha < \beta$, H also has measure zero. Consequently the set of $\omega \in \Omega$ such that $\lim_{t\to\infty} \sup X_t(\omega) > \lim_{t\to\infty} \inf X_t(\omega)$ has measure zero, that is,

$$\lim_{t\to\infty} X_t(\omega) = X_\infty(\omega) \quad \text{exists almost surely.}$$

Also

$$\|X_t\|_1 = E[X_t^+ + X_t^-] = E[X_t] + 2E[X_t^-] \leq E[X_0] + 2E[X_t^-],$$

so by Fatou's lemma

$$E[X_\infty] \leq \sup_t \|X_t\|_1 < \infty \quad \text{and } X_\infty \text{ is, therefore, integrable.}$$ □

Corollary 4.8. *Suppose $\{X_t\}$, $t \in [0, \infty[$, is a supermartingale such that $X_t \geq 0$ a.s. for every $t \in \tau$. Then clearly $X_t^- = 0$ a.s. so the above theorem applies* *and* $\lim_{t\to\infty} X_t(\omega) = X_\infty(\omega)$ *a.s. Furthermore, if* $\mathscr{F}_\infty = \vee_t \mathscr{F}_t$ *and* $A \in \mathscr{F}_t$, *then for* $s > t$

$$\int_A X_t \, dP \geq \int_A X_s \, dP.$$

Letting s tend to infinity, we have by Fatou's lemma, that $X_t \geq E[X_\infty | \mathscr{F}_t]$ *a.s. Therefore,* $\{X_t\}$, $t \in [0, \infty]$ *is a supermartingale.*

Corollary 4.9. *Suppose* $\{X_t\}$, $t \in [0, \infty[$, *is a uniformly integrable supermartingale. Then* $\sup_t \|X_t\|_1 < \infty$, *and, because* $E[X_t^-] \leq \|X_t\|_1$, *the condition of the theorem is satisfied and* $\lim_{t\to\infty} X_t(\omega) = X_\infty(\omega)$ *a.s. By uniform integrability the convergence also takes place in* L^1, *and* $\{X_t\}$, $t \in [0, \infty]$, *is a supermartingale.*

Corollary 4.10. *Suppose* $\{X_t\}$, $t \in [0, \infty[$, *is a uniformly integrable martingale. Then* $\{X_t\}$, $t \in [0, \infty]$, *is a martingale.*

The proof of the following result is similar to that of Theorem 3.20 together with an application of the above Corollary 4.10.

Theorem 4.11. *If* $\mathscr{F}_\infty = \bigvee_t \mathscr{F}_t$ *and Y is an integrable* $\mathscr{F}_\infty$*-measurable random variable, then* $\{E[Y | \mathscr{F}_t]\}$, $t \in [0, \infty]$, *is a uniformly integrable martingale. We can take a right continuous modification* $\{Y_t\}$ *of this martingale, and* $\lim_{t\to\infty} Y_t(\omega) = Y(\omega)$ *a.s. and in* L^1.

Finally we extend the optional stopping theorem to continuous time martingales.

Theorem 4.12 (Optional Stopping). *Suppose* $\{X_t\}$ *is a right continuous supermartingale with respect to a filtration* $\{\mathscr{F}_t\}$, $t \in [0, \infty]$. *If S and T are two stopping times such that* $S \leq T$ *a.s., then the random variables* X_S *and* X_T *are integrable and* $X_S \geq E[X_T | \mathscr{F}_S]$ *a.s.*

PROOF. Suppose n is a positive integer and write Σ_n for the set of all real numbers of the form $2^{-n}k$, $k \in N$, together with ∞. Then $\{X_t\}$, $t \in \Sigma_n$, is a discrete time supermartingale with respect to the σ-fields $\{\mathscr{F}_t\}$, $t \in \Sigma_n$.

For any number $\rho \in [0, \infty[$ define $\rho^{(n)}$ to be the unique number $2^{-n}k$ in Σ_n such that $2^{-n}(k-1) \leq \rho < 2^{-n}k$. Furthermore, define $\infty^{(n)} = \infty$. Then for the stopping times S and T the random variables $S^{(n)}$ and $T^{(n)}$ are stopping times, when they are defined by $S^{(n)}(\omega) = (S(\omega))^{(n)}$, (resp. $T^{(n)}(\omega) = (T(\omega))^{(n)}$). Indeed, $S^{(n)}$ and $T^{(n)}$ are discrete valued stopping times with respect to the filtration $\{\mathscr{F}_t\}$, $t \in \Sigma_n$ and $\lim S^{(n)} = S$ a.s. (resp. $\lim T^{(n)} = T$ a.s.). Therefore, applying theorem 3.32 we have that

$$X_{S^{(n)}} \geq E[X_{T^{(n)}} | \mathscr{F}_{S^{(n)}}] \quad \text{a.s.}$$

Because $S < S^{(n)}$ a.s. $\mathscr{F}_S \subset \mathscr{F}_{S^{(n)}}$, so for any $A \in \mathscr{F}_S$

$$\int_A X_{S^{(n)}} \, dP \geq \int_A X_{T^{(n)}} \, dP. \tag{4.1}$$

Now $S \leq S^{(n+1)} \leq S^{(n)}$ a.s., so $\mathscr{F}_{S^{(n+1)}} \subset \mathscr{F}_{S^{(n)}}$ and working with the discrete parameter set $\Sigma_{n+1}, X_{S^{(n+1)}} \geq E[X_{S^{(n)}} | \mathscr{F}_{S^{(n+1)}}]$. Therefore, $\{X_{S^{(n)}}\}$ is a reverse supermartingale with respect to the decreasing family of σ-fields $\{\mathscr{F}_{S^{(n)}}\}$. However, $E[X_{S^{(n)}}] \leq E[X_0]$ for all n, so applying Theorem 3.28 the set $\{X_{S^{(n)}}\}$ is uniformly integrable.

The same argument applied to $\{X_{T^{(n)}}\}$, where $T \leq T^{(n+1)} \leq T^{(n)}$ and $\mathscr{F}_{T^{(n+1)}} \subset \mathscr{F}_{T^{(n)}}$, shows that $\{X_{T^{(n)}}\}$ is a reverse supermartingale with $E[X_{T^{(n)}}] \leq E[X_0]$, and so the set $\{X_{T^{(n)}}\}$ is uniformly integrable. Because $\{X_t\}$ is right continuous

$$\lim_n X_{S^{(n)}} = X_S \quad \text{a.s.}$$

and $\lim X_{T^{(n)}} = X_T$ a.s. Therefore, X_S and X_T are uniformly integrable, and letting n tend to ∞ in (4.1) we see that

$$\int_A X_S \, dP \geq \int_A X_T \, dP,$$

that is $X_S \geq E[X_t | \mathscr{F}_S]$ a.s. □

Corollary 4.13. *If $\{X_t\}$, $t \in [0, \infty]$, is a right continuous martingale, and S and T are two stopping times such that $S \leq T$ a.s. then $X_S = E[X_T | \mathscr{F}_S]$ a.s.*

Definition 4.14. As in the discrete time case, a nonnegative right continuous supermartingale $\{X_t\}$, $t \in \tau$, is said to be a *potential* if $\lim_{t \to \infty} E[X_t] = 0$.

Then by Corollary 4.8 we know that $\lim_{t \to \infty} X_t(\omega) = X_\infty(\omega)$ almost surely, and by Fatou's lemma $E[X_\infty] = 0$, so $X_\infty(\omega) = 0$ a.s. That is, $\{X_t\}$ is a potential if, and only if,

(1) $X_t(\omega) \geq 0$ a.s., $t \in \tau$.
(2) $\lim_{t \to \infty} X_t(\omega) = 0$ a.s.

The Riesz decomposition for a supermartingale in continuous time is established in the following form:

Theorem 4.15. *Suppose $\{X_t\}$, $t \in \tau$, is a right continuous uniformly integrable supermartingale. Then there is a right continuous uniformly integrable martingale $\{Y_t\}$, and a potential $\{Z_t\}$, such that $X_t = Y_t + Z_t$ a.s. for every $t \in \tau$. This decomposition is unique up to indistinguishable processes, and is called the Riesz decomposition of $\{X_t\}$.*

PROOF. Write Y_t for a right continuous modification of the martingale $E[X_\infty | \mathscr{F}_t]$ and put $Z_t = X_t - Y_t$.

Then Z_t is a right continuous uniformly integrable supermartingale, and

$$\lim_{t\to\infty} Z_t = \lim_{t\to\infty}(X_t - E[X_\infty|\mathcal{F}_t]) = 0 \quad \text{a.s.}$$

Therefore, Z_t is a potential. If $X_t = Y_t^* + Z_t^*$, then $Y_\infty^* = X_\infty = Y_\infty$, so $Y_t^* = E[X_\infty \mathcal{F}|_t] = Y_t$ a.s. for all $t \in \tau$. By right continuity $\{Y_t^*\}$ is indistinguishable from $\{Y_t\}$. □

Theorem 4.16. *Suppose $\{X_t\}$, $t \geq 0$, is a corlol, positive supermartingale. Write*

$$T(\omega) = \inf\{t: X_t(\omega) = 0 \text{ or } t > 0 \text{ and } X_{t-}(\omega) = 0\}.$$

Then for almost every $\omega \in \Omega$

$$X_s(\omega) \text{ is zero on } [T(\omega), \infty[.$$

PROOF. We can suppose the random variable X_∞ is 0 a.s. Consider, for $n \in Z^+$, the stopping times

$$T_n(\omega) = \inf\{t: X_t(\omega) \leq 1/n\}.$$

Clearly $T_{n-1} \leq T_n \leq T$, and $X_{T_n} \leq 1/n$ on $\{T_n < \infty\}$, whilst $X_{T_n} = 0$ on $\{T_n = \infty\}$. By the optional stopping theorem $E[X_{T+t}] \leq E[X_{T_n}] \leq 1/n$ for every rational $t \geq 0$ and every $n \in Z^+$.

Therefore, $X_{T+t} = 0$ a.s., and so the result follows by right continuity. □

Remarks 4.17. The above argument also shows that $\lim T_n = T$.

Furthermore, if $S(\omega) = \inf\{t: X_t(\omega) = 0\}$ then $X_s(\omega)$ is strictly positive on $[0, S(\omega)[$, and every $a < S(\omega)$, $X_s(\omega)$ is bounded below away from zero on $[0, a]$.

Finally we give a characterization of a uniformly integrable martingale.

Lemma 4.18. *Suppose $\{M_t\}$, $t \in [0, \infty]$, is an adapted right continuous process such that for every stopping time T, $E[|M_T|] < \infty$ and $E[M_T] = 0$. Then $\{M_t\}$ is a uniformly integrable martingale.*

PROOF. From Theorem 4.6, $\{M_t\}$ has a corlol version. Consider any time $t \in [0, \infty]$ and any $A \in \mathcal{F}_t$. Define a random variable T by putting $T(\omega) = t$ if $\omega \in A$ and $T(\omega) = \infty$ if $\omega \notin A$. Then T is a stopping time. (In the notation of Definition 5.10 below, $T = t_A$.) Then by hypothesis:

$$E[M_T] = \int_A M_t + \int_{A^c} M_\infty = 0 = E[M_\infty] = \int_A M_\infty + \int_{A^c} M_\infty.$$

Therefore

$$\int_A M_t = \int_A M_\infty \quad \text{for all } A \in \mathcal{F}_t, \qquad \text{so } M_t = E[M_\infty|\mathcal{F}_t] \quad \text{a.s.} \qquad \square$$

CHAPTER 5

Predictable and Totally Inaccessible Stopping Times

Convention 5.1. Throughout this chapter we shall suppose that $(\Omega, \mathscr{F}, P)$ is a complete probability space with a right continuous complete filtration $\{\mathscr{F}_t\}$ for $t \in [0, \infty[$ (so that, in particular, $\mathscr{F}_0$ contains all null sets of $\mathscr{F}$). $\mathscr{F}_\infty$ will denote the σ-field $\bigvee_t \mathscr{F}_t$.

A stopping time was defined in Definition 2.4 as a random variable T with values in $[0, \infty]$ such that $\{T \leq t\} \in \mathscr{F}_t$ for all $t \in [0, \infty[$.

Definition 5.2. A stopping time T is said to be *predictable* if there is a sequence $\{T_n\}$, $n \in N$, of stopping times such that

(1) $\{T_n(\omega)\}$ is almost surely an increasing sequence in $[0, \infty[$ and $\lim_n T_n(\omega) = T(\omega)$ a.s.
(2) on the set $\{T > 0\}$, $T_n(\omega) < T(\omega)$ a.s. for all n.

The sequence $\{T_n\}$ is said to *announce* T.

Remarks 5.3. Predictable stopping times occur naturally in the real world. For example, consider a ship being driven onto a rocky coastline. The time T when the ship is wrecked is a stopping time announced by the family $\{T_n\}$, where T_n is the time it is $1/n$ km. from shore.

Examples 5.4. For any stopping time T and any real number $r > 0$ the random variable $T + r$ is a stopping time. Indeed, $T + r$ is a predictable stopping time because it is announced by the sequence $\{T_n\}$ where $T_n = T + r(1 - 1/n)$. Therefore, by considering a sequence $\{r_n\}$ such that $r_n > 0$ and $\lim r_n = 0$, e.g. $r_n = 1/n$, we see that any stopping time is the limit from above of a decreasing sequence of predictable stopping times. (Clearly, in general $T - r$ is not a stopping time for $r > 0$.)

Definition 5.5. Suppose S and T are stopping times and $S \leq T$ a.s. The (*half open*) *stochastic interval*, denoted by $[\![S, T[\![$, is the set

$$\{(t, \omega) \in [0, \infty[\times \Omega: S(\omega) \leq t < T(\omega)\}.$$

The stochastic intervals $[\![S, T]\!]$, $]\!]S, T]\!]$ and $]\!]S, T[\![$ are defined similarly.

$$[\![T, T]\!] = \{(t, \omega) \in [0, \infty[\times \Omega: T(\omega) = t\}$$

is denoted by $[\![T]\!]$, and is the *graph* of the stopping time T.

If we write $[\![S, T[\![$, etc., it will be understood that $S \leq T$ a.s.

Note that the stochastic integrals and graphs are subsets of $[0, \infty[\times \Omega$ (and not $[0, \infty] \times \Omega$).

Remarks 5.6. We now introduce other classes of stopping times using definitions which involve their relation with predictable stopping times. It perhaps should be mentioned that arbitrary stopping times are sometimes called *optional* stopping times.

Definition 5.7. (1) A stopping time T is said to be *accessible* if there is a countable set $\{T_n\}$, $n \in N$, of predictable stopping times such that

$$[\![T]\!] \subset \bigcup_n [\![T_n]\!],$$

up to an evanescent set, that is $P(\bigcup_n \{\omega: T_n(\omega) = T(\omega)\}) = 1$.

(2) A stopping time T is said to be *totally inaccessible* if, for every predictable stopping time S,

$$[\![T]\!] \cap [\![S]\!] = \varnothing,$$

up to an evanescent set, that is

$$P(\{\omega: T(\omega) = S(\omega) < \infty\}) = 0$$

for every predictable stopping time S.

Notation 5.8. $\mathscr{T}_o$ (resp. $\mathscr{T}_a$, $\mathscr{T}_p$) will denote the set of all (resp. accessible, predictable) stopping times.

Lemma 5.9. *If the stopping time T is both accessible and totally inaccessible then $T = \infty$ a.s.*

PROOF.

$$\{T < \infty\} = \{\omega \in \Omega: (t, \omega) \in [\![T]\!] \text{ for some } t \in [0, \infty]\}.$$

The definitions above imply that $[\![T]\!]$ is evanescent, so $\{T < \infty\}$ has measure zero. □

Definition 5.10. If T is a stopping time and $A \in \mathscr{F}$ the *restriction* of T to A is the random variable T_A defined by

$$T_A(\omega) = \begin{cases} T(\omega), & \text{for } \omega \in A, \\ \infty, & \text{for } \omega \notin A. \end{cases}$$

Recalling Definition 2.8 we prove the following result.

Lemma 5.11. *T_A is a stopping time if and only if $A \in \mathscr{F}_T$.*

PROOF.

$$\{T_A \le t\} = A \cap \{T \le t\}. \qquad \square$$

Definition 5.12. The σ-field $\mathscr{F}_{T-}$ of events *strictly prior* to the stopping time T is the σ-field generated by $\mathscr{F}_0$ and all sets of the form $A \cap \{t < T\}$, where $t \in [0, \infty]$ and $A \in \mathscr{F}_t$.

Theorem 5.13. *Suppose S and T are stopping times. Then*

(1) $\mathscr{F}_{T-} \subset \mathscr{F}_T$,
(2) *T is $\mathscr{F}_{T-}$ measurable,*
(3) *if $T \le S$, $\mathscr{F}_{T-} \subset \mathscr{F}_{S-}$,*
(4) *for every $A \in \mathscr{F}_S$, $A \cap \{S < T\} \in \mathscr{F}_{T-}$.*

PROOF. (1) It is enough to show that for $t \in [0, \infty[$ and $A \in \mathscr{F}_t$ a generating set $A \cap \{t < T\}$ is in $\mathscr{F}_T$. That is, for any $r \in [0, \infty[$ we must show $A \cap \{t < T\} \cap \{T \le r\}$ is in $\mathscr{F}_r$. If $r \le t$ this set is empty, whilst if, $t < r$, $A \cap \{t < T \le r\} \in \mathscr{F}_r$.

(2) $\{T \le t\}$ is clearly an element of $\mathscr{F}_{T-}$.

(3) $A \cap \{t < T\} = A \cap \{t < T\} \cap \{t < S\}$, so any generator of $\mathscr{F}_{T-}$ is a generator of $\mathscr{F}_{S-}$.

(4) $A \cap \{S < T\} = \bigcup_r [A \cap \{S \le r\} \cap \{r < T\}]$, where r runs through all positive rationals. Because $A \in \mathscr{F}_S$, $A \cap \{S \le r\} \in \mathscr{F}_r$ and so (4) is proved. $\square$

Lemma 5.14. *Suppose T is a stopping time and $A \in \mathscr{F}_\infty$. Then*

$$A \cap \{T = \infty\} \in \mathscr{F}_{T-}.$$

PROOF. The sets $B \in \mathscr{F}_\infty$ such that $B \cap \{T = \infty\} \in \mathscr{F}_{T-}$ certainly form a σ-field. Therefore, it is sufficient to show that $A \cap \{T = \infty\} \in \mathscr{F}_{T-}$, when $A \in \mathscr{F}_n$, $n = 1, 2, \ldots$. However,

$$A \cap \{T = \infty\} = \bigcap_{m=n+1}^{\infty} A \cap \{T > m\},$$

and by definition each set $A \cap \{T > m\}$ is in $\mathscr{F}_{T-}$. $\square$

Definition 5.15. If T is any stopping time $\mathscr{S}(T)$ will denote the family of all increasing sequences $\{T_n\}$ of stopping times such that $T_n \leq T$ for all n.

If $\{T_n\} \in \mathscr{S}(T)$ we write

$$K[\{T_n\}] = \{\omega : \lim_n T_n(\omega) = T(\omega) < \infty \text{ and } T_n(\omega) < T(\omega) \text{ for all } n\}.$$

We now derive a decomposition for a stopping time.

Theorem 5.16. *Suppose T is a stopping time. Then there is an essentially unique partition of Ω into two elements A and B of $\mathscr{F}_{T-}$, such that T_A is accessible and T_B is totally inaccessible.*

PROOF. For $\{S_n\} \in \mathscr{S}(T)$ write $A[\{S_n\}] = K[\{S_n\}] \cup \{T = 0\}$, so that $T_{A[\{S_n\}]}$ is predictable and, outside $\{T = 0\}$, $A[\{S_n\}]$ is the set on which $\{S_n\}$ announces T.

Write $S = \lim S_n$. Then $\{S \leq T\} = (\bigcap \{S_n < T\}) \cup \{S = T = 0\}$, and this is in $\mathscr{F}_{T-}$ by Theorem 5.13(4). Therefore, $A[\{S_n\}] = \{S = T\} = \{S \leq T\} \backslash \{S < T\} \in \mathscr{F}_{T-}$. There are at most countably many such sets $A[\{S_n\}]$ of positive measure. Write $A = A(T)$ for their essential union. Then $A \in \mathscr{F}_{T-}$ and A contains $\{T = 0\}$ and $\{T = \infty\}$. Furthermore, $[T_A]$ is contained in a countable union of graphs of predictable stopping times and so T_A is accessible. If B is the complement of A, $B \in \mathscr{F}_{T-}$ and T_B is totally inaccessible. □

The uniqueness is immediate from Lemma 5.9 and the following result.

Lemma 5.17. *Suppose T is a stopping time and $A \in \mathscr{F}_T$. If T is accessible (resp. totally inaccessible) then T_A is accessible (resp. totally inaccessible).*

PROOF. By definition $[\![T_A]\!] \subset [\![T]\!]$. □

Lemma 5.18. *Suppose S and T are two stopping times. If both S and T are predictable (resp. accessible, totally inaccessible), then $S \vee T$ and $S \wedge T$ are predictable (resp. accessible, totally inaccessible).*

PROOF. This is an immediate consequence of the definitions, for example if $\{S_n\}$ announces S and $\{T_n\}$ announces T, then $\{S_n \wedge T_n\}$ announces $S \wedge T$. Also $[\![S \wedge T]\!] \subset [\![S]\!] \cup [\![T]\!]$. □

Theorem 5.19. *Suppose $\{T_n\}$ is an increasing sequence of predictable (resp. accessible) stopping times. Then $T = \lim T_n$ is predictable (resp. accessible).*

PROOF. *Accessible Case.* Write $A = \{\omega \in \Omega : T_n(\omega) < T(\omega) \text{ for all } n\}$ and S_n for the restriction of T_n to A. Then if $R_n = S_n \wedge n$, T_A is announced by the sequence $\{R_n\}$ and so T_A is predictable. If B is the complement of A then for $\omega \in B$, $T(\omega) = T_n(\omega)$ for some n, so

$$[\![T_B]\!] \subset \bigcup_n [\![T_n]\!].$$

Therefore, T_B is accessible and by Lemma 5.18 above $T = T_A \wedge T_B$ is accessible.

Predictable Case. Suppose $\{S_{n,p}\}$, $p \in N$, is a sequence of stopping times which announces T_n. For each $n \in N$ write

$$S_n = \sup_{k \le n, p \le n} S_{k,p}.$$

Then $\{S_n\}$ announces T. □

Theorem 5.20. *Suppose $\{T_n\}$ is a decreasing sequence of predictable (resp. accessible) stopping times and $T = \lim T_n$ a.s. If for almost every $\omega \in \Omega$, $T(\omega) = T_n(\omega)$ for some n then T is predictable (resp. accessible).*

Proof. *Accessible Case.* We have $[\![T]\!] \subset \bigcup_n [\![T_n]\!]$. Because the T_n are accessible the result is immediate from the definition.

Predictable Case. Clearly we can restrict our attention to $\{T < \infty\}$. Suppose that $T < \infty$ and $\{S_{n,p}\}$, $p \in N$, is a sequence of stopping times announcing T_n such that

$$P(\{\omega : d(S_{n,p}(\omega), T_n(\omega)) > 2^{-p}\}) \le 2^{-(n+p)},$$

for every p, where $d(s, t) = (|s - t|)/(1 + |s - t|)$ for $s, t \in [0, \infty[$, $d(\infty, \infty) = 0$ and $d(t, \infty) = d(\infty, t) = 1$. Such a sequence can be obtained by, if necessary, selecting a subsequence from any sequence announcing T_n.

Put $S_p = \inf_n S_{n,p}$. Then the sequence $\{S_p\}$, $p \in N$, is increasing and, by hypothesis, $S_p < T$ on $\{T > 0\}$ for each p. Write $S = \lim_p S_p$. Then for each $p \in N$:

$$\begin{aligned} P(\{\omega : d(S(\omega), T(\omega)) > 2^{-p}\}) &\le P(\{\omega : d(S_p(\omega), T(\omega)) > 2^{-p}\}) \\ &\le \sum_n P(\{\omega : d(S_{n,p}(\omega), T(\omega)) > 2^{-p}\}) \\ &\le \sum_n P(\{\omega : d(S_{n,p}(\omega), T_n(\omega)) > 2^{-p}\}) \le 2^{-p}. \end{aligned}$$

Letting p tend to ∞ we see that

$$P\{S < T\} = 0 \quad \text{and so } S = T.$$ □

Corollary 5.21. *For any stopping time T the family of subsets $A \in \mathscr{F}_T$ such that T_A is predictable is closed under countable unions and intersections.*

Proof. This follows from the previous two results, Theorems 5.19 and 5.20. □

Lemma 5.22. *Suppose S and T are stopping times and $S \le T$. If $S < T$ on $\{0 < T < \infty\}$ then $\mathscr{F}_S \breve{\subset} \mathscr{F}_{T-}$.*

Proof. Recall that $\mathscr{F}_S = \{A \in \mathscr{F}_\infty : A \cap \{S \le t\} \in \mathscr{F}_t$ for all $t \in [0, \infty[\}$. For $A \in \mathscr{F}_S$

$$A = [A \cap \{T = 0\}] \cup [A \cap \{S < T\}] \cup [A \cap \{T = \infty\}].$$

Now

$$A \cap \{T=0\} \in \mathscr{F}_0 \subset \mathscr{F}_{T-} \quad \text{by definition.}$$

$$A \cap \{S<T\} \in \mathscr{F}_{T-} \quad \text{by Theorem 5.13(4).}$$

When $B \in \mathscr{F}_t \subset \mathscr{F}_{t+m}$ we have $B \cap \{t+m<T\} \in \mathscr{F}_{T-}$ for $m \geq 0$, so $B \cap \{T=\infty\} \in \mathscr{F}_{T-}$ for $B \in \mathscr{F}_t$, for any t. Therefore, for $A \in \mathscr{F}_\infty = \bigvee_t \mathscr{F}_t$, $A \cap \{T=\infty\} \in \mathscr{F}_{T-}$. □

Lemma 5.23. *If $\{T_n\}$ is an increasing sequence of stopping times and $T = \lim T_n$, then $\mathscr{F}_{T-} = \bigvee_n \mathscr{F}_{T_n-}$.*

PROOF. From Theorem 5.13(3) $\mathscr{F}_{T-} \supset \bigvee_n \mathscr{F}_{T_n-}$. However, for any $t \in [0, \infty[$ and $A \in \mathscr{F}_t$ $A \cap \{t<T\} = \bigcup_n A \cap \{t<T_n\}$, so any generator of $\mathscr{F}_{T-}$ is in $\bigvee_n \mathscr{F}_{T_n-}$.

Corollary 5.24. *Suppose $\{T_n\}$ is an increasing sequence of stopping times and $T = \lim T_n$. If $T_n < T$ on $\{0<T<\infty\}$ then $\mathscr{F}_{T-} = \bigvee_n \mathscr{F}_{T_n}$.*

PROOF. From Theorem 5.13(1) and Lemma 5.22, $\mathscr{F}_{T_n-} \subset \mathscr{F}_{T_n} \subset \mathscr{F}_{T-}$, so the result follows from the lemma. □

Remark. This result extends Definition 5.2 and is a natural description of $\mathscr{F}_{T-}$.

Theorem 5.25. *Suppose T is a stopping time and $A \in \mathscr{F}_T$.*

If T_A is predictable then $A \in \mathscr{F}_{T-}$. Conversely, if T is predictable and $A \in \mathscr{F}_{T-}$ then T_A is predictable.

PROOF. Suppose T_A is predictable, and let $\{S_n\}$ be a sequence of stopping times announcing T_A. Then

$$A = \{T_A \leq T\} \backslash (A^c \cap \{T=\infty\}).$$

Now $\{T_A \leq T\} = \bigcap_n \{S_n < T\}$, and $\{S_n < T\} \in \mathscr{F}_{T-}$ from Theorem 5.13(3). Also $A \in \mathscr{F}_T \subset \mathscr{F}_\infty$, so $A^c \in \mathscr{F}_\infty$ and, by Lemma 5.14 $A^c \cap \{T=\infty\} \in \mathscr{F}_{T-}$.

Consequently, $A \in \mathscr{F}_{T-}$. Conversely, suppose T is predictable and consider the family of sets A in $\mathscr{F}_T$ which have the property that T_A and T_{A^c} are predictable.

From Corollary 5.21 this family is a σ-field, so it is enough to prove the result when A belongs to a system of generators of $\mathscr{F}_{T-}$. Now if $\{T_n\}$ is a sequence of stopping times announcing T, from Corollary 5.24, $\mathscr{F}_{T-} = \bigvee_n \mathscr{F}_{T_n}$. Therefore, suppose $A \in \mathscr{F}_{T_n}$. Then the sequence of stopping times $\{S_m\}$, where $S_m = (T_{n+m})_A \wedge m$, $m \in N$, announces T_A, so T_A is predictable, and the result follows. □

Theorem 5.26. *Suppose S is a predictable stopping time and T an arbitrary stopping time. For any $A \in \mathscr{F}_{S-}$ the set $A \cap \{S \leq T\}$ belongs to $\mathscr{F}_{T-}$. In particular, the sets $\{S \leq T\}$ and $\{S = T\}$ belong to $\mathscr{F}_{T-}$.*

PROOF. Let $\{S_n\}$ be a sequence of stopping times announcing S and suppose $A \in \mathscr{F}_{S_m}$. Then

$$A \cap \{S \le T\} = [A \cap \{S = 0\}] \cup \left[\bigcap_n (A \cap \{S_{m+n} < T\})\right],$$

$$A \cap \{S = 0\} \in \mathscr{F}_0 \subset \mathscr{F}_{T-} \quad \text{and from Theorem 5.13(4),}$$

$$A \cap \{S_m < T\} \in \mathscr{F}_{T-} \quad \text{as is } \{S_{m+n} < T\} \subset \{S_m < T\}.$$

Therefore, for $A \in \mathscr{F}_{S_m}$ $A \cap \{S \le T\} \in \mathscr{F}_{T-}$. However, the family of sets $\{A\}$ which have the property that $A \cap \{S \le T\} \in \mathscr{F}_{T-}$ is clearly a σ-field, and from Corollary 5.24, $\mathscr{F}_{S-} = \bigvee_m \mathscr{F}_{S_m}$, so $A \cap \{S \le T\} \in \mathscr{F}_{T-}$ for any $A \in \mathscr{F}_{S-}$.

In particular, $\{S \le T\} \in \mathscr{F}_{T-}$. From Theorem 5.4(4) we have for general stopping times S and T that $\{S < T\} \in \mathscr{F}_{T-}$. Therefore, $\{S = T\} = \{S \le T\} \backslash \{S < T\} \in \mathscr{F}_{T-}$. □

Remarks 5.27. From Theorem 2.10(ii) and Theorem 5.13(4) we have that the sets

$$\{S \le T\}, \{S < T\}, \{S = T\}, \{S > T\}, \{S \ge T\}$$

belong to both $\mathscr{F}_S$ and $\mathscr{F}_T$. Reversing the roles of S and T in Theorem 5.13(4) we see that $\{T < S\}$ (and so $\{S \le T\}$), belong to $\mathscr{F}_{S-}$, but not, in general, to $\mathscr{F}_{T-}$. However, the above result shows that when S is predictable $\{T < S\}$ and $\{S \le T\}$ belong to $\mathscr{F}_{T-}$.

Lemma 5.28. *Suppose S and T are predictable stopping times. Then T_A is predictable, where $A = \{T < S\}$.*

PROOF. From the above theorem and the remarks following we see that $A \in \mathscr{F}_{T-}$. Therefore, the result follows from Theorem 5.25. □

Remarks 5.29. We now give some characterizations of accessible and totally inaccessible stopping times.

The first result is an immediate consequence of the definitions.

Lemma 5.30. *A stopping time T is accessible (resp. totally inaccessible) if and only if for every totally inaccessible (resp. accessible) stopping time S*

$$P\{\omega : S(\omega) = T(\omega) < \infty\} = 0.$$

Lemma 5.31. *If $\{T_n\} \in \mathscr{S}(T)$ then the restriction of T to $K[\{T_n\}]$ is accessible.*

PROOF. Write S_n for the restriction of T_n to the set $\{T_n < \lim T_n\}$. Then the stopping time $S = \lim S_n$ is announced by the sequence $\{S_n \wedge n\}$ and so is predictable. However, the graph of $T_{[K\{T_n\}]}$ is contained in $[\![S]\!]$ so $T_{[K\{T_n\}]}$ is accessible. □

Theorem 5.32.

(a) *A stopping time T is accessible if and only if the set $\{0<T<\infty\}$ is the union of a sequence of sets of the form $K[\{S_n\}]$, $\{S_n\}\in\mathscr{S}(T)$.*

(b) *A stopping time T is totally inaccessible if and only if*

$$P\{T=0\}=0$$

and

$$P(K[\{S_n\}])=0$$

for every $\{S_n\}\in\mathscr{S}(T)$.

PROOF. (a) Because for every $\{S_n\}\in\mathscr{S}(T)$ the stopping time $T_{K[\{S_n\}]}$ is accessible the condition is sufficient. Conversely, suppose that T is accessible and let $\{T^m\}$ be a sequence of predictable stopping times such that

$$[\![T]\!]\subset\bigcup_m[\![T^m]\!].$$

For each m let $\{T^m_n\}$ be a sequence which announces T^n and write $S^m_n=T\wedge T^m_n$. Then for m $\{S^m_n\}$, $n=1,2,\ldots$, is a member of $\mathscr{S}(T)$ and the result follows because $\{0<T<\infty\}=\bigcup_m K[\{S^m_n\}]$.

(b) Because $T_{K[\{S_n\}]}$ is accessible for every sequence $\{S_n\}\in\mathscr{S}(T)$ the condition is necessary by Lemma 5.31. It is also sufficient, because by part (a) the accessible part of T is then infinite. □

Theorem 5.33. *Suppose S is an accessible stopping time. Then S is predictable if and only if the set $\{S=T\}\in\mathscr{F}_{T-}$ for every predictable stopping time T.*

PROOF. If S is predictable then $\{S=T\}\in\mathscr{F}_{T-}$ by Theorem 5.26. Conversely, suppose S is accessible and $\{S_n\}$ is a sequence of predictable stopping times such that

$$[\![S]\!]\subset\bigcup_n[\![S_n]\!].$$

Then for each n, by Theorem 5.26 and Remarks 5.27,

$$\{S\le S_n\}\in\mathscr{F}_{S_n-}.$$

Write T_n for the restriction of S_n to $\{S\le S_n\}$, so that by Theorem 5.25 T_n is predictable. For each n the stopping time

$$R_n=T_1\wedge T_2\wedge\cdots\wedge T_n$$

is predictable. The sequence $\{R_n\}$ is decreasing, and for each $\omega\in\Omega$ there is an n such that $R_n(\omega)=S(\omega)$. Therefore, by Theorem 5.20, S is predictable. □

Definition 5.34. The filtration $\{\mathscr{F}_t\}$ is said to be *left quasi-continuous* if for every predictable stopping time T

$$\mathscr{F}_{T-}=\mathscr{F}_T.$$

Remarks 5.35. The word "quasi" is used because equality of $\mathscr{F}_{T-}$ and $\mathscr{F}_T$ is only required for predictable stopping times.

Theorem 5.36. *The following three properties are equivalent*:

(a) *the filtration* $\{\mathscr{F}_t\}$ *is left quasi-continuous,*
(b) *if* $\{T_n\}$ *is any increasing sequence of stopping times then*

$$\mathscr{F}_{(\lim T_n)} = \bigvee_n \mathscr{F}_{T_n},$$

that is the filtration has no times of discontinuity,

(c) *the accessible stopping times are predictable.*

PROOF. (b)$\Rightarrow$(a) follows from Corollary 5.24.

(a)$\Rightarrow$(c) follows from Theorem 5.33 above.

(c)$\Rightarrow$(b) Suppose $\{T_n\}$ is an increasing sequence of stopping times, and write $T = \lim T_n$. Let R (resp. S) denote the accessible (resp. totally inaccessible) part of T. For any set $A \in \mathscr{F}_T$ we can write

$$A = [\{R_A = T\} - (A^c \cap \{T = \infty\})] \cup [\{S_A < \infty\} \cup (A \cap \{T = \infty\})].$$

From Lemma 5.14 the sets $A^c \cap \{T = \infty\}$ and $A \cap \{T = \infty\}$ belong to $\mathscr{F}_{T-}$, and so to $\bigvee_n \mathscr{F}_{T_n}$ by Lemma 5.23. Now by hypothesis R_A is predictable, so by Theorem 5.26

$$\{R_A = T\} \in \mathscr{F}_{T-} = \bigvee_n \mathscr{F}_{T_n-} \subset \bigvee_n \mathscr{F}_{T_n}, \quad \text{again by 5.23.}$$

Consider the set $\{S_A < \infty\}$. Because S_A is totally inaccessible and $\{T_n\} \in \mathscr{S}(S_A)$ it follows from Theorem 5.32(b) that

$$\{S_A < \infty\} = \bigcup_n \{S_A = T_n < \infty\}$$

$$\in \bigvee_n \mathscr{F}_{T_n}.$$

Therefore, $A \in \bigvee_n \mathscr{F}_{T_n}$ and the result is proved. □

CHAPTER 6

The Optional and Predictable σ-Fields

Convention 6.1. Again we shall work with a probability space $(\Omega, \mathscr{F}, P)$ with a right continuous, complete filtration $\{\mathscr{F}_t\}$, $t \in [0, \infty[$.

Definition 6.2. The *optional* (resp. *accessible, predictable*) *σ-field* Σ_o (resp. Σ_a, Σ_p) on $[0, \infty[\times \Omega$ is the σ-field generated by the evanescent sets and all stochastic intervals of the form $[\![T, \infty[\![$ for T an arbitrary (resp. accessible, predictable) stopping time.

Remarks 6.3. Note that this is equivalent to saying Σ_x is generated by the intervals $[\![T, S[\![$, $S, T \in \mathscr{T}_x$ $(x = o, a, p)$. Also $\Sigma_p \subset \Sigma_a \subset \Sigma_o$.

Definition 6.4. A set $A \subset [0, \infty[\times \Omega$ is said to be *progressively measurable*, or *progressive*, if its indicator function I_A is a progressive process. The family of all progressive sets forms a σ-field Σ_π, and a process $\{X_t\}$ is progressive if and only if the map $(t, \omega) \to X_t(\omega)$ is measurable with respect to Σ_π.

Furthermore, because $I_{[\![T,S[\![}$ is adapted and right continuous, the optional σ-field is contained in the progressive σ-field by Theorem 2.3, and so $\Sigma_0 \subset \Sigma_\pi$.

We have the following description:

Lemma 6.5. *Σ_x is generated by the intervals $[\![T, S]\!]$, for $T \in \mathscr{T}_x$ and $S \in \mathscr{T}_o$.*

PROOF. We know from Examples 5.4 that an arbitrary stopping time $S \in \mathscr{T}_o$ is the limit, from above, of the sequence of predictable stopping times $S_n = S + 1/n$. Therefore, $[\![T, S]\!] = \bigcap_n [\![T, S_n[\![$ and $[\![T, S]\!] \in \Sigma_x$ if $T \in \mathscr{T}_x$.

Conversely, if $S \in \mathscr{T}_x$ (and $T \in \mathscr{T}_x$), $[\![S]\!] = \bigcap_n [\![S, S + 1/n]\!]$ and $[\![T, S[\![= [\![T, S]\!] \backslash [\![S]\!]$. Therefore, the kind of generating interval mentioned in the

Remarks 6.3 above is in the σ-field generated by the intervals $[\![T, S]\!]$, $T \in \mathcal{T}_x$. $\square$

Definition 6.6. A stochastic process $\{X_t\}$ defined on $(\Omega, \mathcal{F})$, with values in the measurable space $(E, \mathcal{E})$, is said to be *optional* (resp. *accessible, predictable*) if the map $X: [0, \infty[\times \Omega \to E$ is measurable when $[0, \infty[\times \Omega$ is given the optional (resp. accessible, predictable) σ-field.

Theorem 6.7. *Σ_p is generated by stochastic intervals of the forms $[\![0_A]\!] = \{0\} \times A$, $A \in \mathcal{F}_0$, and $]\!]S, T]\!]$ where S and T are arbitrary stopping times.*

PROOF. 0_A is the stopping time which is 0 on $A \in \mathcal{F}_0$ and ∞ on A^c. 0_A is trivially predictable, so $[\![0_A]\!] \in \Sigma_p$. From Lemma 6.5 above, $[\![0, T]\!] \in \Sigma_p$ and $[\![S + 1/n, \infty[\![\in \Sigma_p$ because $S + 1/n \in \mathcal{T}_p$. Therefore,

$$]\!]S, T]\!] = \left(\bigcup_n [\![S + 1/n, \infty[\![\cap [\![0, T]\!] \right) \in \Sigma_p.$$

Conversely, consider an interval $[\![S, \infty[\![$ for S predictable. We must show this is in the σ-field generated by the above intervals. However, $B = \{S = 0\} \in \mathcal{F}_0$, and if $\{S_n\}$ is a sequence which announces S on $\{S > 0\}$, $[\![S, \infty[\![= [\![0_B]\!] \cup (\bigcap_n]\!]S_n, \infty[\![)$. $\square$

Examples 6.8. We now give some examples of optional, accessible and predictable processes.

(1) Suppose $[\![S, T[\![$ is a stochastic interval and Z is an $\mathcal{F}_S$ measurable random variable. Then the process

$$X_t(\omega) = Z(\omega) I_{[\![S,T[\![}(t, \omega)$$

is optional. If S and T are accessible stopping times then $\{X_t\}$ is accessible, and if Z is $\mathcal{F}_{S-}$ measurable and S and T are predictable then $\{X_t\}$ is predictable. These statements are easily verified by first considering the case $Z = I_A$, where $A \in \mathcal{F}_S$ (resp. $\mathcal{F}_{S-}$). The process $\{X_t\}$ is then the characteristic function of the interval $[\![S_A, T_A[\![$, and the result follows from Lemma 5.11 and 5.17 and Theorem 5.25. The general case follows by approximation.

(2) In the above examples $[\![S, T[\![$ can be replaced by $[\![S, T]\!]$, where $T \in \mathcal{T}_o$ in all cases. (Because $[\![S, T]\!] = \bigcap_n [\![S, T + 1/n[\![$.)

(3) For $S \in \mathcal{T}_o$ and $T \in \mathcal{T}_o$ we know from Theorem 6.7 that $]\!]S, T]\!] \in \Sigma_p$. An approximation argument then shows that if Z is $\mathcal{F}_S$ measurable then the process $Z I_{]\!]S,T]\!]}$ is predictable.

We quote the next result without proof from Dellacherie [22], Chapter I, Theorem 3.2.

Theorem 6.9. *Suppose $(\Omega, \mathscr{F}, P)$ is a complete probability space and $\mathscr{B}([0, t[)$ is the Borel field on $[0, t[$. Write π for the projection map of $[0, t[\times\Omega$ onto Ω.*

If A is a measurable set in the product σ-field $\mathscr{B}([0, t[)\otimes\mathscr{F}$ then the projection $\pi(A)$ is in $\mathscr{F}$.

Definition 6.10. Suppose $A\subset[0, \infty[\times\Omega$. The function $D_A(\omega)=\inf\{t\in[0, \infty[: (t, \omega)\in A\}$ is called the *debut* of A. Here the convention that the infimum of the empty set is ∞ is again used. Using Theorem 6.9 above we show that the debut of a progressive set is a stopping time.

Theorem 6.11. *The debut D_A of a progressively measurable set A is a stopping time.*

PROOF. For each $u>0$, $A\cap[\![0, u]\!]$ is a measurable subset of $[0, u]\times\Omega$ given the σ-field $\mathscr{B}[(0, u])\times\mathscr{F}_u$. Now for any $t\in[0, \infty[$ the set $\{D_A<t\}$ is equal to $\pi(A_t)$, where $A_t=A\cap[\![0, t[\![$. (Note we cannot write $[\![0, t]\!]$ here.) A_t is, therefore, a measurable subset of $([0, t[\times\Omega, \mathscr{B}([0, t[)\otimes\mathscr{F}_t)$. Because each $\mathscr{F}_t$ is complete, from Theorem 6.9 above, $\{D_A<t\}$ is $\mathscr{F}_t$ measurable. For each $s>t$, $\{D_A\leq t\}\in\mathscr{F}_s$, and so $\{D_A\leq t\}\in\mathscr{F}_{t+}=\mathscr{F}_t$, because the filtration is right continuous. □

Corollary 6.12. *Suppose $\{X_t\}$, $t\in[0, \infty[$, is a progressive process with values in the measurable space $(E, \mathscr{E})$. Then for any set $B\in\mathscr{E}$ the random variable*

$$Z(\omega)=\inf\{t\in]0, \infty[: X_t(\omega)\in B\}$$

is a stopping time.

PROOF.

$$A=\{(t, \omega): X_t(\omega)\in B\}$$

is a progressively measurable set, as is $B=A\cap]\!]0, \infty[\![$. Then $Z=D_B$, and so is a stopping time. □

Definition 6.13. Z is called the *first hitting time* of B.

Definition 6.14. For $A\subset[0, \infty[\times\Omega$ the *section* of A by $\omega\in\Omega$ is the set $A(\omega)=\{t\in[0, \infty[; (t, \omega)\in A\}$.

Remarks 6.15. For $A\subset[0, \infty[\times\Omega$ the debut D_A can be defined as

$$D_A(\omega)=\inf\{t\in[0, \infty[: [0, t]\cap A(\omega) \text{ contains at least one point}\}.$$

This can be generalized as follows.

Definition 6.16. For each $n \in \{1, 2, 3, \ldots\}$ the *n-debut* of A, D_A^n, is the random variable

$$D_A^n(\omega) = \inf\{t \in [0, \infty[: [0, t] \cap A(\omega) \text{ contains at least } n \text{ points}\}.$$

Corollary 6.17. *Suppose A is progressively measurable. Then for each $n \in \{1, 2, 3, \ldots\}$ D_A^n is a stopping time.*

PROOF. The proof is by induction. From Theorem 6.11, $D_A^1 = D_A$ is a stopping time. Suppose D_A^n is a stopping time. Then D_A^{n+1} is the debut of the progressive set $A \cap]\!]D_A^n, \infty[\![$ and so D_A^{n+1} is a stopping time. □

The monotone class Theorem 1.22 can be used to give an easy and alternative proof that the debut of an optional set is a stopping time.

Lemma 6.18. *Suppose $A \in \Sigma_o$. The debut D_A is then a stopping time.*

PROOF. We shall apply the monotone class theorem. Let $\mathcal{N}$ be the family of all finite unions, intersections and complements of stochastic intervals of the form $[\![T, \infty[\![$, for $T \in \mathcal{T}_o$. Let $\mathcal{M}$ be the family of all subsets of $[0, \infty[\times \Omega$ whose debut is a stopping time. If $\{A_n\}$ is a monotonic increasing sequence in $\mathcal{M}$ and $A = \bigcup_n A_n$, then $D_A = \bigwedge_n D_{A_n}$, and by Lemma 2.7 this is a stopping time. The proof for monotonic decreasing sequences is similar. Therefore, $\mathcal{M}$ contains the σ-field Σ_o generated by $\mathcal{N}$. □

Notation 6.19. Write $\mathcal{B}_x$ for the Boolean algebra generated by stochastic intervals of the form $[\![S, T[\![$, where $S, T \in \mathcal{T}_x$ $(x = o, a, p)$.

Lemma 6.20. *If $B \in \mathcal{B}_x$ then B is a finite union of the form:*

$$B = [\![S_1, T_1[\![\cup \cdots \cup [\![S_m, T_m[\![, \qquad S_i, T_i \in \mathcal{T}_x, \quad i = 1, \ldots, m.$$

PROOF. $\mathcal{B}_x$ is the family of all finite unions, finite intersections and complements, of stochastic intervals

$$[\![S, T[\![, \qquad S, T \in \mathcal{T}_x.$$

However,

$$[\![S, T[\![^c = [\![0, S[\![\cup [\![T, \infty[\![$$

and

$$[\![S, T[\![\cap [\![U, V[\![= [\![S \vee U, (S \vee U) \vee (T \wedge V)[\![. \qquad \square$$

Lemma 6.21. *Suppose $\{A_n\}$, $n \in N$, is a sequence of elements of $\mathcal{B}_x$, $x = o$, a, p.* Write $A = \bigcap_n A_n$. Then $[\![D_A]\!] \subset A$ and $D_A \in \mathcal{T}_x$.

PROOF. For an interval of the form $C = [\![S, T[\![$, where $S, T \in \mathcal{T}_x$, $D_C = S_{\{S<T\}}$, so using Remarks 5.27 $[\![D_C]\!] \subset [\![S]\!] \subset C$ and (by Lemma 5.17 when $x = a$ and by Theorem 5.25 when $x = p$), $D_C \in \mathcal{T}_x$. If $B = \mathcal{B}_x$, from Lemma

6.20 above $B = C_1 \cup \cdots \cup C_m$, where $C_i = [\![S_i, T_i[\![$, and $D_B = D_{C_1} \wedge \cdots \wedge D_{C_m}$. Therefore $D_B \in \mathcal{T}_x$ and $[\![D_B]\!] \subset B$.

Consider now $A = \bigcap A_n$, $A_n \in \mathcal{B}_x$ and write $\mathcal{D}$ for the set of stopping times in $\mathcal{T}_x$ which are less than D_A. The zero stopping time is in $\mathcal{D}$, so $\mathcal{D}$ is nonempty, and the supremum of a countable family of members of $\mathcal{D}$ is in $\mathcal{D}$ (by Theorem 5.19 when $x = a$ or p). Therefore, there is an increasing sequence of elements $\{T_n\}$ of $\mathcal{D}$ such that $T = \bigvee_n T_n$ is the essential supremum of $\mathcal{D}$ and $T \in \mathcal{T}_x$. Write $B_n = A_n \cap [\![T, \infty[\![$ and $S_n = D_{B_n}$. Now $B_n \in \mathcal{B}_x$ so $S_n \in \mathcal{T}_x$. Because $T \leq D_A$ we see that $A_n \supset B_n \supset A$ and so $\bigcap_n B_n = A$. However, $D_A \geq S_n \geq T$ and $S_n \in \mathcal{D}$, so, because T is the essential supremum of $\mathcal{D}$, $T = S_n$ a.s. Therefore, $[\![T]\!] = [\![S_n]\!] \subset B_n$ for all $n \in N$. That is, $[\![T]\!] \subset \bigcap B_n = A$. Consequently $T \geq D_A$, and so $T = D_A$. □

The following "section theorem" is quoted without proof from Dellacherie [22], Chapter 1, Theorem 37.

Theorem 6.22. *Suppose $(\Omega, \mathcal{F}, P)$ is a probability space and that B is an element of the product σ-field $\mathcal{B} \otimes \mathcal{F}$ on $[0, \infty[\times \Omega$. Then there is an $\mathcal{F}$-measurable random variable T, with values in $[0, \infty]$, such that*

(1) if $T(\omega) < \infty$ then $(T(\omega), \omega) \in B$, and
(2) $P(\{T < \infty\}) = P(\{D_B < \infty\})$.

Remarks 6.23. Although T is not, in general, a stopping time (1) states that the graph $[\![T]\!] = \{(t, \omega) \in [0, \infty[\times \Omega \colon T(\omega) = t\}$ is contained in B.

Write π for the projection map of $[0, \infty[\times \Omega$ onto Ω. Then $\pi(B) = \{D_B < \infty\}$ and (2) states that $\{T < \infty\} = \pi(B)$ a.s. Therefore, up to a P-null set, T is a section of B.

This result is extended to the Σ_x σ-fields as follows:

Theorem 6.24 (Section Theorem). *Suppose $(\Omega, \mathcal{F}, P)$ is a probability space with a filtration $\{\mathcal{F}_t\}$, $t \in [0, \infty[$, and suppose $A \in \Sigma_x$. For any $\varepsilon > 0$ there is a stopping time $S \in T_x$ such that*

(1) $[\![S]\!] \subset A$, and
(2) $P(\{S < \infty\}) \geq P(\pi(A)) - \varepsilon$.

Proof. From the section Theorem 6.22 above there is an $\mathcal{F}$ measurable random variable T such that

(1) if $T(\omega) < \infty$ then $(T(\omega), \omega) \in A$, and
(2) $P(\{T < \infty\}) = P(\pi(A))$.

Define a measure μ on the product σ-field $\mathcal{B} \otimes \mathcal{F}$ by putting

$$\mu(E) = P(\{\omega \colon (T(\omega), \omega) \in E\} \cap \{T < \infty\}) \quad \text{for } E \in \mathcal{B} \otimes \mathcal{F}.$$

This measure has support A and $\mu(A) = P(\pi(A))$.

Now the Boolean algebra $\mathscr{B}_x$ $(x = o, a, p)$ generates Σ_x so from a standard result in measure theory, for any $\varepsilon > 0$ there is a decreasing sequence $\{B_n\}$, $n \in N$, of elements $B_n \in \mathscr{B}_x$ such that, $B = \bigcap_n B_n \subset A$ and $\mu(B) \geq \mu(A) - \varepsilon$. Taking $S = D_B$ we have, from Lemma 6.21 that $[\![S]\!] \subset B \subset A$ and $S \in \mathscr{T}_x$. Because $\mu(B) \leq P(\pi(A \cap B)) \leq P(\pi(B)) = P(\{S < \infty\})$ the result follows. □

The following result provides a check for the indistinguishability of stochastic processes.

Corollary 6.25. *Suppose $\{X_t\}, \{Y_t\}$, $t \in [0, \infty[$, are two Σ_x measurable processes. Then $\{X_t\}$ and $\{Y_t\}$ are indistinguishable if and only if $X_T = Y_T$ a.s. for every $T \in \mathscr{T}_x$ $(x = o, a, p)$.*

Proof. The set $A = \{(t, \omega): X_t(\omega) \neq Y_t(\omega)\}$ is in Σ_x. If $P(\pi(A)) \neq 0$ there is for any $\varepsilon > 0$ an $S \in \mathscr{T}_x$ such that $[\![S]\!] \subset A$ and $P(\{S < \infty\}) \geq P(\pi(A)) - \varepsilon$. Therefore, there is a $t \in [0, \infty[$ such that $X_{S \wedge t}$ differs from $Y_{S \wedge t}$ on a set of positive measure. □

The section theorem also provides the following characterization of stopping times.

Corollary 6.26. *Suppose T is a random variable with values in $[0, \infty]$. Then $T \in \mathscr{T}_x$ $(x = o, a, p)$ if and only if $[\![T]\!] = \{(t, \omega) \in [0, \infty[\times \Omega: T(\omega) = t\} \in \Sigma_x$.*

Proof. The necessity is immediate, because if $T \in \mathscr{T}_x$ then $[\![T]\!] = \bigcap_n [\![T, T + 1/n[\![\in \Sigma_x$. If $[\![T]\!] \in \Sigma_o$ we know from Theorem 6.11 that $D_{[\![T]\!]} = T \in \mathscr{T}_o$. If $[\![T]\!]$ is an accessible (resp. predictable), set, then for each $n \in N$ there is by Theorem 6.24 a stopping time $T_n \in \mathscr{T}_a$ (resp. $\mathscr{T}_p$) such that

(1) $[\![T_n]\!] \subset [\![T]\!]$,
(2) $P(\{T_n < \infty\}) \geq P(\{T < \infty\}) - 2^{-n}$.

Replacing T_n by $T_1 \wedge T_2 \wedge \cdots \wedge T_n$ we can suppose the sequence $\{T_n\}$ is decreasing. Then $T = \lim T_n$ a.s., and for each $\omega \in \Omega$ there is an $n(\omega)$ such that $T_{n(\omega)}(\omega) = T(\omega)$. Therefore, by Theorem 5.20 T is accessible (resp. predictable). □

Theorem 6.27. *Suppose $A \in \Sigma_x$ $(x = o, a, p)$ and $A \subset \bigcup_n [\![S_n]\!]$, where $\{S_n\}$, $n \in N$, is a sequence of stopping times. Then $A = \bigcup_n [\![T_n]\!]$ where $\{T_n\}$, $n \in N$, is a sequence of stopping times, $T_n \in \mathscr{T}_x$ for each n, and $[\![T_n]\!] \cap [\![T_m]\!] = \varnothing$ if $n \neq m$.*

Proof. Write T_n for the stopping time whose graph is $(A - \bigcup_{k<n} [\![S_k]\!]) \cap [\![S_n]\!]$. Then the T_n have disjoint graphs and $A = \bigcup_n [\![T_n]\!]$. This proves the result in the optional case.

Suppose $A \in \Sigma_a$. Then as above $A = \bigcup_n [\![T_n]\!]$ where $T_n \in \mathscr{T}_o$. Write U_n (resp. V_n) for the accessible (resp. totally inaccessible) part of T_n. Then, from Theorem 5.16, $[\![T_n]\!]=[\![U_n]\!]\cup[\![V_n]\!]$, so

$$A-\bigcup_n [\![U_n]\!]=\bigcup_n [\![V_n]\!]\in\Sigma_a.$$

Writing $B=\bigcup_n [\![V_n]\!]$ we have that, if $P(B)>0$, for any $\varepsilon>0$ there is an accessible stopping time W such that $[\![W]\!]\subset B$, $P(\pi(B))\leq P(\{W<\infty\})+\varepsilon$. Because W is accessible $[\![W]\!]\subset\bigcup_m [\![W_m]\!]$, where the W_m are predictable. However, $[\![V_n]\!]\cap[\![W_m]\!]=\emptyset$ for all $n, m\in N$. So $B\cap[\![W_m]\!]=\emptyset$ and therefore, $B\cap[\![W]\!]\subset B\cap(\bigcup_m [\![W_m]\!])=\emptyset$. Therefore, if $P(B)>0$ we would have a contradiction. Consequently, B must be evanescent, and so $A=\bigcup_n [\![T_n]\!]$, where $T_n=U_n$ a.s., and $U_n\in\mathscr{T}_a$.

If $A\in\Sigma_p$ then A is certainly in Σ_a so we can apply the above result to say

$$A=\bigcup_n [\![U_n]\!], \qquad U_n\in\mathscr{T}_a.$$

However, because U_n is accessible its graph is contained in the union of the graphs of a countable number of predictable stopping times. Therefore, $A\subset\bigcup_n [\![R_n]\!]$, where $\{R_n\}$ is a sequence of predictable stopping times. Similarly to the optional case, write T_n for the stopping time whose graph is $(A-\bigcup_{k<n} [\![R_k]\!])\cap[\![R_n]\!]$. Then from Corollary 6.26, T_n is predictable and $A=\bigcup_n [\![T_n]\!]$. Clearly $[\![T_n]\!]\cap[\![T_m]\!]=\emptyset$ if $n\neq m$. □

We now give a function space version of the monotone class theorem. Rather than deriving it as a corollary of Theorem 1.22 it is easier to prove the special case we need.

Theorem 6.28. *Suppose $\mathscr{H}$ is a vector space of real valued bounded functions defined on Ω, containing constants closed under uniform convergence and having the following property:*

Montone Property. *For every increasing sequence $\{f_n\}$ of positive functions the limit $f=\lim_n f_n$ is in $\mathscr{H}$ when f is bounded.*

Suppose $\mathscr{C}$ is a uniformly bounded subset of $\mathscr{H}$ which is closed under multiplication and write $\sigma(\mathscr{C})$ for the σ-field generated by $\mathscr{C}$. Then $\mathscr{H}$ contains all bounded $\sigma(\mathscr{C})$ measurable functions.

PROOF. Without loss of generality we can assume that the functions in $\mathscr{C}$ take their values in $[-1,+1]$. Write Λ for the set of real valued functions $\{\phi\}$ on $[-1,+1]^N$ (that is the countable Cartesian product of $[-1,+1]$), such that $\phi(f_1,\ldots,f_n,\ldots)\in\mathscr{H}$ for every sequence $\{f_n\}$ of elements of $\mathscr{C}$. Then Λ contains the constant functions and the polynomial functions, and Λ is closed under uniform convergence. Consequently, by the Stone–Weierstrass theorem, Λ contains all the continuous functions on $[-1,+1]^N$. Also, because Λ is closed under taking limits of monotone uniformly bounded sequences (by the monotone property), Λ contains all the Borel functions on $[-1,+1]^N$. Therefore, $\mathscr{H}$ contains all bounded $\sigma(\mathscr{C})$ measurable functions. □

Using this result we now show that an optional process differs from a predictable process on quite a "small" set.

Theorem 6.29. *Suppose $\{X_t\}$ is an optional process. Then there is a predictable process $\{Y_t\}$ such that $\{X \neq Y\}$ is contained in the union of the graphs of a sequence of stopping times.*

PROOF. Write $\mathscr{H}$ for the set of optional processes $\{X_t\}$ for each of which there is a predictable process $\{Y_t\}$ such that $\{X \neq Y\}$ is contained in the union of the graphs of a sequence of stopping times. Then $\mathscr{H}$ is a vector space, and $\mathscr{H}$ contains constants. Furthermore, $\mathscr{H}$ is closed under pointwise convergence, because suppose $\{X^n\}$ is a convergent sequence of optional processes in $\mathscr{H}$. For each $\{X_t^n\}$ there is a predictable process $\{Y_t^n\}$ such that $\{X^n \neq Y^n\}$ is contained in the union of the graphs of a sequence of stopping times. Write $X = \lim_n X^n$ and $Y = \lim_n \sup Y^n)I_{\{\lim \sup|Y^n|<\infty\}}$. Then $\{X \neq Y\}$ is contained in the union of the graphs of a sequence of stopping times. Therefore, in particular, $\mathscr{H}$ is closed under uniform convergence and monotone convergence.

Now for any stopping times $S, T \in \mathscr{T}_o$ the process $I_{]\!]S,T]\!]}$ is in $\mathscr{H}$, because from Theorem 6.7, $]\!]S, T]\!]$ is predictable. Consequently from Theorem 6.28 above, $\mathscr{H}$ contains all optional processes. □

The following result gives a characterization of predictable processes.

Theorem 6.30. *An accessible process $\{X_t\}$ is predictable if and only if for every $T \in \mathscr{T}_p$ the random variable $X_t I_{\{T<\infty\}}$ is $\mathscr{F}_{T-}$ measurable.*

Furthermore, if $\{X_t\}$ is predictable then $X_T I_{\{T<\infty\}}$ is $\mathscr{F}_{T-}$ measurable for every stopping time $T \in \mathscr{T}_o$.

PROOF. *Necessity*. Suppose $\mathscr{H}$ is the set of predictable processes such that $X_T I_{\{T<\infty\}}$ is $\mathscr{F}_{T-}$ measurable for every $T \in \mathscr{T}_o$. Then clearly $\mathscr{H}$ satisfies the hypotheses of Theorem 6.28, that is, $\mathscr{H}$ is a monotone class. Consequently, from Theorem 6.7 it is enough to prove the result when $\{X_t\}$ is the indicator function of a stochastic interval of the form $[\![0_A]\!]$, $A \in \mathscr{F}_o$, or $]\!]U, V]\!]$, $U, V \in \mathscr{T}_o$.

Suppose $T \in \mathscr{T}_o$. When $X = I_{[\![0_A]\!]}$, $I_T I_{\{T<\infty\}} = I_{A \cap \{T=0\}}$ which, from Definition 5.12, is $\mathscr{F}_{T-}$ measurable. When $X = I_{]\!]U,V]\!]}$

$$X_T I_{\{T<\infty\}} = (I_{\{U<T\}} - I_{\{V<T\}}) I_{\{T<\infty\}}$$

and by Theorem 5.13(4) this is $\mathscr{F}_{T-}$ measurable.

Sufficiency. The sufficiency is established when $\{X_t\}$ is accessible. From Theorem 6.29 there is a predictable process $\{Y_t\}$ such that $A = \{X \neq Y\}$ is contained in the union of graphs of a sequence of stopping times. However, $A \in \Sigma_a$ so as in Theorem 6.27 there is a sequence of predictable stopping times $\{S_n\}$ with disjoint graphs such that $A \subset \bigcup_n [\![S_n]\!]$. By hypothesis $Z_n =$

$X_{S_n}.I_{\{S_n<\infty\}}$ is $\mathscr{F}_{S_n-}$ measurable, so as in Examples 6.8, $Z_n.I_{[\![S_n]\!]}=X_{S_n}.I_{[\![S_n]\!]}$ is a predictable process. Write $A=\bigcap_n [\![S_n]\!]^c$; then $I_A.Y$ is predictable. Finally

$$X=I_A.Y+\sum_n X_{S_n}.I_{[\![S_n]\!]},$$

so X is predictable. □

Corollary 6.31. *Suppose $\{X_t\}$ and $\{X_t\}$ are two optional (resp. accessible, predictable) processes. Then the following two properties are equivalent:*

(1) *$X_t(\omega)\geq Y_t(\omega)$, except on an evanescent set,*
(2) *if for any $T\in\mathscr{T}_x, x=o$ (resp. $x=a, x=p$), the random variables $X_TI_{\{T<\infty\}}$ and $Y_TI_{\{T<\infty\}}$ are integrable, then*

$$E[X_TI_{\{T<\infty\}}]\geq E[Y_TI_{\{T<\infty\}}].$$

PROOF. That (1) implies (2) is obvious. Suppose that (2) holds and that $\{X<Y\}$ is not evanescent. Then by Theorem 6.24 there is a $T\in\mathscr{T}_x$ $(x=o,a,p)$ such that $P\{T<\infty\}>0$ and $X_TI_{\{T<\infty\}}<Y_TI_{\{T<\infty\}}$. There is certainly a constant $\alpha>0$ such that the measure of the set

$$B=\{T<\infty\}\cap\{|X_T|\leq\alpha\}\cap\{|Y_T|\leq\alpha\}$$

is nonzero. $B\in\mathscr{F}_T$, and in the predictable case $B\in\mathscr{F}_{T-}$ by Theorem 6.30 above. Therefore, $T_B\in\mathscr{T}_x$ $(x=o,a,p)$ and

$$E[X_{T_B}I_{\{T_B<\infty\}}]<E[Y_{T_B}I_{\{T_B<\infty\}}],$$

contradicting (2). □

Finally we describe processes that generate Σ_p and Σ_o.

Theorem 6.32. *The predictable σ-field Σ_p is generated by the family of left continuous adapted stochastic processes.*

PROOF. From Theorem 6.7 we see that Σ_p is generated by all processes of the form $I_{[\![0_A]\!]}, A\in\mathscr{F}_0$ and $I_{]\!]S,T]\!]}, S,T\in\mathscr{T}_o, S\leq T$. These are left continuous and adapted, so it remains to be shown that every left continuous adapted process X is predictable. Write

$$X^n=X_0I_{[0]}+\sum_{k\geq 1}X_{k/n}I_{]\!]k/n,(k+1)/n]\!]}.$$

The process X^n is predictable because it is countable sum of elementary predictable processes, as in Examples 6.8. Because X is left continuous

$$\lim_n X^n(t,\omega)=X(t,\omega)\quad\text{for all }(t,\omega)$$

not in some evanescent set. Therefore, X is predictable. □

Corollary 6.33. *Σ_p is generated by stochastic intervals of the form $[\![0_A]\!]$, $A\in\mathscr{F}_o$, and $[\![s_B,t_B]\!]$, $B\in\mathscr{F}_{s-}$, where s and t are constants.*

PROOF. These intervals are predictable. Conversely, because the random variables $X_{k/n}$ in the above construction are $\mathscr{F}_{k/n-}$ measurable, any left continuous adapted process is the limit of linear combinations of characteristic functions of such intervals. □

Corollary 6.34. *Σ_p is generated by the family of all continuous adapted processes.*

PROOF. An adapted continuous processes is certainly left continuous and so predictable by Theorem 6.32 above. Conversely, $I_{[\![0_A,\infty[\![}$ is continuous and adapted if $A \in \mathscr{F}_o$. An interval $]\!]S, \infty[\![$ is equal to the set $\{X > 0\}$, where X is the continuous adapted process $X_t(\omega) = t - S(\omega) \wedge t$. □

Theorem 6.35. *Σ_o is generated by the family of all adapted processes which are continuous on the right and have limits on the left, that is, all adapted corlol processes.*

PROOF. Σ_o is generated by all corlol adapted processes of the form $I_{[\![S,T[\![}$, $S, T \in \mathscr{T}_o$, $S \leq T$. Consequently, it remains to be shown that every adapted corlol processes is optional.

For such a process $\{X_t\}$ and each integer $k > 0$ define an increasing of stopping times as follows:

$$T_1^k(\omega) = 0 \quad \text{for all } \omega \text{ and all } k,$$

$$T_{n+1}^k(\omega) = \inf\{t > T_n^k(\omega) \colon |X_t(\omega) - X_{T_n^k}(\omega)| \geq 1/k\}.$$

(If this set is empty, or if $T_n^k(\omega) = \infty$, then $T_{n+1}^k(\omega) = \infty$.) Now $\{X_t\}$ is certainly a progressive process, as is $X_{T_n^k} . I_{[\![T_n^k,\infty[\![}$. Therefore, the set

$$B_{n+1}^k =]\!]T_n^k, \infty[\![\cap \{|X_t(\omega) - X_{T_n^k} I_{[\![T_n^k,\infty[\![}| \geq 1/k\}$$

is progressive, and because T_{n+1}^k is the debut of B_{n+1}^k, T_{n+1}^k is a stopping time by Theorem 6.11. Because X is right continuous we have $|X_{T_{n+1}^k} - X_{T_n^k}| \geq 1/k$ on $\{T_{n+1}^k < \infty\}$, and the existence of left-hand limits implies that T_n^k converges to infinity for each integer k. For $k > 0$ write

$$X^k = \sum_{n=1}^{\infty} X_{T_n^k} I_{[\![T_n^k, T_{n+1}^k[\![}.$$

Then X^k is optional because it is a countable sum of elementary optional processes, as in Examples 6.8. Letting k tend to infinity, by right continuity $X_t^k(\omega)$ approaches $X_t(\omega)$, except possibly on an evanescent set. Therefore X is optional. □

Remarks 6.36. It can be shown that an arbitrary right continuous adapted process is optional, but this result will not be needed in the sequel. See Dellacherie [22] Chapter IV, Theorem 27.

Definition 6.37. Suppose $\{X_t\}$, $t \in [0, \infty[$, is a real valued $(\mathscr{B} \otimes \mathscr{F})$ measurable stochastic process. A process $\{X_t^*\}$ is defined by

$$X_t^*(\omega) = \sup_{s \leq t} |X_t(\omega)|.$$

Then $X_\infty^*(\omega) = \sup_t |X_t(\omega)|$ exists, and the process $\{X_t\}$ is said to be *bounded* if $X_\infty^* \in L^\infty(\Omega, \mathscr{F}, P)$. If $\{\mathscr{F}_t\}$ is a filtration on $(\Omega, \mathscr{F}, P)$ then X_t is said to be *locally bounded* if there is an increasing sequence T_n of stopping times such that $\lim_n T_n = \infty$ a.s. and $X_{T_n}^*(\omega) \leq n$ a.s. for each n.

Notation 6.38. Write $B(\mathscr{B} \otimes \mathscr{F})$ for the space of bounded measurable processes, and $B(\Sigma_x)$ $(x = o, a, p)$, for the space of bounded Σ_x measurable processes.

The following projection theorem shows there is a projection map very similar to the conditional expectation operation.

Theorem 6.39 (Projection Theorem). *There is a unique linear order preserving projection* Π_x $(x = o, a, p)$, *of* $B(\mathscr{B} \times \mathscr{F})$ *onto* $B(\Sigma_x)$, *such that for* $X \in B(\mathscr{B} \times \mathscr{F})$ *for every* $T \in \mathscr{T}_x$

$$E[X_T I_{\{T<\infty\}}] = E[(\Pi_x X)_T I_{\{T<\infty\}}]. \tag{6.1}$$

PROOF. First note that if $X, Y \in B(\mathscr{B} \otimes \mathscr{F})$, if $\Pi_x X$ and $\Pi_x Y$ both exist in $B(\Sigma_x)$ and if $X \leq Y$ (except possibly on an evanescent set), then $\Pi_x X \leq \Pi_x Y$. This is because $A = \{\Pi_x X > \Pi_x Y\} \in \Sigma_x$ and if $P(\pi(A)) \neq 0$, by the Section Theorem 6.24, there would be a stopping time $T \in \mathscr{T}_x$ such that $E[(\Pi_x X - \Pi_x Y)_T I_{\{T<\infty\}}] > 0$. So by (6.1) $E[(X - Y)_T I_{\{T<\infty\}}] > 0$, a contradiction. Also, if $\Pi_x X$ and $\tilde{\Pi}_x X$ are two projections of the process $X \in B(\mathscr{B} \times \mathscr{F})$, both satisfying the above conditions, then for every $T \in \mathscr{T}_x$:

$$E[(\Pi_x X)_T I_{\{T<\infty\}}] = E[X_T I_{\{T<\infty\}}] = E[(\tilde{\Pi}_x X)_T I_{\{T<\infty\}}].$$

Therefore, by Corollary 6.31, $\Pi_x X = \tilde{\Pi}_x X$ and so the projection is unique.

Write $\mathscr{H}_x$ for the set of all processes in $B(\mathscr{B} \otimes \mathscr{F})$ which have a projection onto $B(\Sigma_x)$ satisfying (6.1). Then $\mathscr{H}_x$ is a vector space, contains constants and is closed under uniform convergence. By the Lebesgue theorem $\mathscr{H}_x$ is closed under limits of increasing sequences, so $\mathscr{H}_x$ is a monotone class. Consequently, by the monotone class theorem, we need only to prove the existence of Π_x for the elements of a uniformly bounded family of measurable processes which is closed under multiplication and which generates the σ-field Σ_x. The three σ-fields will be treated separately:

(1) *Optional Case* Σ_o. Suppose $X = Z I_{[\![r,s]\!]}$, where Z is $\mathscr{F}$ measurable and bounded, and r, s are real numbers such that $r \leq s$. Write $\{Y_t\}$ for a right continuous modification of the martingale $E[Z|\mathscr{F}_t]$. Then, by Theorem 4.3, Y has left-hand limits, so by Theorem 6.35, Y is optional. Define

$(\Pi_o X)_t = Y_t I_{[\![r,s]\!]}$. Then for $T \in \mathcal{T}_o$

$$E[X_T I_{\{T<\infty\}}] = E[Z I_{r \le T \le s}],$$

and

$$\begin{aligned} E[(\Pi_o X)_T I_{\{T<\infty\}}] &= E[Y_T I_{r \le T \le s} I_{\{T<\infty\}}] \\ &= E[E[Z|\mathcal{F}_T] I_{r \le T \le s}] \quad \text{by Theorem 4.12} \\ &= E[E[Z I_{r \ge T \le s}|\mathcal{F}_T]]. \end{aligned}$$

Consequently, $\Pi_o X$ is an optional process satisfying (6.1). Processes of the form $Z I_{[\![r,s]\!]}$ generate $\mathcal{B} \otimes \mathcal{F}$, so the result follows by the Monotone Class Theorem 6.28.

(2) *Predictable Case* Σ_p. As above, suppose $X = Z I_{[\![r,s]\!]}$ and write $\{Y_t\}$ for the corlol version of the martingale $E[Z|\mathcal{F}_t]$. Then the process $\{W_t\}$, defined by $W_t = Y_{t-}$, is left continuous, and so predictable by Theorem 6.32. Suppose $T \in \mathcal{T}_p$. Then T is announced by an increasing sequence of stopping times $\{T_n\}$ and from Corollary 5.24 $\mathcal{F}_{T-} = \vee_n \mathcal{F}_{T_n}$. Therefore, applying Corollary 3.19 to the uniformly integrable martingale $\{V_n\}$, where $V_n = Y_{T_n} = E[Z|\mathcal{F}_{T_n}]$ we have that $Y_{T-} = \lim_n Y_{T_n} = E[Z|\mathcal{F}_{T-}]$. That is, $W_T = E[Z|\mathcal{F}_{T-}]$ for $T \in \mathcal{T}_p$.

Define $(\Pi_p X)_t = W_t I_{[\![r,s]\!]}$. Then $(\Pi_p X)$ is predictable, and for $T \in \mathcal{T}_p$ announced by the $\{T_n\}$

$$E[X_T I_{\{T<\infty\}}] = E[Z . I_{r \le T \le s}] \quad \text{as above,}$$

and

$$\begin{aligned} E[(\Pi_p X)_T . I_{\{T<\infty\}}] &= E[W_T . I_{r \le T \le s}] \\ &= E[Y_{T-} . I_{r \le T \le s}] = E[E[Z|\mathcal{F}_{T-}] I_{r \le T \le s}] \\ &= E[E[Z . I_{r \le T \le s}|\mathcal{F}_{T-}]], \end{aligned}$$

because $I_{r \le T \le s}$ is $\mathcal{F}_{T-}$ measurable.

(3) *Accessible Case* Σ_a. If Π_a exists then by (6.1) it must satisfy $(\Pi_a X) = \Pi_a(\Pi_o X)$. Consequently we need only establish the existence of Π_a for optional processes. By the monotone class theorem, therefore, we need only consider the processes of the form $X = I_{[\![S,\infty[\![}$, $S \in \mathcal{T}_o$. Suppose S_A (resp. S_B) is the accessible (resp. totally inaccessible) part of S, as in Theorem 5.16. Write $C = [\![S_A, \infty[\![\cup]\!]S_B, \infty[\![$. Then C is the union of an accessible set and a predictable set, so $C \in \Sigma_a$. Furthermore, because $[\![S_B]\!] \cap [\![T]\!]$ is evanescent for every $T \in \mathcal{T}_a$, we see that $C \cap [\![T]\!] = [\![S, \infty[\![\cap [\![T]\!]$. Therefore, $\Pi_a X = I_C$ is an accessible processes satisfying (6.1). □

Definition 6.40. $\Pi_x X$, $x = o$ (resp. a, p), is called the *optional* (resp. *accessible, predictable*) *projection of* X.

Corollary 6.41. *If* $X \in \mathscr{B}(\Sigma_x)$ *then, by uniqueness,* $\Pi_x X = X$ $(x = o, a, p)$.

Remarks 6.42. By the ordering property established at the beginning of the above proof, if $\{X_t^n\}$, $n \in N$, is an increasing sequence of processes, that is $X_t^{n+1}(\omega) \geq X_t^n(\omega)$ except on an evanescent set, then $\{(\Pi_x X)_t^n\}$ is an increasing sequence. Therefore, for a general positive measurable process X, $\Pi_x X$ can be defined as the limit of the increasing sequence of processes $\{\Pi_x(X \wedge n)\}$.

The following two results show that Π_x behaves like a conditional expectation.

Theorem 6.43. *Suppose* X *is a bounded or positive* $\mathscr{B} \otimes \mathscr{F}$ *measurable process, so that* $\Pi_x X$ *exists. Then*

$$\text{for } T \in \mathscr{T}_o\text{:} \quad (\Pi_o X)_T I_{\{T<\infty\}} = E[X_T I_{\{T<\infty\}} | \mathscr{F}_T],$$

$$\text{for } T \in \mathscr{T}_a\text{:} \quad (\Pi_a X)_T I_{\{T<\infty\}} = E[X_T I_{\{T<\infty\}} | \mathscr{F}_T],$$

$$\text{for } T \in \mathscr{T}_p\text{:} \quad (\Pi_p X)_T I_{\{T<\infty\}} = E[X_T I_{\{T<\infty\}} | \mathscr{F}_{T-}].$$

PROOF. Suppose $H \in \mathscr{F}_T$ in the optional and accessible cases, and $H \in \mathscr{F}_{T-}$ in the predictable case. If $T \in \mathscr{T}_x$ then $T_H \in \mathscr{T}_x$. From the projection Theorem 6.39

$$E[X_{T_H} I_{\{T_H<\infty\}}] = E[(\Pi_x X)_{T_H} I_{\{T_H<\infty\}}],$$

so

$$E[X_T I_{\{T<\infty\}} I_H] = E[(\Pi_x X)_T I_{\{T<\infty\}} I_H].$$

Because $(\Pi_x X)_T I_{\{T<\infty\}}$ is $\mathscr{F}_T$ measurable for $x = o$ and a, and $\mathscr{F}_{T-}$ measurable for $x = p$, the result follows. □

Corollary 6.44. *Suppose* $\{X_t\}$ *and* $\{Y_t\}$ *are bounded measurable processes. If* $Y \in B(\Sigma_x \Gamma$ *then* $\Pi_x(XY) = (\Pi_x X)Y$.

PROOF. For $T \in \mathscr{T}_x$ $(x = o, a)$, $(\Pi_x X)_T I_{\{T<\infty\}}$ is $\mathscr{F}_T$ measurable by Theorem 2.31, because an optional process is certainly progressive. For $T \in \mathscr{T}_p$, $(\Pi_x X)_T I_{\{T<\infty\}}$ is $\mathscr{F}_{T-}$ measurable by Theorem 6.30. The result follows from the theorem by uniqueness. □

Definition 6.45. Suppose $\{X_t\}$ is an adapted corlol process. Then $\{X_t\}$ *charges* a stopping time $T \in \mathscr{T}_o$ if $P(\{X_T \neq X_{T-}\} \cap \{T < \infty\}) > 0$, and $\{X_t\}$ has a *jump* at $T \in \mathscr{T}_o$ if $X_T \neq X_{T-}$ a.s. on $\{T < \infty\}$.

The sequence $\{T_n\}$ *exhausts the jumps* of $\{X_t\}$ if $\{X_t\}$ has a jump at T_n for each n, the graphs of the T_n are disjoint, and $\{X_t\}$ does not charge any other stopping time.

Using this definition we have the following description of the jumps of a corlol process.

Theorem 6.46. *Suppose $\{X_t\}$, $t \in [0, \infty[$, is an adapted corlol process. Then there is a sequence $\{T_n\}$ of stopping times which exhaust the jumps of $\{X_t\}$. If $\{X_t\}$ is accessible (resp. predictable) the T_n are accessible (resp. predictable).*

PROOF. Write $\{Y_t\}$ for the process defined by $Y_t = X_{t-}$. Then Y_t is left continuous, and so predictable. Consider the set $A = \{X \neq Y\}$. If we can show that A is contained in the union of graphs of a sequence of stopping times the result will follow from Theorem 6.27. For every positive integer k put $A_k = \{(t, \omega): |X_t(\omega) - Y_t(\omega)| > 1/k\}$.

$A_k \in \Sigma_o$ and $A = \bigcup_k A_k$. Because $\{X_t\}$ is corlol it has no oscillatory discontinuities. Therefore, for almost every $\omega \in \Omega$, the section $A_k(\omega)$ has no finite accumulation point for every $k > 0$. A_k is consequently the union of the graphs of its n-debuts, as n runs through the positive integers, and so A is contained in a countable union of graphs of stopping times. □

Corollary 6.47. *Suppose $\{X_t\}$, $t \in [0, \infty[$, is an adapted corlol process. Then $\{X_t\}$ is accessible if and only if it does not charge any totally inaccessible stopping time.*

PROOF. If $\{X_t\}$ is accessible there is a sequence of accessible stopping times which exhaust the jumps of $\{X_t\}$.

Conversely, if $\{X_t\}$ does not charge any totally inaccessible stopping time the T_n constructed in the theorem must all be accessible. Write $A = \bigcap_n [\![T_n]\!]^c$, so $A \in \Sigma_a$, and $Y_t = X_{t-}$, so that $\{Y_t\}$ is a predictable process. Then $I_A . Y$ is an accessible process, and for each n, $X_{T_n} . I_{[\![T_n]\!]}$ is accessible. However,

$$X = I_A . Y + \sum_n X_{T_n} . I_{[\![T_n]\!]}$$

so X is accessible. □

Finally we show the projections differ on only a small set.

Theorem 6.48. *Suppose X is a bounded measurable process. Then the set $\{\Pi_x X \neq \Pi_y X\}$, $x \neq y$, is a countable union of graphs of stopping times. Furthermore, for $x = 0$ and $y = a$ the stopping times are totally inaccessible.*

PROOF. From Theorem 6.27 we need only show that the set is contained in such a countable union. Further, from the monotone class theorem we need only prove the result for the kind of processes for which the Π_x projections were originally defined.

Consider, therefore, the process $X = Z . I_{[\![r,s]\!]}$, where Z is a bounded $\mathscr{F}$ measurable random variable. By definition, the set $\{\Pi_o X \neq \Pi_p X\}$ is contained in the union of graphs of the sequence of stopping times which exhaust the jumps of the corolol martingale $E[Z | \mathscr{F}_t]$.

For $X = I_{[\![S,\infty[\![}$, $S \in \mathscr{T}_o$, the set where $\{\Pi_o X = X \neq \Pi_a X\}$ is contained in the graph of the totally inaccessible part, $[\![S_B]\!]$, of S.

Finally,

$$\{\Pi_a X \neq \Pi_p X\} \subset \{\Pi_o X \neq \Pi_a X\} \cup \{\Pi_o X \neq \Pi_p X\}. \qquad \square$$

CHAPTER 7

Processes of Bounded Variation

Convention 6.1 of the previous chapter still applies.

Definition 7.1. A $\mathscr{B} \otimes \mathscr{F}$ measurable stochastic process $\{A_t\}$, $t \in [0, \infty[$, with values in $[0, \infty[$, is called an *increasing process* if almost every sample path $A(\omega)$ is right continuous and increasing. (Because increasing functions have left limits an increasing process is, therefore, corlol.) The random variable $A_\infty(\omega) = \lim_{t\to\infty} A_t(\omega)$ exists for an increasing process. Note that, unlike some authors, we do not require $A_0(\omega) = 0$ a.s. However, we shall always follow the convention that $A_{0-}(\omega) = 0$ a.s.

Write $\mathscr{W}^+$ for the set of such processes. $\mathscr{W} = \mathscr{W}^+ - \mathscr{W}^+$ is the set of corlol processes for which almost every sample path is of finite variation on each compact subset of $[0, \infty[$. From Lemma 7.5 below we see that a process is in $\mathscr{W}$ if and only if it is the difference of two members of $\mathscr{W}^+$. Members of $\mathscr{W}$ are the *processes of bounded variation.*

Definition 7.2. $\mathscr{V}^+$ will denote the family of equivalence classes of processes (under the relationship of P-indistinguishability), which admit a representative $\{A_t\}$ which is an increasing process *adapted* to the filtration $\{\mathscr{F}_t\}$, and such that $A_t < \infty$ a.s. for each $t \in [0, \infty[$. The same notation $\{A_t\}$ is used for the equivalence class and a process in the equivalence class. Such a process is called a *version* of $\{A_t\}$. From Theorem 6.35 this version is optional.

$\mathscr{V}_0^+$ will denote those processes $\{A_t\} \in \mathscr{V}^+$ such that $A_0 = 0$.

Definition 7.3. $\mathscr{V} = \mathscr{V}^+ - \mathscr{V}^+$ is the set of (equivalence classes of) processes, each of which is the difference of two elements of $\mathscr{V}^+$. A process is in $\mathscr{V}$ if it is corlol and adapted, and if almost every sample path is of finite variation on each compact subset of $[0, \infty[$. Again, processes in $\mathscr{V}$ are optional.

$\mathscr{V}_0$ will denote those processes $\{A_t\} \in \mathscr{V}$ such that $A_0 = 0$.

Definition 7.4. If $\{A_t\} \in \mathscr{W}$ then $\{D_t\} \in \mathscr{W}^+$, where $D_t(\omega) = \int_0^t |dA_s(\omega)|$ is the *total variation* of $\{A_t\}$. $\{D_t\}$ is the unique process in $\mathscr{W}^+$ such that, for almost every ω, the measure $dD_t(\omega)$ on $[0, \infty[$ is the absolute value of the signed measure $dA_t(\omega)$.

Note that $D_0 = |A_0|$.

Lemma 7.5. *If $\{A_t\} \in \mathscr{W}$ there is a unique decomposition $A_t = B_t - C_t$, where $\{B_t\}, \{C_t\} \in \mathscr{W}^+$ and $B_t + C_t = D_t = \int_0^t |dA_s|$.*

If $\{A_t\}$ is an optional (resp. a predictable), process then $\{B_t\}$ and $\{C_t\}$ are optional (resp. predictable).

PROOF. Write $B_t = (A_t + D_t)/2$ and $C_t = (D_t - A_t)/2$. Then $\{B_t\}$ and $\{C_t\}$ are increasing processes, and are the unique processes satisfying $A_t = B_t - C_t$, $D_t = B_t + C_t$. For almost every ω:

$$D_t(\omega) = |A_0(\omega)| + \lim_{n \to \infty} \sum_{k=1}^{n} |A_{tk/n} - A_{t(k-1)/n}|$$

so if $\{A_t\}$ is optional then $\{D_t\}$ is adapted and corlol. Therefore $\{D_t\}$, and so $\{B_t\}$ and $\{C_t\}$, belong to $\mathscr{V}^+$.

Suppose $\{A_t\}$ is predictable. Then the process $\{\bar{D}_t\}$, where $\bar{D}_t = D_{t-}$, is left continuous and adapted, and so predictable by Theorem 6.32. Now the process $\{\Delta A_t\}$, where $\Delta A_t = A_t - A_{t-}$, is predictable, so the process $\{\Delta D_t\}$ is predictable because $\Delta D_t = |\Delta A_t|$. Consequently, because $D_t = \bar{D}_t + \Delta D_t$, $\{D_t\}$ is predictable so, finally, $\{B_t\}$ and $\{C_t\}$ are predictable. □

Definition 7.6. $\mathscr{A}^+$ denotes the set of *integrable* increasing processes, that is, the set of processes $\{A_t\} \in \mathscr{V}^+$ such that $E[A_\infty] < \infty$.

$\mathscr{A} = \mathscr{A}^+ - \mathscr{A}^+$ denotes the set of processes of *integrable variation*, that is, the set of processes $\{A_t\} \in \mathscr{V}$ such that $E[D_\infty] = E[\int_0^\infty |dA_s|] < \infty$.

$\mathscr{A}_0^+$ and $\mathscr{A}_0$ consist of processes in $\mathscr{A}^+$ (resp. $\mathscr{A}$) which are 0 a.s. when $t = 0$.

The following result shows how Stieltjes integrals with respect to an increasing deterministic function can be defined in terms of a related Lebesgue integral.

Theorem 7.7. *Suppose $\alpha: [0, \infty[\to [0, \infty]$ is an increasing, right continuous deterministic function. For $t \in [0, \infty[$ define $\gamma(t) = \inf\{s: \alpha(s) > t\}$. Then γ is an increasing, right continuous function and $\gamma(t) < \infty$ if and only if $t < \alpha(\infty) = \lim_{s \to \infty} \alpha(s)$. Furthermore, $\alpha(s) = \inf\{t: \gamma(t) > s\}$. If α is finite then for every* *positive Borel function f on* $[0, \infty[$:

$$\int_0^\infty f(t)\, d\alpha(t) = \int_{\alpha(0)}^{\alpha(\infty)} f(\gamma(t))\, dt = \int_{\alpha(0)}^{\infty} I_{\{\gamma<\infty\}}(t) f(\gamma(t))\, dt.$$

PROOF. The relationship between α and γ is mostly easily verified by observing that the graph of γ is obtained from the graph of α by reflection in the diagonal. Preserving right continuity, intervals where α is constant correspond to jumps of γ, and jumps of α correspond to intervals where γ is constant.

To establish the identity of the integrals it is sufficient to consider f of the form $f(t) = I_{[0,s]}(t)$ where $s \in [0, \infty[$. Then

$$\int_0^\infty I_{[0,s]}(t)\, d\alpha(t) = \alpha(s) - \alpha(0),$$

and $\int_{\alpha(0)}^{\alpha(\infty)} I_{[0,s]}(\gamma(t))\, dt$ is the length of the interval $\{t : t \geq \alpha(0) \text{ and } \gamma(t) \leq s\}$. However, $\sup\{t : \gamma(t) \leq s\} = \inf\{t : \gamma(t) > s\} = \alpha(s)$, so the integral equals $\alpha(s) - \alpha(0)$. Because simple functions are dense the result follows. □

Remarks 7.8. Note that $\gamma(\alpha(t)) = t$ if, and only if, α is not right constant at t. Similarly, $\alpha(\gamma(t)) = t$ if, and only if, γ is not right constant at t. If α is continuous, and so has no jumps, there are no intervals on which γ is right constant, so applying the theorem above to $f(t) = g(\alpha(t))$ we have the following result:

Corollary 7.9. *Suppose* $\alpha : [0, \infty[\to [0, \infty]$ *is an increasing, continuous deterministic function. Then for every positive Borel function* g

$$\int_0^\infty g(\alpha(t))\, d\alpha(t) = \int_{\alpha(0)}^{\alpha(\infty)} g(t)\, dt.$$

Definition 7.10. Suppose $\{A_t\}$ is either an increasing process or a process of bounded variation, that is, the difference of two increasing processes, and $\{X_t\}$ is a real valued $\mathscr{B} \otimes \mathscr{F}$ measurable process on $[0, \infty[\times \Omega$. For $t \in [0, \infty[$ consider the Stieltjes integral

$$X \,.\, A_t(\omega) = \int_{[0,t]} X_s(\omega)\, dA_s(\omega),$$

whenever it exists. (The integral may take the value $+\infty$ or $-\infty$.) If $X \,.\, A_t(\omega)$ exists for all $t \in [0, \infty[$ for almost every ω a process $\{X \,.\, A_t\}$ is defined and $X \,.\, A_0 = X_0 A_0$.

Remarks 7.11. If $\{A_t\} \in \mathscr{V}$ there is a version $\{A_t\}$ which is corlol and adapted. Therefore, $\{A_t\}$ is optional, so if $\{X_t\}$ is optional there is an optional version of $\{X \,.\, A_t\}$. Similarly, if $\{X_t\}$ and $\{A_t\}$ are predictable there is a predictable version of $\{X \,.\, A_t\}$.

Definition 7.12. If $\{A_t\}$ is an increasing process a positive measure μ_A can be defined on $([0,\infty[\times\Omega, \mathscr{B}\otimes\mathscr{F})$ by putting $\mu_A(X)=E[X\,.\,A_\infty]=E[\int_{[0,\infty[} X_s\,dA_s]$ for each bounded $\mathscr{B}\otimes\mathscr{F}$-measurable process X.

We have the following characterization for finite measures of this form.

Theorem 7.13. *Suppose μ is a positive measure on $([0,\infty[\times\Omega, \mathscr{B}\otimes\mathscr{F})$. For μ to be of the form μ_A, where $\{A_t\}\in\mathscr{A}^+$, it is necessary and sufficient that*

(1) *μ has finite mass,*
(2) *the evanescent sets have μ-measure zero,*
(3) *$\mu([0,t]\times H)=\mu(E[I_H|\mathscr{F}_t].\,I_{[\![0,t]\!]}$ for every $t\in[0,\infty[$ and $H\in\mathscr{F}$.*

$\{A_t\}$ is then unique up to indistinguishability.

Proof. *Necessity.* Suppose $A\in\mathscr{A}^+$ and $\mu=\mu_A$. Then $\mu([0,\infty[\times\Omega)=E[A_\infty]<\infty$. If $E\subset[0,\infty[\times\Omega$ is evanescent then $\mu(E)\le E[A_\infty I_{\pi(E)}]=0$. Finally, because $\{A_t\}$ is adapted for $t\in[0,\infty[$ and $H\in\mathscr{F}$

$$\begin{aligned}\mu_A([0,t]\times H)&=E\left[\int_{[0,t]} I_H\,dA_s\right]\\&=E[I_HA_t]=E[E[I_H\,.\,A_t|\mathscr{F}_t]]\\&=E[E[I_H|\mathscr{F}_t]\,.\,A_t]=\mu_A(E[I_H|\mathscr{F}_t]\,.\,I_{[\![0,t]\!]}).\end{aligned}$$

Sufficiency. Given μ satisfying (1) and (2) for each $t\in[0,\infty[$ define a measure m_t on $(\Omega,\mathscr{F})$ by putting $m_t(H)=\mu([0,t]\times H)$. Then, for each t, m_t is a positive bounded measure on $(\Omega,\mathscr{F})$ and, from property (2), m_t is absolutely continuous with respect to the underlying measure P on $(\Omega,\mathscr{F})$. Write $\hat{A}_t$ for a version of the Radon–Nikodym derivative dm_t/dP. Because μ is positive, $\hat{A}_s\le\hat{A}_t$ if $s\le t$. Define $\hat{\hat{A}}_t=\sup_r\hat{A}_r$ where the supremum is taken over all rational numbers $r\le t$. Then $\hat{\hat{A}}_t$ is also a version of dm_t/dP, and if $\{t_n\}$ is a sequence of converging to t from above, by Lebesgue's theorem, $\|\hat{\hat{A}}_{t_n}-\hat{\hat{A}}_t\|_1$ converges to zero in $L^1(\Omega,\mathscr{F},P)$ and so $\hat{\hat{A}}_{t_n}$ converges almost surely to $\hat{\hat{A}}_t$. Finally, therefore, write $A_t=\hat{\hat{A}}_{t+}$, so $\{A_t\}$ is a right continuous increasing process. By construction, for $t\in[0,\infty[$ and $H\in\mathscr{F}$,

$$\mu_A([0,t]\times H)=E\left[\int_{[0,t]} I_H\,dA_s\right]=E[I_HA_t]=\mu([0,t]\times H).$$

Clearly sets of the form $[0,t]\times H$ generate the product σ-field $\mathscr{B}\otimes\mathscr{F}$, and so the measures are determined by their values on such sets. Therefore, $\mu=\mu_A$.

We must show that if $\mu=\mu_A$ satisfies (3) then $\{A_t\}$ is adapted. However, for $t\in[0,\infty[$ and $H\in\mathscr{F}$:

$$\begin{aligned}\mu_A([0,t]\times H)&=E[I_HA_t]\\&=\mu_A(E[I_H|\mathscr{F}_t]\,.\,I_{[\![0,t]\!]})\end{aligned}$$

$$= E[E[I_H | \mathscr{F}_t] . A_t]$$
$$= E[E[E[A_t | \mathscr{F}_t] I_H | \mathscr{F}_t]]$$
$$= E[E[A_t | \mathscr{F}_t] I_H].$$

That is, $\int_H A_t \, dP = \int_H E[A_t | \mathscr{F}_t] \, dP$ for all $H \in \mathscr{F}$, so $\{A_t\}$ is adapted.

Uniqueness. If $\{B_t\} \in \mathscr{A}^+$ is another such process, then B_t is also a version of dm_t/dP, so $A_t = B_t$ a.s. However, because $\{A_t\}$ and $\{B_t\}$ are both right continuous they are indistinguishable by Lemma 2.21. □

Definition 7.14. Suppose $\{A_t\} \in \mathscr{V}^+$. For $t \in [0, \infty[$ and $\omega \in \Omega$ define

$$C_t(\omega) = \inf\{s : A_s(\omega) > t\}.$$

Now

$$\{C_t < s\} = \bigcup_n \{A_{s-1/n} > t\},$$

and

$$\{A_{s-1/n} > t\} \in \mathscr{F}_{s-1/n} \subset \mathscr{F}_s,$$

because $\{A_t\}$ is adapted. Therefore, $\{C_t < s\} \in \mathscr{F}_s$, $\{C_t \leq s\} \in \mathscr{F}_{s+} = \mathscr{F}_s$, by right continuity. Consequently, C_t is a stopping time. The definition is analogous to that of Theorem 7.7 and $\{C_t\}$ is called the *time change* associated with $\{A_t\}$.

Theorem 7.15. *Suppose $\{X_t\}$ and $\{Y_t\}$ are two $\mathscr{B} \otimes \mathscr{F}$-measurable processes such that for every stopping time T*

$$E[X_T . I_{\{T<\infty\}}] = E[Y_T . I_{\{T<\infty\}}].$$

Then for every process $\{A_t\} \in \mathscr{V}^+$ and every stopping time T

$$E\left[\int_0^T X_s \, dA_s\right] = E\left[\int_0^T Y_s \, dA_s\right].$$

PROOF. First suppose $T = \infty$ a.s. and write $\{C_t\}$ for the time change associated with $\{A_t\}$. Then by Theorem 7.7 and Fubini's theorem:

$$E\left[\int_0^\infty X_s \, dA_s\right] = E\left[\int_0^\infty X_{C_s} . I_{\{C_s<\infty\}} \, ds\right]$$
$$= \int_0^\infty E[X_{C_s} . I_{\{C_s<\infty\}}] \, ds.$$

Similarly

$$E\left[\int_0^\infty Y_s \, dA_s\right] = \int_0^\infty E[Y_{C_s} . I_{\{C_s<\infty\}}] \, ds,$$

so, because $\{C_s\}$ is a stopping time for all s, the result is proved when $T=\infty$ a.s.

The result for a general stopping time is obtained by applying the above result to the process $\{B_t\}\in\mathscr{V}^+$ where $B_t=A_{t\wedge T}$. □

Corollary 7.16. *Suppose $\{A_t\}\in\mathscr{V}^+$ and $\{M_t\}$ is a positive martingale which is right continuous and uniformly integrable. Then for every stopping time T.*

$$E\left[\int_0^T M_s\,dA_s\right]=E[M_TA_T].$$

PROOF. Write

$$X_t=M_t\,.\,I_{[\![0,T]\!]}(t)$$

and

$$Y_t=M_T\,.\,I_{[\![0,T]\!]}(t).$$

Then for any stopping time S;

$$\begin{aligned}E[X_S\,.\,I_{\{S<\infty\}}]&=E[M_S\,.\,I_{[\![0,T]\!]}(S)]\\&=E[M_S\,.\,I_{\{S\le T\}}I_{\{S<\infty\}}],\end{aligned}$$

and by Doob's Theorem 4.12

$$\begin{aligned}E[Y_S\,.\,I_{\{S<\infty\}}]&=E[M_T\,.\,I_{[\![0,T]\!]}(S)]\\&=E[E[M_T\,.\,I_{\{S\le T\}}\,.\,I_{\{S<\infty\}}\,|\,\mathscr{F}_S]]\\&=E[M_S\,.\,I_{\{S\le T\}}\,.\,I_{\{S<\infty\}}].\end{aligned}$$

Consequently, by Theorem 7.15

$$\begin{aligned}E\left[\int_0^T X_s\,dA_s\right]&=E\left[\int_0^T M_s\,dA_s\right]\\&=E\left[\int_0^T Y_s\,dA_s\right]=E\left[M_T\,.\int_0^T dA_s\right]\\&=E[M_TA_T].\end{aligned}$$

□

Remarks 7.17. We have seen that to every increasing function $\alpha(t)$ defined on $[0,\infty[$ there corresponds a measure $d\alpha(t)$. In the terminology of measure theory, (see [2] or [62]), α is continuous if and only if $d\alpha$ is a diffuse measure (that is, the support of $d\alpha$ has no atoms). If $d\alpha$ is purely atomic (that is, the support of $d\alpha$ consists of a countable number of atoms), then α is said to be purely discontinuous. Every such measure on $[0,\infty[$ can be written in a unique way as the sum of a diffuse and purely atomic measure. This corresponds to the unique decomposition of an increasing function into the sum of a continuous and purely discontinuous function. For increasing processes we have the following analogous result.

Theorem 7.18. *Suppose $\{A_t\}$ is an increasing process. Then A_t has a unique decomposition as $A_t^c + A_t^d$, where $\{A_t^c\}$ is an increasing continuous process, and $\{A_t^d\}$ is an increasing purely discontinuous process.*

If $\{A_t\}$ is adapted then it is optional and $\{A_t^d\}$ is optional. If $\{A_t\}$ is accessible (resp. predictable) then $\{A_t^d\}$ is accessible (resp. predictable). If $\{A_t\}$ is adapted $\{A_t^c\}$ is predictable.

PROOF. Only the second part remains to be proved.

Suppose, from Theorem 6.46, that $\{T_n\}$ is a sequence of stopping times which exhaust the jumps of A.

Write

$$B_t^n = (A_{T_n} - A_{T_n-}) \,.\, I_{[\![T_n,\infty[\![}(t),$$

$$A_t^d = \sum_n B_t^n.$$

For each n $\{B_t^n\}$ is an increasing process.

$\{A_t^d\}$ is, therefore, an increasing process, the right continuity of the trajectories following from the definition. If $A_t^c = A_t - A_t^d$, then $\{A_t^c\}$ is an increasing continuous process.

Because $\{A_t\}$ is corlol, if it is adapted it is optional. In all three cases $\{A_t^c\}$ is then continuous and adapted, and so predictable. Consequently, if $\{A_t\}$ is Σ_x measurable $(x = o, \alpha, p)$ then, because $A_t^d = A_t - A_t^c$, $\{A_t^d\}$ is Σ_x measurable. □

Theorem 7.19. *Suppose $\{A_t\}$ is an increasing, purely discontinuous Σ_x measurable process $(x = o, a, p)$. Then there is a sequence $\{\alpha_n\}$ of positive numbers and a sequence $\{T_n\}$ of stopping times $T \in \mathcal{T}_x$ such that $A_t = \sum_n \alpha_n I_{[\![T_n,\infty[\![}(t)$.*

PROOF. Consider the predictable case $(x = p)$; the other cases are proved similarly. By Theorem 6.46 there is a sequence $\{S_n\}$, $S_n \in \mathcal{T}_p$, of stopping times which exhausts the jumps of $\{A_t\}$. Therefore,

$$A_t = \sum_n (A_{S_n} - A_{S_n-}) \,.\, I_{[\![S,\infty[\![},$$

and it is enough to prove the result when the process is of the form $(A_S - A_{S-}) \,.\, I_{[\![S,\infty[\![}$, where S is a predictable stopping time. By Theorem 6.30 the random variable $X = (A_S - A_{S-}) \,.\, I_{\{S<\infty\}}$ is $\mathcal{F}_{S-}$ measurable and so there is an increasing sequence $\{X_n\}$ of simple $\mathcal{F}_{S-}$ measurable functions such that $X = \lim X_n$ a.s. Consequently, there is a sequence $\{\alpha_n\}$ of positive numbers and a sequence $\{H_n\}$ of elements of $\mathcal{F}_{S-}$ such that

$$X = \sum_n \alpha_n \,.\, I_{H_n}.$$

Writing $T_n = S_{H_n}$, so that $T_n \in \mathcal{T}_p$, we see that $A_t = \sum_n \alpha_n \,.\, I_{[\![T_n,\infty[\![}$. □

Remark 7.20. It is a consequence of the theorem below that if the process $\{A_t\} \in \mathscr{V}^+$ is Σ_x measurable the corresponding measure μ_A is determined by its restriction to Σ_x.

Theorem 7.21. *Suppose $\{A_t\} \in \mathscr{V}^+$ is Σ_x measurable $(x = o, a, p)$. If $\{X_t\}$ and $\{Y_t\}$ are two bounded $\mathscr{B} \otimes \mathscr{F}$ measurable processes whose projections $\{\Pi_x X_t\}$ and $\{\Pi_x Y_t\}$ are indistinguishable then $\mu_A(X) = E[X \,.\, A_\infty] = E[Y \,.\, A_\infty] = \mu_A(Y)$.*

PROOF. *Optional Case.* By uniqueness, $\Pi_o X = \Pi_o Y$ if and only if

$$E[X_T \,.\, I_{\{T<\infty\}}] = E[Y_T \,.\, I_{\{T<\infty\}}],$$

for every $T \in \mathscr{T}_o$. Therefore, because $\{A_t\}$ is adapted the result follows from Theorem 7.15.

Accessible and Predictable Cases. Because Π_a and Π_p factor through Π_o we can suppose $\{X_t\}$ and $\{Y_t\}$ are optional. By the above decomposition $\{A_t\}$ can be written

$$A_t = A_t^c + \sum_n \alpha_n I_{[T_n, \infty[}(t),$$

where $\{T_n\}$, $T_n \in \mathscr{T}_x$ $(x = a$ or $p)$, is a sequence of stopping times which exhausts the jumps of A_t. It is, therefore, enough to establish the result for each term in this decomposition. The sets $\{X \neq \Pi_x X\}$ and $\{Y \neq \Pi_x Y\}$, $x = a$ (resp. p) are countable unions of graphs of stopping times by Theorem 6.48. Therefore, because $\Pi_x X = \Pi_x Y$ (except on an evanescent set), the set $\{X \neq Y\}$ is also such a countable union. For the continuous part A_t^c we have $E[X \,.\, A_\infty^c] = E[Y \,.\, A_\infty^c]$. If $B_t^n = I_{[\![T_n, \infty[\![}(t)$, $T_n \in \mathscr{T}_x$, $x = a$ (resp. p) then $E[X \,.\, B_\infty^n] = E[X_{T_n} \,.\, I_{\{T_n<\infty\}}]$ and

$$E[Y \,.\, B_\infty^n] = E[Y_{T_n} \,.\, I_{\{T_n<\infty\}}] \quad \text{but if } \Pi_x X = \Pi_x Y, \quad x = a \text{ (resp. } p),$$

the two right-hand sides are equal and the result follows. □

Corollary 7.22. *Suppose $\{A_t\} \in \mathscr{V}^+$ is Σ_x measurable $(x = o, a, p)$. If $\{X_t\}$ and $\{Y_t\}$ are two bounded $\mathscr{B} \otimes \mathscr{F}$ measurable processes such that $\Pi_x X = \Pi_x Y$ then for every pair of stopping times $S, T \in \mathscr{T}_o$, $S \leq T$, we have*

$$E\left[\int_{]\!]S,T]\!]} X_t \, dA_t | \mathscr{F}_S\right] = E\left[\int_{]\!]S,T]\!]} Y_t \, dA_t | \mathscr{F}_S\right].$$

PROOF. We must show that for any $H \in \mathscr{F}_S$ the two random variables have the same integral over H. However, these integrals can be written as $E[I_{]\!]S_H, T_H]\!]} X \,.\, A_\infty]$ and $E[I_{]\!]S_H, T_H]\!]} Y \,.\, A_\infty]$. Because $I_{]\!]S_H, T_H]\!]}$ is predictable

$$\Pi_x(I_{]\!]S_H, T_H]\!]} X) = I_{]\!]S_H, T_H]\!]} \Pi_x X = I_{]\!]S_H, T_H]\!]} \Pi_x Y, \qquad x = o, a, p.$$

Therefore, the two integrals are equal by Theorem 7.21 above. □

We next show that the property described in Theorem 7.21 characterizes measures obtained from Σ_x measurable processes.

Corollary 7.23. *Suppose μ_A is the measure generated by the increasing process $\{A_t\}$. Then $\{A_t\}$ is Σ_x measurable if, and only if, μ_A satisfies the following condition "C": C: if $\{X_t\}$ and $\{Y_t\}$ are two bounded $\mathscr{B} \otimes \mathscr{F}$ measurable processes such that $\Pi_x X = \Pi_x Y$ $(x = o, a, p)$, then $\mu_A(X) = \mu_A(Y)$.*

PROOF. The necessity of condition C is established in the above theorem. (1) $x = o$. Because $\{A_t\}$ is corlol we need only show it is adapted. As in Theorem 7.13, this is the case if for all $H \in \mathscr{F}$ and $t \in [0, \infty[$:

$$E[I_H . A_t] = E[E[I_H | \mathscr{F}_t] . A_t].$$

However, the processes $X_t = I_H . I_{[\![0,t]\!]}$ and $Y_t = E[I_H | \mathscr{F}_t] . I_{[\![0,t]\!]}$ have the same Π_o projection, namely $M_t . I_{[\![0,t]\!]}$, and so the result follows. Here, M_s is the right continuous version of the martingale $E[I_H | \mathscr{F}_s]$.

(2) $x = a$. To prove A_t accessible, using Corollary 6.47, we must show that A_t does not charge any totally inaccessible stopping time. Suppose S is totally inaccessible. Then, as in Theorem 6.39,

$$\Pi_a(I_{[\![S]\!]}) = 0.$$

Consequently,

$$E[A_S - A_{S-}] = \mu_A(I_{[\![S]\!]})$$

$$= \mu_A(\Pi_a I_{[\![S]\!]}) = 0.$$

(3) $x = p$. If $\{A_t\}$ satisfies condition C when $x = p$ it certainly satisfies condition C when $x = a$, because $\Pi_p = \Pi_p \Pi_a$. Therefore, by the argument above, $\{A_t\}$ is accessible. From Theorem 6.30, because $A_\infty(\omega) = \lim_t A_t(\omega)$, to show $\{A_t\}$ predictable we must show that A_T is $\mathscr{F}_{T-}$ measurable for every $T \in \mathscr{T}_p$. That is, for every $H \in \mathscr{F}$ we must show that

$$E[I_H . A_T] = E[E[I_H | \mathscr{F}_{T-}] A_T],$$

because then

$$E[I_H A_T] = E[E[A_T | \mathscr{F}_{T-}] I_H],$$

so that

$$A_T = E[A_T | \mathscr{F}_{T-}].$$

However, for $T \in \mathscr{T}_p$ and $H \in \mathscr{F}$ consider the processes $X_t = I_H I_{]\!]0,T]\!]}$ and $Y_t = E[I_H | \mathscr{F}_{T-}] I_{]\!]0,T]\!]}$. Then, if $\{\bar{M}_t\}$ is the left continuous version of the martingale $E[I_H | \mathscr{F}_t]$, from Theorem 6.39 we have that

$$\Pi_p X_t = \Pi_p Y_t = \bar{M}_t I_{]\!]0,T]\!]}.$$

Consequently

$$\mu_A(X) = E[I_H A_T] = \mu_A(Y) = E[E[I_H | \mathcal{F}_{T-}] A_T],$$

and the result follows. □

Remarks 7.24. If $\{X_t\}$ is a bounded $\mathcal{B} \otimes \mathcal{F}$ measurable process and $\{A_t\}$ a $\mathcal{B} \otimes \mathcal{F}$ measurable process of integrable variation, an inner product (see Definition 7.10), $\langle X, A \rangle = E[\int_0^\infty X_s(\omega)\, dA_s(\omega)]$ can be defined. It is of interest to investigate the map which is the adjoint of the projection Π_x. Consider first an increasing process.

Theorem 7.25. *Suppose $\{A_t\}$ is an increasing $\mathcal{B} \otimes \mathcal{F}$ measurable process such that $E[A_\infty] < \infty$. Then there is a unique Σ_x $(x = o, a, p)$, measurable increasing process $\{\Pi_x^* A_t\} \in \mathcal{A}^+$ such that for every bounded $\mathcal{B} \otimes \mathcal{F}$ measurable process $\{X_t\}$*

$$\langle \Pi_x X, A \rangle = \langle X, \Pi_x^* A \rangle = \langle \Pi_x X, \Pi_x^* A \rangle.$$

Proof. Given the increasing $\mathcal{B} \otimes \mathcal{F}$ measurable process $\{A_t\}$ we can define a measure μ_x on $\mathcal{B} \otimes \mathcal{F}$ by putting

$$\mu_x(X) = \langle \Pi_x X, A \rangle,$$

for every bounded $\mathcal{B} \otimes \mathcal{F}$ measurable process $\{X_t\}$. Then for any $r \in [0, \infty[$ and $H \in \mathcal{F}$ the processes $X_t = I_H I_{[\![0,r]\!]}$ and $Y_t = E[I_H | \mathcal{F}_r] I_{[\![0,r]\!]}$ have the same Π_x projection. Therefore, μ_x satisfies the conditions of Theorem 7.13 and so is generated by a unique process $\{\Pi_x^* A_t\} \in \mathcal{A}^+$. However, from Corollary 7.23 above we see that $\{\Pi_x^* A_t\}$ is Σ_x measurable. □

Definition 7.26. $\{\Pi_x^* A_t\}$, $x = o$ (resp. a, resp. p), is called the *dual optional* (resp. *accessible, predictable*) *projection of* $\{A_t\}$.

Remarks 7.27. $\{\Pi_x^* A_t\}$ should not be confused with $\{\Pi_x A_t\}$, which may not even be an increasing process.

The map Π_x^* $(x = o, a, p)$, can be extended to the $\mathcal{B} \otimes \mathcal{F}$ measurable processes of integrable variation, each of which is the difference of two increasing processes.

Corollary 7.28. *If $\{A_t\} \in \mathcal{A}^+$ is Σ_x measurable then $\Pi_x^* A_t = A_t$.*

Proof. If $\{A_t\}$ is Σ_x measurable then, by Corollary 7.23

$$\langle X, A \rangle = \langle \Pi_x X, A \rangle = \langle X, \Pi_x^* A \rangle$$

for all bounded $\mathcal{B} \times \mathcal{F}$-measurable processes $\{X_t\}$. □

Theorem 7.29. *Suppose $\{A_t\}$ is a continuous $\mathcal{B} \otimes \mathcal{F}$ measurable process of integrable variation. Then $\{\Pi_x^* A_t\}$ is a continuous process, and*

$$\Pi_o^* A_t = \Pi_a^* A_t = \Pi_p^* A_t.$$

Proof. Suppose $T \in \mathcal{T}_o$ and $X_t = I_{[\![T]\!]}(t)$. Then $\{X_t\}$ is optional so

$$\begin{aligned}\langle X, A\rangle &= \langle \Pi_o X, A\rangle \\ &= \langle X, \Pi_o^* A\rangle = E[A_T - A_{T-}] \\ &= 0 = E[\Pi_o^* A_T - \Pi_o^* A_{T-}].\end{aligned}$$

Therefore, $\{\Pi_o^* A_t\}$ has no jumps and so is continuous. Therefore, $\{\Pi_o^* A_t\}$ is predictable, and $\Pi_o^* A_t = \Pi_a^* A_t = \Pi_p^* A_t$. □

Theorem 7.30. *Suppose $\{A_t\} \in \mathcal{A}_0$. Then the following conditions are equivalent*

(1) *$\{A_t\}$ is a martingale,*
(2) *$\{\Pi_p^* A_t\}$ is indistinguishable from 0 (the zero process),*
(3) *the restriction of μ_A to Σ_p is zero.*

Proof. Consider a process $\{X_t\}$ of the form $X_t = I_{]\!]S,T]\!]}$ for $S, T \in \mathcal{T}_o$, or $X_t = I_{[\![0_F]\!]}$ for $F \in \mathcal{F}_0$, so that $\{X_t\}$ is predictable. Then $\mu_A(]\!]S, T]\!]) = \langle X, A\rangle = \langle \Pi_p X, A\rangle = \langle X, \Pi_p^* A\rangle$, and, similarly, $\mu_A(\{0\}\times F) = \langle I_{[\![0_F]\!]}, \Pi_p^* A\rangle$.

The equivalence of (2) and (3) is immediate because Σ_p is generated by stochastic intervals of the form $]\!]S, T]\!]$, $S, T \in \mathcal{T}_o$, and $[\![0_F]\!]$, $F \in \mathcal{F}_0$. From Corollary 6.33, Σ_p is generated by stochastic intervals of the form $[\![0_F]\!]$, $F \in \mathcal{F}_0$, and $]\!]s_B, t_B]\!] =]s, t]\times B$ for $B \in \mathcal{F}_s$ and s and t constants.

Now $\mu_A([\![0_F]\!]) = 0$ because $A_0 = 0$ a.s. and $\{A_t\}$ is right continuous. Furthermore, $\mu_A(]\!]s_B, t_B]\!]) = E[I_B(A_t - A_s)]$ so $E[A_t|\mathcal{F}_s] = A_s$ if, and only if, the restriction of μ_A to Σ_p is zero. Therefore, (1) is equivalent to (3). □

Corollary 7.31. *If $\{A_t\} \in \mathcal{A}$ then the process $B_t = A_t - \Pi_p^*(A_t)$ is a martingale.*

Proof. $\{A_t\}$ is adapted and $\Pi_p^*(B_t) = 0$ by Corollary 7.28. □

Corollary 7.32. *If a predictable process $\{A_t\} \in \mathcal{A}$ is a martingale then for all t $A_t = A_0$ a.s.*

Proof. $\{A_t - A_0\} \in \mathcal{A}_0$ and is a martingale. Therefore, from the theorem, $\Pi_p^*(A_t - A_0) = A_t - A_0$ is indistinguishable from the zero process. □

Definition 7.33. Two processes of integrable variation are said to be *associated* if they have the same dual predictable projection.

Rephrasing the uniqueness stated in Theorem 7.25 we can say that, if $\{A_t\}$ is an increasing $\mathcal{B} \otimes \mathcal{F}$ measurable process such that $E[A_\infty] < \infty$, then $\{A_t\}$ is associated with one, and only one, predictable process.

Lemma 7.34. *Suppose $\{A_t\}$ is an increasing $\mathcal{B} \otimes \mathcal{F}$ measurable process and $\{\Pi_x^* A_t\}$ $(x = o, a, p)$, is its dual Σ_x projection. If $\{X_t\}$ is a bounded $\mathcal{B} \otimes \mathcal{F}$*

measurable process and S, T are stopping times, $S \leq T$, then:

$$E\left[\int_S^T \Pi_x X_t \, dA_t \middle| \mathscr{F}_S\right] = E\left[\int_S^T X_t \, d\Pi_x^* A_t \middle| \mathscr{F}_S\right] = E\left[\int_S^T (\Pi_x X_t) \, d\Pi_x^* A_t \middle| \mathscr{F}_S\right].$$

PROOF. The result follows from the definition of $\{\Pi_x^* A_t\}$ by considering the process $I_{]\!]S_B, T_B]\!]} \cdot X_t$ for any $B \in \mathscr{F}_S$, as in Corollary 7.22. □

Corollary 7.35. *If $\{A_t\}$ and $\{B_t\} \in \mathscr{A}_0$ are associated, then $\{M_t\}$ is a martingale if $M_t = A_t - B_t$. Conversely, if $A_t - B_t$ is a martingale then $\{A_t\}$ and $\{B_t\}$ are associated.*

PROOF. Take $X_t = 1$ a.s. for all t, and $S = s \leq t = T$ in the above theorem. Then, because $\Pi_p^* A_t = \Pi_p^* B_t$

$$E\left[\int_s^t d\Pi_p^* A \middle| \mathscr{F}_s\right] = E\left[\int_s^t dA \middle| \mathscr{F}_s\right]$$

$$= E\left[\int_s^t d\Pi_p^* B \middle| \mathscr{F}_s\right] = E\left[\int_s^t dB \middle| \mathscr{F}_s\right].$$

That is,

$$E[A_t - A_s | \mathscr{F}_s] = E[B_t - B_s | \mathscr{F}_s],$$

so

$$E[A_t - B_t | \mathscr{F}_s] = A_s - B_s.$$

The converse follows from applying Theorem 7.30 to $A_t - B_t$. □

Notation 7.36. If $\mathscr{C}$ is some family of processes (for example $\mathscr{A}$ or $\mathscr{A}^+$), then $\mathscr{C}_{\text{loc}}$ will denote the family of processes which are locally in $\mathscr{C}$. That is, $\{Y_t\} \in \mathscr{C}_{\text{loc}}$ if there is an increasing sequence of stopping times $\{T_n\}$ such that $\lim_n T_n = \infty$ a.s. and such that each stopped process $\{Y_t^{T_n}\} = \{Y_{t \wedge T_n}\}$ is in $\mathscr{C}$.

Remarks 7.37. Most of the results proved above for processes in $\mathscr{A}$ or $\mathscr{A}^+$ have "local" versions valid for processes in $\mathscr{A}_{\text{loc}}$ or $\mathscr{A}_{\text{loc}}^+$. In particular, the dual predictable projections can be extended to processes which satisfy "locally" the conditions of Theorem 7.25. That is, suppose $\{A_t\}$ is an increasing $\mathscr{B} \otimes \mathscr{F}$ measurable process for which there is an increasing sequence $\{T_n\}$ of stopping times such that $\lim_n T_n = \infty$ a.s. and $E[A_\infty^{T_n}] = E[A_{T_n}] < \infty$ for each n. Then the dual Σ_x projection can be defined for each stopped process $\{A_t^{T_n}\}$, and by "pasting" together the pieces the dual Σ_x projection $\{\Pi_x^* A_t\}$ of the process $\{A_t\}$ is a Σ_x measurable process in $\mathscr{A}_{\text{loc}}^+$. Here for example, is the "local" version of Theorem 7.13.

Theorem 7.38. *Suppose μ is a positive measure on $([0, \infty[\times \Omega, \mathscr{B} \times \mathscr{F})$. For μ to be of the form μ_A where $\{A_t\} \in \mathscr{A}_{\text{loc}}^+$, it is necessary and sufficient that*

(1) *there is an increasing sequence of stopping times* $\{T_n\}$ *such that* $\lim_n T_n = \infty$ *and* $\mu([\![0, T_n]\!]) < \infty$ *for each* n,
(2) *the evanescent sets have* μ *measure zero,*
(3) $\mu([0, t] \times H) = \mu(E[I_H | \mathscr{F}_t] \,.\, I_{[\![0,t]\!]})$ *for every* $t \in [0, \infty[$ *and* $H \in \mathscr{F}$.

PROOF. For each n consider the measure μ_n on $([0, \infty[\times \Omega, \mathscr{B} \times \mathscr{F})$ defined by $\mu_n(B) = \mu(B \cap [\![0, T_n]\!])$, $B \in \mathscr{B} \times \mathscr{F}$. Then μ_n satisfies the conditions of the Theorem 7.13, and so is of the form $\mu_n = \mu_{A(n)}$, where $\{A(n)_t\} \in \mathscr{A}^+$ and $\mathscr{A}(n)_t$ is constant after T_n. If $m > n$, μ_m is similarly of the form $\mu_{A(m)}$ for $\{A(m)_t\} \in \mathscr{A}^+$. However, by uniqueness $A(m)_t(\omega) = A(n)_t(\omega)$ if $t \leq T_n(\omega)$, so a process $\{A_t\} \in \mathscr{A}^+_{\text{loc}}$ is obtained by "pasting" the processes $A(n)$ together. □

The following result will be used in Theorem 8.22.

Lemma 7.39. *Suppose* $A \in \mathscr{V}$ *is either predictable or has bounded jumps. Then* $A \in \mathscr{A}_{\text{loc}}$.

PROOF. We can suppose $A_0 = 0$. From Lemma 7.5, if A is predictable then the increasing process $\int_0^t |dA_s|$ is predictable. Write

$$S_k = \inf\left\{t : \int_0^t |dA_s| \geq k\right\}.$$

Because $A_0 = 0$, S_k is almost surely positive. Write $D_t = \int_0^t |dA_s|$ and $B_k = \{(t, \omega): D_t \geq k\}$. Then S_k is the debut of $B_k \in \Sigma_p$, and because $\{D_t\}$ is right continuous, $[\![S_k]\!] \subset B_k$. In fact $[\![S_k]\!] = B_k \backslash]\!]S_k, \infty[\![\in \Sigma_p$, and so S_k is predictable by Corollary 6.26.

Suppose $\{S_{km}\}$ is a sequence of stopping times which announce S_k. Then the variation of A on $[\![0, S_{km}]\!]$ is less than k, and so integrable. Define $T_n = \max_{k \leq n, m \leq n} S_{km}$. Then $\lim T_n = \infty$ and $E \int_0^{T_n} |dA_s| \leq n$, so $A \in \mathscr{A}_{\text{loc}}$.

If A has bounded jumps one can immediately take $T_n = S_n$. □

CHAPTER 8

The Doob–Meyer Decomposition

In Chapter 4 we showed in Theorem 4.15 that a uniformly integrable supermartingale can be expressed as the sum of a martingale and a potential; this was called its Riesz decomposition. Of course, a decreasing adapted process is a particular case of a supermartingale and we now describe a more explicit decomposition.

Definition 8.1. Suppose $\{X_t\}$, $t \in \tau = [0, \infty]$, is a right continuous uniformly integrable supermartingale. Then $\{X_t\}$ is said to have a *Doob–Meyer* decomposition if there is a right continuous martingale $\{M_t\}$ and an adapted increasing process $\{A_t\} \in \mathscr{V}_0^+$ such that $X_t = M_t - A_t$ a.s., $t \in \tau$. That is, $M_t = X_t + A_t$ a.s. for all $t \in \tau$.

Remarks 8.2. We know from Theorem 4.15 that $\{X_t\}$ has a Riesz decomposition $X_t = Y_t + Z_t$ a.s., $t \in \tau$, where $\{Y_t\}$ is a uniformly integrable martingale and $\{Z_t\}$ is a potential. Therefore, $\{X_t\}$ has a Doob–Meyer decomposition if, and only if, its potential $\{Z_t\}$ has a Doob–Meyer decomposition.

Suppose $\{Z_t\}$ has a Doob–Meyer decomposition

$$Z_t = M_t - A_t \quad \text{a.s.,} \qquad t \in \tau.$$

Then

$$E[Z_t] = E[M_t] - E[A_t]$$

and

$$\lim_{t \to \infty} E[Z_t] = 0.$$

Because $\{A_t\}$ is an increasing process the random variable $A_\infty(\omega) = \lim_{t \to \infty} A_t(\omega)$ certainly exists, almost surely, and

$$\lim_{t \to \infty} E[A_t] = E[A_\infty] = E[M_0] \quad \text{so } \{A_t\} \in \mathscr{A}_0^+.$$

In fact, because $\lim_{t\to\infty} Z_t(\omega)=0$ a.s., $\lim_{t\to\infty} M_t(\omega)=M_\infty(\omega)=A_\infty(\omega)$ a.s. Furthermore, $E[M_t]\le E[Z_0]+E[A_\infty]$ so $\{M_t\}$ is a uniformly integrable martingale, and $M_t=E[A_\infty|\mathscr{F}_t]$. Therefore,

$$Z_t=E[A_\infty|\mathscr{F}_t]-A_t \quad \text{a.s.}$$

Definition 8.3. In this situation we say that the potential $\{Z_t\}$ is generated by the increasing integrable process $\{A_t\}$.

Lemma 8.4. *In the above Doob–Meyer decomposition the increasing process $\{A_t\}$ can be taken to be predictable.*

Proof. If $\{A_t\}$ is not predictable consider the dual predictable projection $B_t=\Pi_p^* A_t$ of $\{A_t\}$. Certainly $\{A_t\}$ and $\{B_t\}$ are associated, so $\{\Pi^*(A_t-B_t)\}$ is indistinguishable from the zero process, and, from Corollary 7.35, $\{A_t-B_t\}$ is a martingale. Therefore,

$$E[A_\infty-B_\infty|\mathscr{F}_t]=A_t-B_t$$

and we can write

$$Z_t=E[B_\infty|\mathscr{F}_t]-B_t \quad \text{a.s.}$$

That is, $\{A_t\}$ and $\{B_t\}$ generate the same potential. □

Remarks 8.5. When $\{X_t\}$ is a uniformly integrable martingale the set of random variables $\{X_T\}$, for T any stopping time, is uniformly integrable. This follows from Doob's optional stopping Theorem 4.12 because $X_T=E[X_\infty|\mathscr{F}_T]$ a.s. However, this is not true in general when $\{X_t\}$ is a uniformly integrable supermartingale. (See Johnson and Helms [51] for an example.)

Definition 8.6. A right continuous uniformly integrable supermartingale $\{X_t\}$ is said to be of *class* D if the set of random variables $\{X_T\}$, for T any stopping time, is uniformly integrable.

Theorem 8.7. *If the potential $\{Z_t\}$ has a Doob–Meyer decomposition*

$$\begin{aligned} Z_t&=M_t-A_t \quad \textit{a.s.}\\ &=E[A_\infty|\mathscr{F}_t]-A_t \quad \textit{a.s.}\end{aligned}$$

with $\{A_t\}\in\mathscr{A}_0^+$ then $\{Z_t\}$ is of class D. *Furthermore, the increasing process $\{A_t\}$ can be taken as predictable and for every stopping time T*

$$Z_T=E[A_\infty|\mathscr{F}_T]-A_T \quad \textit{a.s.}$$

Proof. In the above decomposition $M_T=E[A_\infty|\mathscr{F}_T]$ for any stopping time T, so the set of random variables $\{M_T\}$ is uniformly integrable. Because $\{A_t\}\in\mathscr{A}_0^+$ is an increasing process

$$0\ge -A_T\ge -A_\infty \quad \text{a.s.}$$

for any stopping time T, so the set of random variables $\{-A_T\}$ is uniformly integrable. Therefore, $\{Z_t\}$ is of class D.

We have noted above that the increasing process in the Doob–Meyer decomposition can be taken to be predictable. By Theorem 4.12, for any stopping time T

$$Z_T = E[A_\infty|\mathcal{F}_T] - A_T \quad \text{a.s.} \qquad \square$$

Remark 8.8. The main result of this chapter is the following theorem, which is a converse of the above, and shows that if a potential is of class D then it has a Doob–Meyer decomposition.

Theorem 8.9. *Suppose $\{Z_t\}$ is a potential of class* D. *Then there is a unique predictable integrable increasing process $\{A_t\} \in \mathcal{A}_0^+$ such that $\{Z_t\}$ is the potential generated by $\{A_t\}$. That is,*

$$Z_t = E[A_\infty|\mathcal{F}_t] - A_t \quad a.s., \qquad t \in \tau.$$

PROOF. *Uniqueness.* If $\{A_t\} \in \mathcal{A}_0^+$ and $\{B_t\} \in \mathcal{A}_0^+$ are two predictable processes occurring in Doob–Meyer decompositions of $\{Z_t\}$, then $\{A_t - B_t\}$ is a predictable process in $\mathcal{A}_0$ which is also a martingale. Therefore, by Corollary 7.32 $\{A_t - B_t\}$ is indistinguishable from the zero process.

Existence. The existence of a Doob–Meyer decomposition for a potential of class D will be established by a series of lemmas. The first step is to construct a measure on the predictable σ-field Σ_p.

From Theorem 6.7 we know that Σ_p is generated by stochastic intervals of the form $[\![0_A]\!]$, $A \in \mathcal{F}_0$ and $]\!]S, T]\!]$, where S and T are arbitrary stopping times. Write $\mathcal{C}_p$ for the Boolean algebra generated by intervals of this form. Because

$$]\!]S, T]\!]^c =]\!]0, S]\!] \cup]\!]T, \infty[\![$$

and

$$]\!]S, T]\!] \cap]\!]U, V]\!] =]\!]S \vee U, (S \vee U) \vee (T \wedge V)]\!],$$

every element of $\mathcal{C}_p$ is a finite union of intervals of the above form. In fact, for any $C \in \mathcal{C}_p$ define S_1 as the debut of C, T_1 the debut of $]\!]S_1, \infty[\![\cap C^c$, S_2 the debut of $]\!]T_1, \infty[\![\cap C$, and so on. Then C can be written in a unique way as

$$C = [\![0_F]\!] \cup]\!]S_1, T_1]\!] \cup \cdots \cup]\!]S_n, T_n]\!], \quad \text{where } F = \{S_1 = 0\} \in \mathcal{F}_0.$$

Define

$$\bar{C} = [\![0_F]\!] \cup [\![S_1, T_1]\!] \cup \cdots \cup [\![S_n, T_n]\!].$$

A function μ can be defined on $\mathcal{C}_p$ by putting

$$\mu(C) = E[Z_{S_1} - Z_{T_1}] + \cdots + E[Z_{S_n} - Z_{T_n}]$$

for $C \in \mathcal{C}_p$.

Clearly if $C_1, C_2 \in \mathscr{C}_p$ and $C_1 \cap C_2 = \varnothing$ then $\mu(C_1 \cup C_2) = \mu(C_1) + \mu(C_2)$. Because $\{Z_t\}$ is a supermartingale μ is positive, and because $\{Z_t\}$ is a potential μ is bounded. To allow us to say that μ can be extended to the σ-algebra Σ_p the next two lemmas show that μ satisfies continuity conditions.

Lemma 8.10. *Suppose $C \in \mathscr{C}_p$. Then for $\varepsilon > 0$ there is an element $D \in \mathscr{C}_p$ such that $\bar{D} \subset C$ and $\mu(C) \leq \mu(D) + \varepsilon$.*

PROOF. It is sufficient to consider the case when $C =]\!]S, T]\!]$, where $S < T$ on $\{S < \infty\}$. Write S_n for the restriction of $S + 1/n$ to the set $\{S + 1/n < T\}$ and T_n for the restriction of T to the same set. Then $S_n > S$ on $\{S < \infty\}$ and, certainly in general, $S_n \geq S$. Furthermore, $S = \lim_n S_n$. Similarly, $T_n = T$ on $\{S_n < \infty\}$ and $T_n \geq T$. Also, $T = \lim_n T_n$. Therefore, for all n, $[\![S_n, T_n]\!] \subset]\!]S, T]\!]$. Because $\{Z_t\}$ is right continuous $\lim_n Z_{S_n} = Z_S$ a.s. and $\lim_n Z_{T_n} = Z_T$ a.s. $\{Z_t\}$ is of class D, so these random variables are uniformly integrable and these limits hold in $L^1(P)$. Consequently $\lim_n E[Z_{S_n} - Z_{T_n}] = E[Z_S - Z_T]$, and the result is proved by taking $D = [\![S_n, T_n]\!]$ for n large enough. □

Lemma 8.11. *Suppose $\{C_n\}$ is a decreasing sequence of elements of $\mathscr{C}_p$ such that $\bigcap_n C_n = \varnothing$. Then $\lim_n \mu(C_n) = 0$.*

PROOF. Suppose $\varepsilon > 0$. From Lemma 8.10 for each n there is a set $D_n \in \mathscr{C}_p$ such that $\bar{D}_n \subset C_n$ and $\mu(C_n) \leq \mu(D_n) + 2^{-n}\varepsilon$. Write

$$\Delta_n = D_1 \cap D_2 \cap \cdots \cap D_n \subset D_n.$$

Then

$$\begin{aligned} C_n \backslash \Delta_n &= C_n \cap \Delta_n^c \\ &= C_n \cap (D_1 \cap D_2 \cap \cdots \cap D_n)^c \\ &= C_n \cap (D_1^c \cup D_2^c \cup \cdots \cup D_n^c) = \bigcup_{k=1}^n (C_n \cap D_k^c) \subset \bigcup_{k=1}^n (C_k \cap D_k^c), \end{aligned}$$

because $C_n \subset C_{n-1} \subset \cdots \subset C_1$. Therefore,

$$\begin{aligned} \mu(C_n \backslash \Delta_n) &= \mu(C_n) - \mu(\Delta_n) \\ &\leq \sum_{k=1}^n (\mu(C_k) - \mu(D_k)) \leq \varepsilon \sum_{k=1}^n 2^{-k} \\ &\leq \varepsilon. \end{aligned}$$

The sequence $\{\bar{\Delta}_n\}$ is decreasing and $\bigcap_n \bar{\Delta}_n \subset \bigcap C_n = \varnothing$. Write S_n for the debut of $\bar{\Delta}_n$. Then, because the sections of $\bar{\Delta}_n$ are closed (this is precisely why we need them closed), the S_n are increasing and $\lim_n S_n = +\infty$ a.s. Now

$$\mu(\Delta_n) \leq E[Z_{S_n} - Z_\infty] = E[Z_{S_n}].$$

Because $\{Z_t\}$ is a potential of class D, $E[Z_{S_n}]\to 0$ so $\lim_n \mu(\Delta_n)=0$. Consequently, $\lim \mu(C_n)\le\varepsilon$ for every $\varepsilon>0$, and so $\lim \mu(C_n)=0$. □

Remarks 8.12. The function μ can, therefore, be extended in a unique way to a measure on the σ-field generated by $\mathscr{C}_p$. That is, we have a measure, again denoted by μ, on the predictable σ-field Σ_p.

Lemma 8.13.

(a) $\mu([\![0]\!])=0$ *and* $\mu([\![0,\infty]\!])<\infty$,
(b) *for every predictable evanescent set* H, $\mu(H)=0$.

PROOF. (a) is obvious from the definition of μ.

(b) Because H is evanescent its debut D_H is almost surely infinite. Therefore, $A=\{D_H<\infty\}$ is of measure zero, and so belongs to $\mathscr{F}_0$. Certainly $H\subset[\![0_A]\!]\cup]\!]0_A,\infty[\![$ so, because 0_A is a.s. infinite, $\mu(H)=0$. □

COMPLETION OF PROOF OF THEOREM 8.9. The measure μ, defined on Σ_p, can be immediately extended to a measure $\bar{\mu}$ defined on the σ-field $\mathscr{B}\otimes\mathscr{F}$ of measurable sets by putting

$$\bar{\mu}(X)=\mu(\Pi_p X)$$

for every bounded measurable process $X\in B(\mathscr{B}\otimes\mathscr{F})$. Clearly $\bar{\mu}$ has finite mass, because μ is bounded, so from Theorem 7.13 μ is generated by an integrable increasing process $\{A_t\}\in\mathscr{A}_0^+$ and from Corollary 7.23, $\{A_t\}$ is predictable. For any two stopping times S, T, with $S\le T$, by definition:

$$\mu(]\!]S,T]\!])=E[Z_S-Z_T]$$

$$=E[A_T-A_S].$$

Therefore, for any $t\in[0,\infty[$ and $H\in\mathscr{F}_t$ put $S=t_H$ and $T=\infty$ so that

$$E[I_H(A_\infty-A_t)]=E[I_H Z_t].$$

That is,

$$Z_t=E[A_\infty-A_t|\mathscr{F}_t]\quad\text{a.s.}$$

for all t, so Z is the potential generated by the increasing predictable process $\{A_t\}\in\mathscr{A}_0^+$, and

$$Z_t=E[A_\infty|\mathscr{F}_t]-A_t$$

is the Doob–Meyer decomposition of Z. □

Remarks 8.14. For a general right continuous uniformly integrable supermartingale $\{X_t\}$, the potential in its Riesz decomposition is of class D if, and only if $\{X_t\}$ is of class D. We, therefore, have the following result:

Theorem 8.15. *Suppose $\{X_t\}$ is a right continuous supermartingale of class D. Then there exists a unique predictable increasing process $\{A_t\} \in \mathscr{A}_0^+$ such that the process $M_t = X_t + A_t$ is a uniformly integrable martingale.*

To justify the following remarks and definition, first note the following result.

Theorem 8.16. *Suppose $\{Y_t\}$ is a right continuous uniformly integrable martingale, and $\{T_n\}$ is an increasing sequence of stopping times. Then from Corollary* 3.19 *and Theorem* 3.32

$$\lim Y_{T_n} = E\left[Y_{\lim T_n} \middle| \bigvee_n \mathscr{F}_{T_n} \right]$$

almost surely, and in $L^1(P)$.

In particular, if T is a predictable stopping time announced by the sequence $\{T_n\}$, then Corollary 5.24, *states that $\mathscr{F}_{T-} = \bigvee_n \mathscr{F}_{T_n}$, so that*

$$Y_{T-} = \lim Y_{T_n} = E[Y_T | \mathscr{F}_{T-}] \quad a.s.$$

Remarks 8.17. Suppose $\{X_t\}$ is a right continuous uniformly integrable supermartingale and that

$$X_t = Y_t + Z_t$$

is its Riesz decomposition. If T is a predictable stopping time announced by the sequence $\{T_n\}$, then, because Y is a right continuous uniformly integrable martingale, the above result states that

$$Y_{T-} = E[Y_T | \mathscr{F}_{T-}].$$

Because the sequence $E[Z_{T_n}]$ is decreasing, by Doob's optional stopping theorem, the random variable $Z_{T-} = \lim Z_{T_n}$ is integrable. Consequently, X_{T-} is integrable. However,

$$X_{T_n} \geq E[X_T | \mathscr{F}_{T_n}],$$

again by Doob's theorem, so in the limit

$$X_{T-} \geq E[X_T | \mathscr{F}_{T-}],$$

because

$$E[X_T | \mathscr{F}_{T-}] = \lim_n E[X_T | \mathscr{F}_{T_n}]$$

by the martingale convergence theorem. Therefore,

$$E[X_{T-}] = E[X_T],$$

if, and only if, $X_{T-} = E[X_T | \mathscr{F}_{T-}]$.

Supermartingales which satisfy this identity for every predictable stopping time will play a significant role in the sequel.

Definition 8.18. Suppose $\{X_t\}$ is a right continuous uniformly integrable supermartingale. Then $\{X_t\}$ is said to be *regular* if, for every predictable stopping time T,

$$E[X_{T-}] = E[X_T].$$

In fact, if X is regular then

$$X_{T-} = E[X_T | \mathscr{F}_{T-}] \quad \text{a.s.}$$

for every stopping time T.

Remarks 8.19. Clearly a right continuous uniformly integrable martingale is regular. Therefore, a right continuous uniformly integrable supermartingale is regular if, and only if, its potential in its Riesz decomposition is regular. If $\{X_t\}$ is a positive, or bounded, supermartingale, then $\{X_t\}$ is regular if, and only if, its predictable projection equals its left continuous version, because then

$$(\Pi_p X)_t = X_{t-},$$

so for every $T \in \mathscr{T}_p$

$$E[X_{T-}] = E[(\Pi_p X)_T] = E[X_T].$$

In particular, $\{X_t\}$ is regular if $\{X_t\}$ is continuous.

Lemma 8.20. *Suppose $\{X_t\}$ is a right continuous supermartingale of class* D, *and $\{X_t\}$ is the predictable increasing process in its Doob–Meyer decomposition. Then $\{A_t\}$ is continuous if and only if $\{X_t\}$ is regular.*

PROOF. $X_t = M_t - A_t$ where $\{A_t\} \in \mathscr{A}_0^+$ and $\{M_t\}$ is a uniformly integrable martingale. For every predictable stopping time T

$$M_T = X_T + A_T \quad \text{and} \quad M_{T-} = X_{T-} + A_{T-}.$$

However,

$$E[M_T] = E[M_{T-}]$$

so

$$E[A_T - A_{T-}] = E[X_T - X_{T-}].$$

From Theorem 6.46, because $\{A_t\}$ is predictable it is continuous if, and only if, it does not charge any predictable stopping time, so the result follows. □

Remarks 8.21. In particular, if $\{X_t\}$ is a continuous supermartingale of class D, then it is certainly regular, and the increasing process in its Doob–Meyer decomposition is continuous.

By localization we now obtain the Doob–Meyer decomposition of a general supermartingale.

Theorem 8.22. *Suppose* $\{X_t\}$, $t \in [0, \infty]$, *is a right continuous supermartingale. Then* $\{X_t\}$ *has a unique decomposition of the form*

$$X_t = M_t - A_t.$$

Here $\{A_t\}$ *is an increasing predictable process,* $A_0 = 0$ *a.s., and there is an increasing sequence* $\{T_n\}$ *of stopping times such that* $\lim T_n = \infty$ *a.s. and each stopped process* $M_t^{T_n}$ *is a uniformly integrable martingale.*

Remark 8.23. In the terminology of Definition 9.2 of the following chapter $M \in \mathcal{M}_{\text{loc}}$, the set of local martingales. From Lemma 7.39, A is locally integrable.

Definition 8.24. The above decomposition is called the *Doob–Meyer* decomposition of the supermartingale $\{X_t\}$.

PROOF. *Uniqueness.* This is similar to the proof of uniqueness of Theorem 8.9 above.

Suppose

$$X_t = M_t - A_t = N_t - B_t$$

are two such decompositions. Then there is an increasing sequence $\{S_n\}$ of stopping times such that $\lim S_n \to \infty$ and $M^{S_n} - N^{S_n} = A^{S_n} - B^{S_n}$ is both a martingale and a predictable process of integrable variation. Therefore, by Corollary 7.32

$$M_t^{S_n} - N_t^{S_n} = A_t^{S_n} - B_t^{S_n} = 0.$$

Letting $n \to \infty$ we see that $M = N$ and $A = B$.

Existence. Suppose there are two stopping times S, T and two increasing predictable processes A, B zero at $t = 0$, such that

$$X^S - A = M$$

and

$$X^T - B = N$$

are both local martingales. Then

$$X^S - A^S = M^S,$$

so that M^S is also a local martingale and A^S is predictable, and the above uniqueness argument applied to X^S implies that

$$A = A^S.$$

That is, A is stopped at S. Now,

$$X^{S \wedge T} - A^T = M^T$$

and

$$X^{S \wedge T} - B^S = N^S.$$

Both M^T and N^S are local martingales, and A^T and B^S are predictable. Again, applying the uniqueness argument to $X^{S\wedge T}$ we see that $A^T = B^S$, that is, $A = B$ up to the time $T \wedge S$.

From these remarks we see that it is enough to establish the existence, for each positive integer n, of an increasing predictable process A^n, which is zero at $t=0$, such that $X_t^n + A_t^n$ is a local martingale. Here $X_t^n = X_{t\wedge n}$.

For each $t \in [0, \infty[$ we should then have

$$A_t^n = A_{t\wedge n}^{n+1},$$

and an increasing predictable process A can be defined by putting

$$A_t = \lim_n A_t^n.$$

Clearly $A_{T\wedge n} = A_t^n$, $A_0 = 0$ a.s., and $X + A$ is a local martingale.

Therefore, the result will be proved if we can obtain a Doob–Meyer decomposition for a supermartingale $\{X_t\}$ which is stopped at time $t = n$. In this case consider the nonnegative process

$$Y_t = X_t - E[X_n | \mathscr{F}_t] \geq 0 \quad \text{a.s.},$$

so that Y_t is a potential. For each positive integer n write

$$S_n = \inf\{t \colon Y_t \geq n\} \wedge n.$$

Note that $P\{S_n = 0\}$ can be positive. However, as $n \to \infty$ the sequence of stopping times S_n tends to $+\infty$. The potential Y^{S_n} is bounded above by n on $[\![0, S_n[\![$, and so by Y_{S_n} on $[\![S_n, \infty[\![$, and so by $n \vee Y_{S_n}$ everywhere. However by Theorem 4.12 this random variable is integrable, so the potential Y^{S_n} is of class D. Therefore, by Theorem 8.9 above, Y^{S_n} has a Doob–Meyer decomposition, that is there is a unique predictable increasing process $A^n \in \mathscr{A}_0^+$ such that

$$Y_t^{S_n} + A_t^n = E[A_\infty^n | \mathscr{F}_t],$$

a uniformly integrable martingale. By the remarks above we can, therefore define a predictable process $A \in (\mathscr{A}_0^+)_{\text{loc}}$ such that

$$A^{S_n} = A^n \quad \text{for every } n,$$

and $Y + A$ is then a local martingale reduced by the increasing sequence of stopping times $\{S_n\}$. □

CHAPTER 9

The Structure of Square Integrable Martingales

Convention 9.1. We assume, as in previous chapters, that we are working on a probability space $(\Omega, \mathscr{F}, P)$ which has a right continuous, complete, filtration $\{\mathscr{F}_t\}$, $t \in [0, \infty[$. Furthermore, indistinguishable processes will be indentified, so that when we speak of a process we really mean an equivalence class of indistinguishable process.

Definition 9.2. $\mathscr{M}$ will denote the set of uniformly integrable martingales on $(\Omega, \mathscr{F}, P)$ with respect to the filtration $\{\mathscr{F}_t\}$. That is, if $\{M_t\} \in \mathscr{M}$ then $\{M_t\}$ is a martingale and the set of random variables $\{M_t\}$, $t \in [0, \infty[$, is uniformly integrable.

Following Notation 7.36 $\mathscr{M}_{\text{loc}}$ will denote the set of processes which are locally in $\mathscr{M}$, and if $\mathscr{C}$ is any class of processes $\mathscr{C}_0$ will denote the set of $X \in \mathscr{C}$ such that $X_0 = 0$ a.s. Therefore, $\mathscr{M}_0$ will denote the set of martingales $\{M_t\} \in \mathscr{M}$ such that $M_0 = 0$ a.s. and we shall write $\mathscr{L}$ for the set $(\mathscr{M}_0)_{\text{loc}} = (\mathscr{M}_{\text{loc}})_0$. As in Definition 6.37, if $\{X_t\}$ is any real, measurable process, the process $\{X_t^*\}$ is defined by

$$X_t^* = \sup_{s \leq t} |X_t|.$$

Lemma 9.3. *Every martingale is a local martingale.*

Proof. Suppose $\{M_t\}$ is a martingale. For each positive integer $n \in N$ write $M_s^n = E[M_n | \mathscr{F}_s]$. Therefore, $\{M_t^n\} \in \mathscr{M}$ and $\{M_t\} \in \mathscr{M}_{\text{loc}}$. □

Lemma 9.4. *Suppose $\{M_t\} \in \mathscr{M}_{\text{loc}}$. Then $\{M_t\}$ is in $\mathscr{M}$ if, and only if, it is of class* D.

PROOF. If $\{M_t\} \in \mathcal{M}$ then for every stopping time T

$$M_T = E[M_\infty | \mathcal{F}_T],$$

where $M_\infty \in L^1(\Omega, \mathcal{F}, P)$. Therefore, the set of random variables $\{M_T\}$ is uniformly integrable, and so $\{M_t\}$ is of class D.

Conversely, suppose $\{M_t\} \in \mathcal{M}_{\text{loc}}$ is of class D. Then $\{M_t\}$ is corlol and the set of random variables $\{M_t\}$, $t \geq 0$, is uniformly integrable. To show that $\{M_t\} \in \mathcal{M}$ we must establish that $M_s = E[M_t | \mathcal{F}_s]$ whenever $s \leq t < \infty$. Let $\{T_n\}$ be an increasing sequence of stopping times such that $\{M_t^{T_n}\} \in \mathcal{M}$. Then for each n: $M_s^{T_n} = E[M_t^{T_n} | \mathcal{F}_s]$. However, because $\{M_t\}$ is of class D the sequences $\{M_s^{T_n}\}$ and $\{M_t^{T_n}\}$ converge both almost surely and in L^1 to, respectively, M_s and M_t. The result follows. □

Definition 9.5. For $M \in \mathcal{M}$ and $p \in [1, \infty]$ write

$$\|M\|_{\mathcal{H}^p} = \|M_\infty^*\|_p.$$

Here $\| \ \|_p$ denotes the norm in $L^p = L^p(\Omega, \mathcal{F}, P)$. Then $\mathcal{H}^p$ is the space of martingales such that

$$\|M\|_{\mathcal{H}^p} < \infty.$$

Identifying indistinguishable martingales $\mathcal{H}^p$ is a Banach space under this norm.

Lemma 9.6. (1) If $p' \leq p$, $\mathcal{H}^p \subset \mathcal{H}^{p'}$,
(2) *if* $1 < p < \infty$ *and* $M \in \mathcal{M}$

$$\|M_\infty\|_p \leq \|M_\infty^*\|_p \leq q\|M_\infty\|_p, \quad \textit{where } p^{-1} + q^{-1} = 1,$$

(3) *if* $p = \infty$, $\|M_\infty\|_\infty = \|M_\infty^*\|_\infty$.

PROOF. (1) If $p' \leq p$, then using Hölders inequality

$$\|M_\infty^*\|_{p'} \leq \|M_\infty^*\|_p.$$

(2) The first inequality is trivial and the second is the result of Theorem 4.2(5).

(3) Now

$$|M_t| \leq |M_\infty^*| \quad \text{a.s.,}$$

so

$$|M_\infty| \leq |M_\infty^*| \quad \text{a.s.}$$

and

$$\|M_\infty\|_\infty \leq \|M_\infty^*\|_\infty.$$

Conversely, for any $\varepsilon > 0$ define

$$T_\varepsilon = \inf\{t \geq 0 : |M_t| \geq \|M_\infty^*\| - \varepsilon\}.$$

Now

$$P\{|M_\infty^*| > \|M_\infty^*\|_\infty - \varepsilon\} > 0,$$

so

$$P\{\exists t\colon |M_t| > \|M_\infty^*\|_\infty - \varepsilon\} > 0.$$

Therefore,

$$\|M_{T_\varepsilon}\|_\infty \geq \|M_\infty^*\|_\infty - \varepsilon.$$

However,

$$M_{T_\varepsilon} = E[M_\infty | \mathscr{F}_{T_\varepsilon}] \quad \text{a.s.},$$

so

$$\|M_{T_\varepsilon}\|_\infty \leq \|M_\infty\|_\infty.$$

Because ε is arbitrary $\|M_\infty\| \geq \|M_\infty^*\|_\infty$, and the result follows. □

Lemma 9.7. $\mathscr{H}^p \subset \mathscr{M}$.

PROOF. If M_∞^* is integrable the martingale $\{M_t\}$ is of class D, and the result follows from Lemma 9.4. □

Remarks 9.8. From Lemma 9.6(2) we see that if $1 < p \leq \infty$ the norm $\|M_\infty\|_p$ is equivalent to the norm $\|M\|_{\mathscr{H}^p}$, and so $\mathscr{H}^q$ can be identified with the Banach space $L^p(\Omega, \mathscr{F}_\infty, P)$ by the map which associates to $\{M_t\} \in \mathscr{H}^p$ its terminal value $M_\infty \in L^p$.

The space $\mathscr{H}^1$ requires more careful analysis, and will not be discussed here.

Sometimes (for example, [61]), the space $\mathscr{H}^p$, $1 < p < \infty$, is defined as those martingales for which

$$\sup_t E[M_t^p] < \infty.$$

However, from Corollary 1.19 we would then know that $\{M_t\}$ is uniformly integrable, that M_∞ exists and $M_\infty = \lim_{t \to \infty} M_t$ a.s., and that $\{M_t^p\}$, $t \in [0, \infty]$ is a submartingale. Therefore

$$E[M_t^p] \leq E[M_\infty^p],$$

so the definitions are equivalent.

Theorem 9.9. *Suppose $1 < p \leq \infty$, and that $\{M_t^n\}$ is a sequence of martingales which converge to the martingale $\{M_t\}$ in the norm of $\mathscr{H}^p$. Then there is a subsequence $\{M_t^{n_k}\}$ such that, for almost every $\omega \in \Omega$, $M_t^{n_k}(\omega)$ converges uniformly to $M_t(\omega)$ on $[0, \infty]$.*

PROOF. By definition

$$\lim_n \|M^n - M\|_{\mathscr{H}^p} = 0.$$

This implies that

$$\lim_n \|(M^n - M)^*\|_p = 0.$$

Therefore, there is a subsequence $\{n_k\}$ such that

$$\sup_t |M_t^{n_k}(\omega) - M_t(\omega)| = 0 \quad \text{a.s.} \qquad \square$$

Remark 9.10. The limit in $\mathscr{H}^p$ of a sequence of continuous martingales is, therefore, a continuous martingale, and the jumps of a limit are the limits of the jumps of an approximating sequence.

Definition 9.11. Two local martingales M, N in $\mathscr{M}_{\text{loc}}$ are *orthogonal* if their product $MN = \{M_t N_t\}$ is in $\mathscr{L}$.

We shall then write $M \perp N$. Note that, in particular, orthogonality implies that

$$M_0 N_0 = 0 \quad \text{a.s.}$$

Lemma 9.12. *Suppose $M, N \in \mathscr{H}^2$ are orthogonal. Then for every stopping time $T \in \mathscr{T}_o$ the random variables M_T, N_T are orthogonal in L^2 and $MN \in \mathscr{H}_0^1$.*

Conversely, if $M_0 N_0 = 0$ a.s. and the random variables M_T, N_T are orthogonal in L^2 for every $T \in \mathscr{T}_o$ then M, N are orthogonal.

Proof. Because M_∞^* and N_∞^* are in L^2 the product $M_\infty^* N_\infty^*$ is in L^1. Now

$$(MN)_\infty^* = \sup_t |M_t N_t|$$

$$\leq M_\infty^* N_\infty^*,$$

so $MN \in \mathscr{H}_0^1$ if M and N are orthogonal. Therefore, for any $T \in \mathscr{T}$

$$E[M_T N_T] = E[M_0 N_0] = 0.$$

Conversely, for any $T \in \mathscr{T}_o$, $M_T N_T \in L^1$ so $E[|M_T N_T|] < \infty$, and $E[M_T N_T] = 0$. Therefore, by Lemma 4.18, $\{M_t N_t\}$ is a uniformly integrable martingale. $M_0 N_0 = 0$ a.s. so certainly $MN \in \mathscr{L}$. $\square$

Remark 9.13. Recall Notation 3.9 and 7.35, that if X is a process and $T \in \mathscr{T}$ then X^T is the process X stopped at T: $X_t^T = X_{t \wedge T}$.

Definition 9.14. A subspace $\mathscr{K} \subset \mathscr{H}^2$ is said to be *stable* if:

(1) it is closed under the L^2 topology,
(2) it is closed under stopping, that is $T \in \mathscr{T}$ and $M \in \mathscr{K}$ imply $M^T \in \mathscr{K}$,
(3) if $M \in \mathscr{K}$ and $A \in \mathscr{F}_0$ then $I_A M \in \mathscr{K}$.

Theorem 9.15. *Suppose $\mathscr{K}$ is a stable subspace of $\mathscr{H}^2$, and write $\mathscr{K}^\perp$ for the set of martingales $N \in \mathscr{H}^2$ such that $E[M_\infty N_\infty] = 0$ for all $M \in \mathscr{K}$. Then $\mathscr{K}^\perp$ is a stable subspace, and if $M \in \mathscr{K}$, $N \in \mathscr{K}^\perp$ then M and N are orthogonal.*

PROOF. Consider $M \in \mathscr{K}$, $N \in \mathscr{K}^\perp$ and $T \in \mathscr{T}_o$. Then $E[L_\infty N_\infty] = 0$ for all $L \in \mathscr{K}$ and $\mathscr{K}$ is closed under stopping. Taking $L = M^T \in \mathscr{K}$:

$$E[M_T N_\infty] = 0.$$

Thus,

$$E[M_T N_\infty] = E[E[M_T N_\infty | \mathscr{F}_T]]$$
$$= E[M_T E[N_\infty | \mathscr{F}_T]] = E[M_T N_T] = 0.$$

Taking $T = 0$ and $A \in \mathscr{F}_0$

$$E[I_A M_0 N_0] = 0 \quad \text{so } M_0 N_0 = 0 \quad \text{a.s.},$$

and we see M and N are orthogonal. Furthermore,

$$E[I_A M_T N_T] = E[M_\infty (I_A N^T)_\infty] = 0,$$

which implies that $I_A N^T \in \mathscr{K}^\perp$ for any $N \in \mathscr{K}^\perp$, $T \in \mathscr{T}$ and $A \in \mathscr{F}_0$. Consequently, $\mathscr{K}^\perp$ is stable. □

Remark 9.16. The proof above does not use the closedness of $\mathscr{K}$ or the vector space operations.

Corollary 9.17. *Suppose $\mathscr{K} \subset \mathscr{H}^2$ is a stable subspace. Then every element $M \in \mathscr{H}^2$ has a unique decomposition*

$$M = N + N',$$

where $N \in \mathscr{K}$ and $N' \in \mathscr{K}^\perp$.

PROOF. Suppose $\mathscr{K}_\infty$ is the closed subspace of L^2 generated by the random variables $\{M_\infty : M \in \mathscr{K}\}$. $\mathscr{K}_\infty^\perp$ is defined similarly. Clearly, for any $M \in \mathscr{H}^2$ M_∞ has a unique decomposition

$$M_\infty = N_\infty + N'_\infty,$$

where $N_\infty \in \mathscr{K}$ and $N'_\infty \in \mathscr{K}^\perp$. Then N (resp. N') is the corlol version of the martingale defined by

$$N_t = E[N_\infty | \mathscr{F}_t] \quad (\text{resp. } N'_t = E[N'_\infty | \mathscr{F}_t]).$$ □

Remark 9.18. If X_t, $t \in [0, \infty]$, is any process we shall follow the convention introduced in Chapter 7 that $X_{0-} = 0$ a.s.

Definition 9.19. $\mathscr{H}^{2,c} \subset \mathscr{H}^2$ will denote the space of continuous square integrable martingales. Clearly $\mathscr{H}^{2,c}$ is closed, by Theorem 9.9, and is closed under stopping, so $\mathscr{H}^{2,c}$ is stable. Also, because of the convention that $M_{0-} = 0 = M_0$ a.s. $\mathscr{H}^{2,c} \subset \mathscr{H}_0^2$.

Definition 9.20. $\mathscr{H}^{2,d}$ is the stable subspace orthogonal to $\mathscr{H}^{2,c}$. Martingales in $\mathscr{H}^{2,d}$ are said to be *purely discontinuous*.

We shall determine the structure of $\mathscr{H}^{2,d}$ by studying certain simple subspaces.

Definition 9.21. Suppose $T \in \mathscr{T}$ is a stopping time. $\mathscr{M}(T)$ will denote the space of martingales in $\mathscr{H}^{2,d}$ which are continuous outside the graph of T. Note that $\mathscr{M}(T)$ is a stable subspace.

Lemma 9.22. *For $T = 0$, $\mathscr{M}(0)$ is the space of constant martingales.*

PROOF. If $H \in \mathscr{M}(0)$, then $M_t = H_t - H_0$ is a continuous martingale, but also it is purely discontinuous. Therefore, $E[M_t^2] = E[M_0^2] = 0$ so $H_t = H_0$ a.s. for all t. □

Remarks 9.23. Because of the simple structure of $\mathscr{M}(0)$ it will be assumed below that $T > 0$ a.s. Again, following the convention that $H_{0-} = 0$, because $H \in \mathscr{M}(T)$ is continuous outside the graph of T, $H_{0-} = 0 = H_0$, so $\mathscr{M}(T) \subset \mathscr{H}_0^{2,d}$.

The structure theorems below are generalizations of the following classical result established by Paul Lévy for independent increment processes: suppose $\{X_t\}$ is a real corlol process with independent increments. Then in general $\sum_{0<s\le t} \Delta X_s$ is not convergent, and the sample paths of $\{X_t\}$ are not of bounded variation. However, for any $\varepsilon > 0$ consider the sum of the jumps of size between ε and 1:

$$A_t^\varepsilon = \sum_{0<s\le t} \Delta X_s I_{\{\varepsilon \le |\Delta X_s| \le 1\}}.$$

Then it can be shown that A^ε is adapted and of integrable variation on any finite interval. However, A^ε is not predictable because its jumps are all totally inaccessible. Its dual predictable projection $\Pi_p^* A_t^\varepsilon$ is of the form $c_\varepsilon t$. As $\varepsilon \to 0$ neither c_ε nor A^ε has a limit, but

$$A_t^\varepsilon - \Pi_p^* A_t^\varepsilon$$

does have a limit in L^2, and this enabled Lévy to describe the structure of independent increment processes.

Theorem 9.24. *Suppose M is a martingale which is also a process of integrable variation. Then*

$$M_t = M_0 + A_t - \Pi_p^* A_t.$$

Here

$$A_t = \sum_{0<s\le t} \Delta M_s \in \mathscr{A}^+$$

and $\Pi_p^* A_t$ is continuous.

PROOF. Consider the process

$$B_t = M_t - M_0 - A_t,$$

where A is given above. Then $B \in \mathscr{A}_0$ and B is continuous. Therefore, by Corollary 7.28 $\Pi_p^* B = B$, so from Theorem 7.30

$$\Pi_p^* B = \Pi_p^*(M - M_0) - \Pi_p^* A = -\Pi_p^* A,$$

and

$$\begin{aligned} M_t &= M_0 + A_t + B_t \\ &= M_0 + A_t - \Pi_p^* A_t. \end{aligned}$$

Clearly $\Pi_p^* A_t = -B_t$ is continuous. □

Lemma 9.25. *For M a martingale of integrable variation and any bounded corlol martingale N*

$$E[M_\infty N_\infty] = E\Big[\sum_{s \geq 0} \Delta M_s \Delta N_s\Big].$$

PROOF. $\{M_t - M_0\} \in \mathscr{A}_0$ and is a martingale, so from Theorem 7.30(3) the restriction to Σ_p of the measure associated with $M_t - M_0$ is zero. The process N_{t-} is predictable so consequently

$$E\Big[\int_{[0,\infty[} N_{s-}\, dM_s\Big] = 0.$$

However, the constant process N_∞ has optional projection N_t, so by Theorem 7.21

$$E[M_\infty N_\infty] = E\Big[\int_{[0,\infty[} N_\infty\, dM_s\Big] = E\Big[\int_{[0,\infty[} N_s\, dM_s\Big], \quad \text{because } M_{0-} = 0.$$

Subtracting,

$$\begin{aligned} E[M_\infty N_\infty] &= E\Big[\int_{[0,\infty[} \Delta N_s\, dM_s\Big] \\ &= E\Big[\sum_{s \geq 0} \Delta M_s \Delta N_s\Big]. \end{aligned}$$

(Note that if $M_0 \neq 0$, $N_0 \neq 0$ this sum includes the term $\Delta M_0 \Delta N_0$.) □

Corollary 9.26. *For any M, N as above*

$$L_t = M_t N_t - \sum_{s \leq t} \Delta M_s \Delta N_s \in \mathscr{M}_0.$$

PROOF. Applying the lemma to the martingale N_t^T, stopped at an arbitrary stopping time $t \in \mathscr{T}$

$$E[M_\infty N_T] = E\Big[\sum_{s \leq T} \Delta M_s \Delta N_s\Big].$$

However,

$$E[M_\infty N_T] = E[M_T N_T],$$

so $E[L_T] = 0$ for any $T \in \mathscr{T}$. Therefore, by Lemma 4.18, $L \in \mathscr{M}_0$. □

Corollary 9.27. *A martingale of integrable variation M is orthogonal to every bounded continuous martingale N.*

PROOF. If N is continuous, $\Delta N_s \equiv 0$ so $MN \in \mathscr{M}_0 \subset \mathscr{L}$. □

Theorem 9.28. *Suppose T is a totally inaccessible stopping time and $\Phi \in L^2(\mathscr{F}_T, P)$. Write*

$$A_t = \Phi I_{t \geq T} \in \mathscr{A}_0.$$

Then

$$B_t = \Pi_p^* A_t \quad \textit{is continuous, and}$$

$$M_t = A_t - B_t$$

is a square integrable martingale in $\mathscr{M}(T) \subset \mathscr{H}_0^{2,d}$.

PROOF. It is sufficient to consider the case when $\Phi \geq 0$ a.s. Note that T totally inaccessible implies $T > 0$ a.s.

From Corollary 7.31

$$M_t = A_t - B_t \quad \text{is a martingale.}$$

The measure on Σ_p associated with the dual predictable projection B_t coincides, by definition, with the measure μ_A associated with A_t. However, the support of μ_A is the graph of the totally inaccessible stopping time T. Therefore, B_t does not charge any predictable stopping time, and so is continuous (Theorem 6.46).

Now $A_\infty = \Phi \in L^2 \subset L^1$ and $E[B_\infty] = E[A_\infty]$, so $B \in L^1$. To show M_t is square integrable we must show $B_\infty \in L^2$. From Lemma 8.4, $\{A_t\}$ and $\{B_t\}$ generate the same potential, that is

$$Z_t = E[A_\infty | \mathscr{F}_t] - A_t = E[B_\infty | \mathscr{F}_t] - B_t \quad \text{a.s.}$$

for all t.

Write N_t for the corlol martingale $E[B_\infty | \mathscr{F}_t]$. Then the predictable projection of N_t is just the left continuous martingale N_{t-}. Because B_t is continuous it is predictable, so the predictable projection of the potential Z_t is Z_{t-}.

From Corollary 7.16

$$E\left[\int_0^\infty N_t \, dB_t\right] = E\left[\int_0^\infty E[B_\infty | \mathscr{F}_t] \, dB_t\right] = E[B_\infty^2],$$

so

$$2E\left[\int_0^\infty (N_t - B_t)\, dB_t\right] = 2E\left[\int_0^\infty Z_t\, dB_t\right] = E[B_\infty^2].$$

By Corollary 7.28 and Theorem 7.25

$$E[B_\infty^2] = 2E\left[\int_0^\infty Z_{t-}\, dB_t\right] = 2E\left[\int_0^\infty Z_{t-}\, dA_t\right].$$

But

$$E\left[\int_0^\infty Z_{t-}\, dA_t\right] \le E\left[\int_0^\infty (Z_{t-} + A_{t-})\, dA_t\right] \le E[\sup_t (Z_{t-} + A_{t-}) A_\infty]$$

and

$$\|\sup_t |E[A_\infty|\mathcal{F}_t]|\,\|_2 \le 2\|A_\infty\|_2$$

by Theorem 4.2(5).

Therefore, $B_\infty \in L^2$ and $M \in \mathcal{M}(T) \subset \mathcal{H}_0^{2,d}$. □

Theorem 9.29. *Suppose $T > 0$ a.s. is a predictable stopping time and $\Phi \in L^2(\mathcal{F}_T, P)$ is such that*

$$E[\Phi|\mathcal{F}_{T-}] = 0 \quad \text{a.s.}$$

Then $M_t = A_t = \Phi I_{t \ge T}$ is a square integrable martingale in $\mathcal{M}(T) \subset \mathcal{H}_0^{2,d}$.

PROOF. Because $\Phi I_{\{T=\infty\}}$ is $\mathcal{F}_{T-}$ measurable we can suppose that $\Phi = 0$ a.s. on the set $\{T = \infty\}$. By Theorem 6.30 if $\{X_t\}$ is any bounded predictable process then $X_T I_{\{T<\infty\}}$ is $\mathcal{F}_{T-}$ measurable. Therefore,

$$E\left[\int_0^\infty X_t\, dA_t\right] = E[X_T(A_T - A_{T-})]$$

$$= E[X_T E[\Phi|\mathcal{F}_{T-}]] = 0,$$

so the restriction of μ_A to Σ_p is zero. Therefore, by Theorem 7.30 $\Pi_p^* A_t$ is the zero process and $M_t = A_t$ is a martingale. Clearly

$$\|M_\infty^*\|_2 = \|A_\infty\|_2 = \|\Phi\|_2 < \infty \quad \text{and} \quad M_t \in \mathcal{M}(T).$$ □

Corollary 9.30. *Suppose as above that either*

(i) *T is a totally inaccessible stopping time and $\Phi \in L^2(\mathcal{F}_T, P)$, or*
(ii) *$T > 0$ a.s. is a predictable stopping time, $\Phi \in L^2(\mathcal{F}_T, P)$ and $E[\Phi|\mathcal{F}_{T-}] = 0$ a.s.*

Write

$$M_t = A_t - \Pi_p^* A_t,$$

where

$$A_t = \Phi I_{t \geq T}.$$

Then, for every martingale $N \in \mathcal{H}^2$ the process

$$L_t = M_t N_t - \Delta M_T \Delta N_T I_{t \geq T}$$

is in $\mathcal{M}_0$. Therefore, M_t is orthogonal to every martingale in $\mathcal{H}^2$ which is continuous at T.

In particular,

$$M_t^2 - (\Delta M_T)^2 I_{t \geq T} \in \mathcal{M}_0$$

and

$$EM_\infty^2 = E[(\Delta M)_T^2].$$

PROOF. In both cases M_t is a martingale of integrable variation, so if N_t is a bounded corlol martingale

$$E[M_\infty N_\infty] = E[\Delta M_T \Delta N_T]$$

$$= E[\Phi \Delta N_T].$$

From Remark 9.10 following Theorem 9.9, the result follows for general $N \in \mathcal{H}^2$ by approximation.

Suppose S is any stopping time. Then $N_\infty^S = N_S$, so applying the above equality to the martingale N_t^S we have

$$E[M_S N_S] = E[\Delta M_T \Delta N_T I_{T \leq S}],$$

that is

$$E[L_S] = 0.$$

Therefore, as in Lemma 4.18, $\{L_t\}$ is a martingale, which is uniformly integrable. Further $L_0 = 0$ a.s. □

Corollary 9.31. *Suppose T is as above and $N \in \mathcal{H}^2$. Then the projection of N onto $\mathcal{M}(T)$ is the martingale*

$$M_t = A_t - \Pi_p^* A_t,$$

where, as above,

$$A_t = \Phi I_{t \geq T}$$

and

$$\Phi = \Delta N_T.$$

PROOF. By construction,

$$\Delta M_T = \Phi = \Delta N_T,$$

and from the above corollary every martingale in $\mathcal{M}(T)$ is orthogonal to every martingale which is continuous at T. Therefore, $N_t - M_t$ is continuous at T and so orthogonal to $\mathcal{M}(T)$. Consequently, M_t is the projection of N_t on $\mathcal{M}(T)$. $\square$

Definition 9.32. When M_t is as in Corollary 9.30 we say it is a *compensated jump martingale.*

Remarks 9.33. In Theorem 9.24 it was shown that a martingale of integrable variation is the sum of a series of compensated jump martingales. This result is now extended to martingales in $\mathcal{H}^{2,d}$. However, unlike the martingales of integrable variation considered in Theorem 9.24, where the process A_t obtained by summing all the jumps can be defined and the dual predictable projection $\Pi_p^* A_t$ obtained, for square integrable martingales the compensated jump martingale for each jump must first be defined and the convergence of these then discussed in $\mathcal{H}^2$. This procedure is motivated by the work of Lévy referred to earlier.

Theorem 9.34. *Suppose $M \in \mathcal{H}_0^{2,d}$. Then M is the sum of a series of compensated jump martingales. Furthermore, M is orthogonal to every martingale $N \in \mathcal{H}^2$ which does not charge a common jump time with M. In particular, M is orthogonal to every continuous martingale in $\mathcal{H}^2$.*

PROOF. M_t is adapted and corlol, so from Theorem 6.46 $\{(t, \omega): \Delta M_t(\omega) \neq 0\}$ is contained in the union of graphs of a countable sequence of stopping times $\bigcup_n [\![S_n]\!]$.

By replacing, if necessary, $[\![S_n]\!]$ by $[\![S_n]\!] \backslash \bigcup_{m<n} [\![S_m]\!]$, we may suppose that the $[\![S_n]\!]$ are disjoint. From Theorem 5.16 each $[\![S_n]\!]$ may be written as $[\![S_n^i]\!] \cup [\![S_n^a]\!]$, where S_n^i is totally inaccessible and S_n^a is accessible. The graphs $[\![S_n^i]\!]$ are disjoint. Each $[\![S_n^a]\!]$ is contained in a countable union of graphs of predictable stopping times. Therefore, by the above procedure, $\bigcup [\![S_n^a]\!]$ can be represented as the union of a sequence of disjoint graphs of predictable stopping times.

Therefore, we can suppose that $\{(t, \omega): \Delta M_t(\omega) \neq 0\}$ is contained in a set of the form $\bigcup_n [\![T_n]\!]$, where each T_n is either totally inaccessible or predictable, and their graphs are disjoint.

For each $n \in Z^+$ consider

$$A_t^n = \Delta M_{T_n} I_{t \geq T_n},$$

$$M_t^n = A_t^n - \Pi_p^* A_t^n.$$

From the above results, M_t^n is a martingale in $\mathcal{H}_0^{2,d}$ which is continuous except at the stopping time T_n, and its jump at T_n is ΔM_{T_n}. Write

$$B^k = M^1 + \cdots + M^k.$$

Then $M-B^k$ is continuous at $T_1, \ldots, T_k$ and so orthogonal to $M^1, \ldots, M^k$ and to their sum B^k. Therefore

$$\begin{aligned} E[M_\infty^2] &= E[(B_\infty^k)^2] + E[(M-B^k)_\infty^2] \\ &= \sum_{n=1}^{k} E[(M_\infty^n)^2] + E[(M-B^k)_\infty^2] \\ &= \sum_{n=1}^{k} E[\Delta M_{T_n}^2] + E[(M-B^k)_\infty^2]. \end{aligned}$$

Clearly $\lim_k \sum_{n=1}^k E[(M_\infty^n)^2]$ is finite, so the sequence $B_t^k = M_t^1 + \cdots + M_t^k$ of partial sums converges in $\mathcal{H}^2$ to a martingale B_t. Because $M-B^k$ is orthogonal to B^k for each k, $M-B$ is orthogonal to B. From Theorem 9.9 there is a subsequence of $\{B^k\}$ whose sample paths converge uniformly, almost surely, to B. Therefore, $M-B$ is continuous. However $M \in \mathcal{H}_0^{2,d}$, so M is orthogonal to every continuous martingale. Consequently, $M-B$ is orthogonal to itself, and so 0.

That is, $M = B = \lim_k \sum_{n=1}^k M^n$. □

Corollary 9.35. *If $M \in \mathcal{H}^{2,d}$ is not zero at $t=0$, the above result is still valid because, following the convention that $M_{0-}=0$ a.s. $\Delta M_0 = M_0$ so we can write $T_0 = 0$ and*

$$\Delta M_{T_0} I_{t \geq T_0} = M_t^0.$$

Therefore,

$$(M - M_0) \in \mathcal{H}_0^{2,d},$$

so

$$M - M_0 = B.$$

That is

$$M = \lim_k \sum_{n=0}^{k} M^n.$$

Corollary 9.36. *The decomposition of the above theorem can be applied to any $M \in \mathcal{H}^2$, so that $M = B^k + (M - B^k)$. As in the theorem, $B = \lim B^k$ exists in $\mathcal{H}^2$, and $M-B$ is continuous. Clearly $M-B$ is the projection of M onto $\mathcal{H}_0^{2,c}$, and B is the projection of M onto $\mathcal{H}^{2,d}$. That is, $M = M^c + M^d$, where $M^c = M - B \in \mathcal{H}_0^{2,c}$ and $M^d = B \in \mathcal{H}^{2,d}$. This decomposition is unique.*

Corollary 9.37. *For any $M \in \mathcal{H}^2$*

$$E\left[\sum_s \Delta M_s^2\right] \leq E[M_\infty^2].$$

Equality holds here if, and only if, $M \in \mathcal{H}^{2,d}$. (Here again we use the convention that $\Delta M_0 = M_0$.)

PROOF. Following the proof of the above theorem we have

$$E[M_\infty^2] = \sum_{n=1}^{k} E[\Delta M_{T_n}^2] + E[(M - B^k)_\infty^2].$$

For almost every ω: $\sum_n \Delta M_{T_n}^2(\omega) = \sum_s \Delta M_s^2(\omega)$. So in the limit:

$$E[M_\infty^2] = E\Big[\sum_s \Delta M_s^2\Big] + E[(M - B)_\infty^2].$$

The result follows. □

Corollary 9.38. *For any t and $M \in \mathscr{H}^2$ $\sum_{s \le t} \Delta M_s^2 < \infty$ a.s.*

Corollary 9.39. *If M and N are in $\mathscr{H}^2$*

(i) $\sum_s |\Delta M_s \Delta N_s| \le (\sum_s \Delta M_s^2)^{1/2} (\sum_s \Delta N_s^2)^{1/2}$,
(ii) $E[\sum_s |\Delta M_s \Delta N_s|] \le \|M\|_2 \|N\|_2$.

PROOF. From Corollary 9.37 the right-hand side of (i) is in $L^1(\Omega)$ and is, therefore, almost surely finite. The result follows.

Taking expectations, part (ii) is also immediate from (i) and Corollary 9.37. □

Theorem 9.40. *Suppose M and N belong to $\mathscr{H}^2$ and one of them, say M, belongs to $\mathscr{H}^{2,d}$. Then*

(i) $E[M_\infty N_\infty] = E[\sum_s \Delta M_s \Delta N_s]$,
(ii) $L_t = M_t N_t - \sum_{s \le t} \Delta M_s \Delta N_s$

belongs to $\mathscr{H}_0^1$.

PROOF. Suppose first that both M and N belong to $\mathscr{H}^{2,d}$. Then, from Corollary 9.37

$$E[M_\infty^2] = E\Big[\sum_s \Delta M_s^2\Big],$$

$$E[N_\infty^2] = E\Big[\sum_s \Delta N_s^2\Big]$$

and

$$E[(M + N)_\infty^2] = E\Big[\sum_s (\Delta M_s + \Delta N_s)^2\Big].$$

Part (i) follows by subtraction. Applying part (i) to the martingales M^T, N^T, stopped at $T \in \mathscr{T}$, we have $E[L_T] = 0$. Therefore, as in Lemma 4.18, L_t is a martingale. Furthermore,

$$|L_t| \le M^* N^* + \sum_s |\Delta M_s \Delta N_s|,$$

which by Corollary 9.39 is in L^1, so $L \in \mathscr{H}_0^1$.

In general, suppose $N \in \mathscr{H}^2$. Then $N = N^c + N^d$, where $N^c \in \mathscr{H}_0^{2,c}$ and $N^d \in \mathscr{H}^{2,d}$. By orthogonality, $M_t N_t^c$ is a martingale which is zero at $t=0$. Therefore, $E[M_\infty N_\infty^c]=0$ and $M_t N_t^c$ is bounded in L^1. As $MN = MN^c + MN^d$ the result follows. □

Theorem 9.41. *If $M \in \mathscr{H}^2 \cap \mathscr{V}$ then $M \in \mathscr{H}^{2,d}$.*

PROOF. If $M \in \mathscr{H}^2 \cap \mathscr{A}$ then by Lemma 9.25

$$E[M_\infty N_\infty] = E\Big[\sum_s \Delta M_s \Delta N_s\Big],$$

for any bounded martingale N.

Both sides of the above identity are continuous in N under the $\mathscr{H}^2$ norm, and so the above identity is valid for $N \in \mathscr{H}^2$. Therefore, in particular,

$$E[M_\infty^2] = E\Big[\sum_s \Delta M_s^2\Big],$$

for $M \in \mathscr{H}^2 \cap \mathscr{A}$, and by continuity, for $M \in \mathscr{H}^2 \cap \mathscr{V}$. Therefore, by Corollary 9.37 $M \in \mathscr{H}^{2,d}$. □

CHAPTER 10

Quadratic Variation Processes

Suppose $M \in \mathscr{H}^2$. Then $M_\infty^* \in L^2(\Omega)$ and $\{M_t^2\}$ is a submartingale such that for all $t \geq 0$, $M_t^2 \leq M_\infty^{*2} \in L^1(\Omega)$. The right continuous version of the process $X_t = E[M_\infty^2 | \mathscr{F}_t] - M_t^2$ is, therefore, a supermartingale of class D. Clearly $X_t \geq 0$ a.s. and $X_\infty = 0$, so X is a potential of class D.

Definition 10.1. $\langle M, M \rangle$ is the unique predictable increasing process in $\mathscr{A}^+$ given by the Doob–Meyer decomposition of

$$X_t = E[M_\infty^2 | \mathscr{F}_t] - M_t^2.$$

(See Theorem 8.9.)

$\langle M, M \rangle$ is called the *predictable quadratic variation of M.*

$M_t^2 - \langle M, M \rangle_t$ is a martingale, and $\langle M, M \rangle_0 = M_0^2$.

Definition 10.2. Suppose $M \in \mathscr{H}^2$ and $M = M^c + M^d$ is its decomposition as in Corollary 9.36 into a continuous martingale and a sum of compensated jump martingales. $[M, M]_t$ is the optional increasing process $\langle M^c, M^c \rangle_t + \sum_{s \leq t} \Delta M_s^2$. $[M, M]_t$ is called the *optional quadratic variation of M.* From Corollary 9.37, $[M, M] \in \mathscr{A}^+$ and from Corollary 9.38 $[M, M]_t < \infty$ a.s.

Remarks 10.3. For $M = M^c + M^d \in \mathscr{H}^2$:

$$M_t^2 = M_t^{c2} + 2M_t^c M_t^d + M_t^{d2}.$$

Because M^c and M^d are orthogonal $M^c M^d$ is a martingale. From Theorem 9.40 $M_t^{d2} - \sum_{s \leq t} \Delta M_s^2$ is a martingale in $\mathscr{H}_0^1$, and from the remarks preceding Definition 10.1 $M_t^{c2} - \langle M^c, M^c \rangle_t$ is in $\mathscr{H}_0^1$. Therefore $M_t^2 - [M, M]_{tt} \in \mathscr{H}_0^1$.

Lemma 10.4. *For* $M \in \mathcal{H}^2$, $\langle M, M\rangle$ *is the dual prediction projection of* $[M, M]$.

Proof. From the definitions $\langle M, M\rangle - [M, M]$ is a martingale, so the result follows from Theorem 7.25 and Corollary 7.35. □

Lemma 10.5. $\langle M^c, M^c\rangle$ *is continuous.*

Proof. This follows from Remarks 8.21. □

Definition 10.6. Suppose $M, N \in \mathcal{H}^2$. Then

$$\langle M, N\rangle = \tfrac{1}{2}(\langle M+N, M+N\rangle - \langle M, M\rangle - \langle N, N\rangle).$$

$\langle M, N\rangle$ is the unique predictable process of integrable variation such that

$$MN - \langle M, N\rangle \in \mathcal{H}_0^1 \quad \text{and} \quad M_0N_0 = \langle M, N\rangle_0.$$

Definition 10.7. Suppose $M, N \in \mathcal{H}^2$. Define

$$[M, N] = \tfrac{1}{2}([M+N, M+N] - [M, M] - [N, N]).$$

Then

$$[M, N] \in \mathcal{A}, \qquad MN - [M, N] \in \mathcal{H}_0^1 \quad \text{and} \quad M_0N_0 = [M, N]_0 = \Delta M_0 \Delta N_0.$$

Remarks 10.8. From the definitions

$$[M, N]_t = \langle M^c, N^c\rangle_t + \sum_{s \le t} \Delta M_s \Delta N_s.$$

Furthermore, M and N are orthogonal if and only if $\langle M, N\rangle$ is the zero process. From the definition, $M^2 - \langle M, M\rangle$ is a martingale in $\mathcal{H}_0^1$, so by Doob's Theorem 4.12 for any $T \in \mathcal{T}$,

$$E[\langle M, M\rangle_\infty - \langle M, M\rangle_T | \mathcal{F}_T] = E[M_\infty^2 | \mathcal{F}_T] - M_T^2.$$

Note that for any $\alpha \in R$, $\langle M, \alpha M\rangle = \langle \alpha M, M\rangle = \alpha\langle M, M\rangle$.

Lemma 10.9. *Suppose* $M, N \in \mathcal{H}^2$ *and* $T \in \mathcal{T}$. *Then* $\langle M, N\rangle^T = \langle M, N^T\rangle$.

Proof. By Doob's Theorem 4.12

$$(M.N)^T - \langle M, N\rangle^T \in \mathcal{H}_0^1$$

and

$$M_t.N_t^T - M_t^T.N_t^T = E[(M_\infty - M_T)N_T | \mathcal{F}_t] \in \mathcal{H}_0^1.$$

Therefore,

$$M.N^T - \langle M, N\rangle^T \in \mathcal{H}_0^1.$$

By definition

$$M.N^T - \langle M, N^T\rangle \in \mathcal{H}_0^1 \quad \text{so} \quad \langle M, N^T\rangle - \langle M, N\rangle^T \in \mathcal{H}_0^1.$$

Therefore, by Corollary 7.32

$$\langle M, N^T\rangle = \langle M, N\rangle^T. \qquad \square$$

Corollary 10.10. *Under the above hypotheses*

$$[M, N]^T = [M, N^T].$$

PROOF. This follows from the definition and the lemma because for any $t \geq 0$

$$[M, N]_t^T = \langle M^c, M^c\rangle_t^T + \sum_{s \leq T \wedge t} \Delta M_s \Delta N_s = [M, N^T]_t. \qquad \square$$

Theorem 10.11. *Suppose $M, N \in \mathcal{H}^2$ and $H, K \in B(\mathcal{B} \otimes \mathcal{F})$. (See Notation 6.38.) Then, almost surely,*

$$\int_0^\infty |H_s|\,|K_s|\,|d\langle M, N\rangle_s| \leq \left(\int_0^\infty H_s^2\, d\langle M, M\rangle_s\right)^{1/2} \left(\int_0^\infty K_s^2 d\langle N, N\rangle_s\right)^{1/2},$$

$$\int_0^\infty |H_s|\,|K_s|\,|d[M, N]_s| \leq \left(\int_0^\infty H_s^2\, d[M, M]_s\right)^{1/2} \left(\int_0^\infty K_s^2\, d[N, N]_s\right)^{1/2}.$$

PROOF. Take $s, t \in [0, \infty]$, $s < t$ and s and t rational. Then for any rational λ, because $\langle M + \lambda N, M + \lambda N\rangle$ is increasing,

$$\langle M + \lambda N, M + \lambda N\rangle_t - \langle M + \lambda N, M + \lambda N\rangle_s \geq 0 \quad \text{a.s.}$$

Writing $\Delta_s^t X = X_t - X_s$ for any process X, this implies that

$$\Delta_s^t\langle M, M\rangle + 2\lambda\, \Delta_s^t\langle M, N\rangle + \lambda^2 \Delta_s^t\langle N, N\rangle \geq 0 \quad \text{a.s.}$$

for all real λ, by continuity. Because the processes $\langle M, M\rangle$, etc., are right continuous by definition, we can conclude that there is a single set $\Lambda \subset \Omega$ of measure zero, such that for $\omega \notin \Lambda$

$$|\Delta_s^t\langle M, N\rangle(\omega)| \leq (\Delta_s^t\langle M, M\rangle(\omega))^{1/2} (\Delta_s^t\langle N, N\rangle(\omega))^{1/2},$$

for all real $s, t \in [0, \infty]$, $s < t$. Furthermore, we have for $\omega \in \Omega$:

$$\begin{aligned} |\Delta\langle M, N\rangle_0(\omega)| &= |M_0(\omega) N_0(\omega)| \\ &\leq |\Delta\langle M, M\rangle_0(\omega)|^{1/2} |\Delta\langle N, N\rangle_0(\omega)|^{1/2}. \end{aligned}$$

Consider a finite subdivision of $[0, \infty]$, say $0 = t_0 < t_1 < \cdots < t_{n+1} = \infty$, and a finite number of bounded random variables

$$\begin{aligned} &H_0, H_{t_i}, \qquad 0 \leq i \leq n, \\ &K_0, K_{t_i}, \qquad 0 \leq i \leq n. \end{aligned}$$

Put

$$H_t = H_0 I_{\{t=0\}} + \sum_i H_{t_i} I_{]t_i, t_{i+1}]}(t),$$

$$K_t = K_0 I_{\{t=0\}} + \sum_i K_{t_i} I_{]t_i, t_{i+1}]}(t).$$

Write the above inequality for $s=t_i$, $t=t_{i+1}$, multiply by $|H_{t_i}K_{t_i}|$, sum and apply the Schwarz inequality. Then almost surely:

$$\left|\int_0^\infty H_sK_s\,d\langle M,N\rangle_s\right| \leq \int_0^\infty |H_sK_s|\,|d\langle M,N\rangle_s|$$

$$= |H_0K_0|\,|\Delta\langle M,N\rangle_0| + \sum_{i=0}^n |H_{t_i}K_{t_i}|\,|\Delta_{t_i}^{t_{i+1}}\langle M,N\rangle|$$

$$\leq |H_0^2\,\Delta\langle M,N\rangle_0|\,|K_0^2\,\Delta\langle N,N\rangle_0|$$

$$+\sum_0^n (H_t^2\,\Delta_{t_i}^{t_{i+1}}\langle M,M\rangle)^{1/2}(K_{t_i}^2\,\Delta_{t_i}^{t_{i+1}}\langle N,N\rangle)^{1/2}$$

$$\leq \left\{H_0^2\,\Delta\langle M,M\rangle_0 + \sum_0^n H_{t_i}^2\,\Delta_{t_i}^{t_{i+1}}\langle M,M\rangle\right\}^{1/2}$$

$$\cdot\left\{K_0^2\,\Delta\langle N,N\rangle_0 + \sum_0^n K_{t_i}^2\,\Delta_{t_i}^{t_{i+1}}\langle N,N\rangle\right\}$$

$$= \left(\int_0^\infty H_s^2\,d\langle M,M\rangle_s\right)^{1/2}\left(\int_0^\infty K_s^2\,d\langle N,N\rangle_s\right)^{1/2}$$

The left continuous step processes of the same form as H and K are an algebra, and they generate the product σ-field $\mathscr{B}\otimes\mathscr{F}$ on $[0,\infty]\times\Omega$. A monotone class theorem argument (see Theorem 6.28) then establishes the validity of the above inequality for all bounded processes H and K. That is, for $H, K \in B(\mathscr{B}\otimes\mathscr{F})$

$$\left|\int_0^\infty H_sK_s\,d\langle M,N\rangle_s\right| \leq \left(\int_0^\infty H_s^2\,d\langle M,M\rangle_s\right)^{1/2}\left(\int_0^\infty K_s^2\,d\langle N,N\rangle_s\right)^{1/2} \quad \text{a.s.}$$

This almost proves the first inequality, except that the modulus on the left-hand side of the inequality is outside the integral. However, suppose J_s is the measurable process with values in $\{-1,1\}$ such that $|d\langle M,N\rangle_s| = J_s\,d\langle M,N\rangle_s$. The desired result follows by applying the above inequality to the bounded processes H_s and K_sJ_s.

The proof of the second inequality involving the process $[M,N]$ is identical. □

Corollary 10.12. *If* $1<p<\infty$ *and* $1/p+1/q=1$ *then*

$$E\left[\int_0^\infty |H_s|\,|K_s|\,|d\langle M,N\rangle_s|\right]$$

$$\leq \left\|\left(\int_0^\infty H_s^2\,d\langle M,M\rangle_s\right)^{1/2}\right\|_p \left\|\left(\int_0^\infty K_s^2\,d\langle N,N\rangle\right)^{1/2}\right\|_q$$

and

$$E\left[\int_0^\infty |H_s|\,|K_s|\,|d[M,N]_s|\right]$$
$$\leq \left\|\left(\int_0^\infty H_s^2\,d[M,M]_s\right)^{1/2}\right\|_p \cdot \left\|\left(\int_0^\infty K_s^2\,d[N,N]_s\right)^{1/2}\right\|_q.$$

Remark 10.13. Truncation and a monotone class argument enables the above inequalities to be extended to nonbounded integrands.

Definition 10.14. Suppose $M \in \mathcal{H}^2_{\text{loc}}$. That is, there is an increasing sequence of stopping times $\{T_n\}$, such that $\lim_n T_n = \infty$ a.s. and $M(n) \in \mathcal{H}^2$, where

$$M(n)_t = M_t^{T_n} = M_{T_n \wedge t}.$$

Because the decomposition of Corollary 9.36

$$M(n) = M(n)^c + M(n)^d$$

is unique, if $T_n \leq T_m$ then $M(n)_t^c = M(m)_t^c$ and $M(n)_t^d = M(m)_t^d$, for $(t, \omega) \in [\![0, T_n]\!]$. Furthermore, the predictable quadratic variation process is unique, so

$$\langle M(n), M(n)\rangle_t = \langle M(m), M(m)\rangle_t \quad \text{for } (t, \omega) \in [\![0, T_n]\!].$$

We can, therefore, define the predictable quadratic variation process of $M \in \mathcal{H}^2_{\text{loc}}$ as the unique process $\langle M, M\rangle \in \mathcal{A}^+_{\text{loc}}$ such that

$$\langle M, M\rangle_t^{T_n} = \langle M(n), M(n)\rangle_t.$$

Similarly, the optional quadratic variation process of $M \in \mathcal{H}^2_{\text{loc}}$ is the process $[M, M] \in \mathcal{A}^+_{\text{loc}}$ such that

$$[M, M]_t^{T_n} = [M(n), M(n)]_t$$
$$= \langle M(n)^c, M(n)^c\rangle_t + \sum_{s \leq t \wedge T_n} \Delta M(n)^2.$$

Lemma 10.15. *If M is a continuous local martingale then $M \in \mathcal{H}^2_{\text{loc}}$.*

PROOF. Write

$$T_n = \inf\{t \colon |M_t| \geq n\}.$$

Then $\lim_n T_n = \infty$ a.s. and $M^{T_n} \in \mathcal{H}^2$ because $|M_{t \wedge T_n}| \leq n$. □

Remark 10.16. If M is corlol, but not continuous, M_t is bounded by n on the interval $[\![0, T_n[\![$, but one knows nothing about the jump at T_n. (Here T_n is as in the above lemma.)

CHAPTER 11

Stochastic Integration with Respect to Martingales and Local Martingales

The construction of the Ito integral for stochastic integrals with respect to Brownian motion will be followed below. However, in the Brownian motion case, because almost all sample paths are continuous, the predictable role of the integrands is not obvious.

Definition 11.1. Suppose f is a left continuous simple function on $[0, \infty]$. That is, there is a partition

$$0 = t_0 < t_1 < t_2 < \cdots < t_n < t_{n+1} = \infty,$$

and $f(t) = f_i$ for $t \in]t_i, t_{i+1}]$, where f_i is a real constant. Suppose g is a corlol function. Then one can define the integrals

$$\int_0^\infty f\,dg = f(0)g(0) + \sum_0^n f_i(g(t_{i+1}) - g(t_i)),$$

$$\int_0^t f\,dg = \int_0^\infty f I_{[0,t]}\,dg$$

$$= f(0)g(0) + \sum_0^n f_i(g(t_{i+1} \wedge t) - g(t_i \wedge t)).$$

Clearly $\int_0^t f\,dg$ is corlol, and its jump at t is $f(t)\Delta g(t)$.

Write Λ for the space of bounded simple predictable processes. That is, $H \in \Lambda \subset B(\Sigma_p)$ if there is a partition $\{t_i\}$ of $[0, \infty]$ as above and a family $\{H_i\}$ of random variables such that $H_0 = H_0$, $H_t = H_i$ for $t \in]t_i, t_{i+1}]$. Here all the H_i are bounded, H_0 is $\mathcal{F}_0$ measurable and H_i is $\mathcal{F}_{t_i}$ measurable.

Suppose $M \in \mathcal{H}^2$, so M is corlol. Then the integral $(H \cdot M)_t = \int_0^t H_s\,dM_s$ exists for $H \in \Lambda$.

Lemma 11.2. *$H \cdot M \in \mathscr{H}^2$ and*

$$E[(H \cdot M)_\infty^2] = E\left[\int_0^\infty H_s^2 \, d\langle M, M\rangle_s\right]$$
$$= E\left[\int_0^\infty H_s^2 \, d[M, M]_s\right].$$

PROOF. From the above definition

$$(H \cdot M)_t = H_0 M_0 + \sum_0^n H_i (M_{t_{i+1} \wedge t} - M_{t_i \wedge t}).$$

By the optional stopping theorem, for $s \leq t$:

$$E[(H \cdot M)_t | \mathscr{F}_s] = (H \cdot M)_s.$$

For $i < j$, so that $i + 1 \leq j$,

$$E[H_i H_j (M_{t_{i+1} \wedge t} - M_{t_i \wedge t})(M_{t_{j+1} \wedge t} - M_{t_j \wedge t})]$$
$$= E[E[H_i H_j (M_{t_{i+1} \wedge t} - M_{t_i \wedge t})(M_{t_{j+1} \wedge t} - M_{t_j \wedge t}) | \mathscr{F}_{t_j}]$$
$$= 0.$$

Therefore,

$$E[(H \cdot M)_t^2] = E\left[\sum_0^n H_i^2 (M_{t_{i+1} \wedge t}^2 - M_{t_i \wedge t}^2)\right]$$
$$= E\left[\sum_0^n H_i^2 (\langle M, M\rangle_{t_{i+1} \wedge t} - \langle M, M\rangle_{t_i \wedge t})\right]$$
$$= E\left[\int_0^t H_s^2 \, d\langle M, M\rangle_s\right]$$
$$\leq E\left[\int_0^\infty H_s^2 \, d\langle M, M\rangle_s\right] < \infty,$$

because H is bounded and $M \in \mathscr{H}^2$. Therefore, by Lebesgue's theorem, letting $t \to \infty$:

$$E[(H \cdot M)_\infty^2] = E\left[\int_0^\infty H_s^2 \, d\langle M, M\rangle_s\right].$$

The final equality follows because $\langle M, M\rangle - [M, M]$ is a martingale of integrable variation. (See Lemma 10.4.) □

The result below follows Ito's construction of a stochastic integral by considering convergence in a certain L^2 space.

Theorem 11.3. *Write $L^2(\langle M\rangle)$ for the space of predictable processes $\{H\}$ such that*

$$\|H\|_{\langle M\rangle}^2 = E\left[\int_0^\infty H_s^2 \, d\langle M, M\rangle_s\right] < \infty.$$

Then the map $H \to H \cdot M$ of Λ into $\mathcal{H}^2$ extends in a unique manner to a linear isometry of $L^2(\langle M \rangle)$ into $\mathcal{H}^2$. This map is again denoted by $H \to H \cdot M$.

PROOF. $L^2(\langle M \rangle)$ is the L^2 space, with respect to a certain measure, of the predictable σ-field. Therefore, Λ is dense in $L^2(\langle M \rangle)$. The map $H \to H \cdot M$ of Λ into $\mathcal{H}^2$ is an isometry by Lemma 11.2 above, if Λ is given the semi norm $\|\cdot\|_{\langle M \rangle}$. Therefore, this map extends in a unique manner to an isometry of $L^2(\langle M \rangle)$ into $\mathcal{H}^2$.

The image of $H \in L^2(\langle M \rangle)$ under this map will be written $H \cdot M$. □

Lemma 11.4. *For $H \in L^2(\langle M \rangle)$ the processes $\Delta(H \cdot M)_t$ and $H_t \Delta M_t$ are indistinguishable.*

PROOF. For a step function $H \in \Lambda$, clearly $\Delta(H \cdot M)_t = H_t \Delta M_t$. For a general process $H \in L^2(\langle M \rangle)$ suppose $\{H^n\} \in \Lambda$ is a sequence that approximates H in the $\|\cdot\|_{\langle M \rangle}$ seminorm. Then

$$E[(H^n \cdot M)^2_\infty] \text{ converges to } E[(H \cdot M)^2_\infty],$$

so by Lemma 9.6,

$$(H^n \cdot M)^* \text{ converges to } (H \cdot M)^* \text{ in } L^2(\Omega).$$

From Theorem 9.9 there is, therefore, a subsequence $\{n_k\}$ such that $(H^{n_k} \cdot M)_t$ converges to $(H \cdot M)$ uniformly in t for almost every ω. In particular,

$$\Delta(H^{n_k} \cdot M)_t \text{ converges to } \Delta(H \cdot M)_t \quad \text{a.e.}$$

That is,

$$H_t^{n_k} \cdot \Delta M_t \text{ converges to } \Delta(H \cdot M)_t \quad \text{a.e.}$$

However, because convergence in $L^2(\langle M \rangle)$ is convergence in L^2 certainly

$$H_t^{n_k} \text{ converges to } H_t \quad \text{a.e.}$$

Therefore, $H_t \cdot \Delta M_t = \Delta(H \cdot M)_t$ except on an evanescent set. □

Example 11.5. Suppose T is any stopping time and $H = I_{[\![0,T]\!]}$, so H is predictable, then

$$\|H\|^2_{\langle M \rangle} = E[\langle M, M \rangle_T - \langle M, M \rangle_0]$$
$$< \infty,$$

so $H \in L^2(\langle M \rangle)$. By considering approximations from above of T by simple stopping times we see that

$$(H \cdot M)_t = \int_0^t I_{[\![0,T]\!]} \, dM$$
$$= M_t^T.$$

Remark 11.6. The above construction of Ito defines the stochastic integral in terms of an isometry from $L^2(\langle M\rangle)$ into $\mathscr{H}^2$. The following result, due to Kunita and Wantabe [57], characterizes the stochastic integral in terms of the process $H \cdot M$.

Theorem 11.7. *Suppose* $H \in L^2(\langle M\rangle)$:

(a) *Then for every* $N \in \mathscr{H}^2$

$$E\left[\int_0^\infty |H_s|\,|d\langle M, N\rangle_s|\right] < \infty,$$

$$E\left[\int_0^\infty |H_s|\,|d[M, N]_s|\right] < \infty.$$

(b) *The stochastic integral* $L = H \cdot M$ *is characterized as the unique element of* $\mathscr{H}^2$ *such that for every* $N \in \mathscr{H}^2$

$$E[L_\infty N_\infty] = E\left[\int_0^\infty H_s\, d\langle M, N\rangle_s\right]$$

$$= E\left[\int_0^\infty H_s\, d[M, N]_s\right].$$

(c) *Furthermore, for every* $N \in \mathscr{H}^2$

$$\langle L, N\rangle = H \cdot \langle M, N\rangle,$$

$$[L, M] = H \cdot [M, N],$$

the integrals on the right being Stieltjes integrals.

Proof. The inequalities of (a) are immediate consequences of the Kunita–Watanabe inequalities, Corollary 10.12.

Consider the linear functional on $L^2(\langle M\rangle)$ defined by

$$H \to E\left[(H \cdot M)_\infty N_\infty - \int_0^\infty H_s\, d\langle M, N\rangle_s\right].$$

This is continuous, again the Kunita–Watanabe inequalities. Further, it is zero on Λ, and so it is zero on $L^2(\langle M\rangle)$ by continuity. The second identity of part (b) follows because $\langle M, N\rangle - [M, N]$ is a martingale of integrable variation.

To prove the final identities, note that the process

$$J_t = L_t N_t - \int_0^t H_s\, d\langle M, N\rangle_s$$

is dominated by $L^* N^* + \int_0^\infty |H_s|\,|d\langle M, N\rangle_s|$ which is in L^1. Recall that $\langle M, N^T\rangle = \langle M, N\rangle^T$ for any stopping time T, and apply the identity of part (b) to N^T to see that

$$E[J_T] = 0.$$

Therefore, J_t is a martingale. However, $\langle L, N\rangle_t$ is the unique predictable process of integrable variation such that $L_t N_t - \langle L, N\rangle_t$ is a martingale. Therefore,

$$\langle L, N\rangle = H \cdot \langle M, N\rangle. \tag{11.1}$$

Decompose M and N into the sum of their continuous and totally discontinuous parts:

$$M = M^c + M^d, \qquad N = N^c + N^d.$$

Then if $K \in \mathcal{H}_0^{2,c}$

$$\langle H \cdot M^d, K\rangle = H \cdot \langle M^d, K\rangle = 0,$$

by the above. $H \cdot M^c$ is certainly a continuous martingale, so the decomposition of $H \cdot M$ into continuous and purely discontinuous martingales is given by:

$$(H \cdot M)^c = H \cdot M^c,$$
$$(H \cdot M)^d = H \cdot M^d.$$

Therefore by Remarks 10.8

$$[L, N]_t = \langle L^c, N^c\rangle_t + \sum_{s \le t} \Delta L_s \Delta N_s$$

$$= H \cdot \langle M^c, N^c\rangle_t + \sum_{s \le t} H_s \Delta M_s \Delta N_s$$

by Lemma 11.4 and (11.1), and this is

$$= \int_0^t H_s \, d[M, N]_s.$$

Note that the first identity in (b) uniquely characterizes the stochastic integral $L = H \cdot M$. This is because the right-hand side is a continuous linear functional in N (given H and M), whilst the left-hand side is just the inner product of L and N in the Hilbert space $\mathcal{H}^2$. Consequently, given H and M there is a unique element $L = H \cdot M \in \mathcal{H}^2$ which gives this linear functional. □

Let us explicitly note the following result obtained in the course of the above proof.

Lemma 11.8. *For $H \in L^2(\langle M\rangle)$ the continuous and purely discontinuous parts of the martingale $H \cdot M$ are $H \cdot M^c$ and $H \cdot M^d$, respectively.*

Corollary 11.9. *For $H \in L^2(\langle M\rangle)$ and K a bounded predictable process*

$$(KH) \cdot M = K \cdot (H \cdot M).$$

PROOF. For any $N \in \mathscr{H}^2$

$$E[((KH)\cdot M)_\infty N_\infty] = E[(KH)\cdot\langle M, N\rangle_\infty]$$
$$= E[K \cdot \langle H\cdot M, N\rangle_\infty]$$
$$= E[(K\cdot(H\cdot M))_\infty N_\infty],$$

and the result follows by Theorem 11.7(b). □

Theorem 11.10. *Suppose $\mathscr{S} \subset \mathscr{H}^2$ is a stable subspace. (See Definition 9.14.) Then for $M \in \mathscr{S}$ and $H \in L^2(\langle M\rangle)$ $H\cdot M \in \mathscr{S}$.*

Conversely, if for each $M \in \mathscr{S}$ and $H \in L^2(\langle M\rangle)$ we have $H\cdot M \in \mathscr{S}$ then $\mathscr{S}$ is stable.

PROOF. The converse is proved by observing that if $H = I_{[\![0,T]\!]}$ then $H \in L^2(\langle M\rangle)$ and $H\cdot M = M^T$. Also, if $A \in \mathscr{F}_0$ then $H = I_{[0,\infty[\times A} \in L^2(\langle M\rangle)$ and $H\cdot M = I_A M$.

To prove the direct statement of the theorem we must show that $H\cdot M \in \mathscr{S}^{\perp\perp}$. Suppose $N \in \mathscr{S}^\perp$. Then $M_0N_0 = 0$ and $\langle M, N\rangle = 0$. Therefore, $(H\cdot M)_0 N_0 = H_0M_0N_0 = 0$ and $\langle H\cdot M, N\rangle = H\cdot\langle M, N\rangle = 0$, by Theorem 11.7 above. Therefore

$$E[(H\cdot M)_\infty N_\infty] = 0, \quad \text{so } H\cdot M \in \mathscr{S}^{\perp\perp}.$$ □

Theorem 11.11. *Consider $M \in \mathscr{H}^2$. The stable subspace generated by M is the set $\{H\cdot M, H \in L^2(\langle M\rangle)\}$ of stochastic integrals.*

For $N \in \mathscr{H}^2$ the projection of N on this subspace is $D\cdot M$, where D is a predictable density of $\langle M, N\rangle$ with respect to $\langle M, M\rangle$.

PROOF. The map $H \to H\cdot M$ of $L^2(\langle M\rangle)$ into $\mathscr{H}^2$ is an isometry. Write $\mathscr{K}$ for its image. Then $\mathscr{K}$ is closed, stable, and continued in every stable subspace containing M by Theorem 11.10 above. Therefore $\mathscr{K}$ is the stable subspace generated by M.

For $N \in \mathscr{H}^2$, write $N = N_1 + N_2$, where N_1 is the projection of N on $\mathscr{K}$ and $N_2 \in \mathscr{K}^\perp$. Then $\langle M, N\rangle = \langle M, N_1 + N_2\rangle = \langle M, N_1\rangle$, because $\langle M, N_2\rangle = 0$. Because $N_1 \in \mathscr{K}$ it can be written as $N_1 = D\cdot M$, where $D \in L^2(\langle M\rangle)$. Therefore $\langle M, N_1\rangle = \langle D\cdot M, M\rangle = D\cdot\langle M, M\rangle$, and so D is a predictable density of $\langle M, N\rangle$ with respect to $\langle M, M\rangle$.

If D' is another such density then $D = D'$ "almost surely $\langle M, M\rangle$". Therefore,

$$E\left[\int (D_s - D'_s)^2\, d\langle M, M\rangle_s\right] = 0 \quad \text{and so } D' \in L^2(\langle M\rangle) \text{ and } D\cdot M = D'\cdot M.$$ □

Remark 11.12. The following result relates stochastic and Stieltjes integrals.

Theorem 11.13. *Suppose $M \in \mathscr{H}^2$ is also a process of integrable variation. Furthermore, suppose $H \in L^2(\langle M\rangle)$ is also such that $E[\int_0^\infty |H_s||dM_s|] < \infty$.*

Then the stochastic integral $H \cdot M$ and the Stieltjes stochastic integral $\int H\,dM$ (see Definition 7.10) are indistinguishable processes.

PROOF. Consider first the integrable variation martingale $M-M_0 \in \mathscr{A}_0$. Then from Theorem 7.30(3) the restriction to Σ_p of its associated measure μ is zero. H is μ integrable and predictable, so the measure $H\mu$ is bounded, and zero on Σ_p. However, $H\mu$ is the measure associated with the Stieltjes integral $\int H\,d(M-M_0) \in \mathscr{A}_0$. Therefore, by Theorem 7.30(1) $\int H\,dM$ is a martingale of integrable variation.

From Lemma 9.25, for every bounded martingale N

$$E\Big[N_\infty\Big(\int_0^\infty H\,dM\Big)\Big] = E\Big[\sum_s H_s \Delta M_s \Delta N_s\Big].$$

However, from Theorem 9.41, M is a purely discontinuous martingale so

$$[M, N]_t = \sum_{s \leq t} \Delta M_s \Delta N_s$$

and for every $N \in \mathscr{H}^2$ (in particular for bounded N), by Theorem 11.7(b)

$$\begin{aligned} E[N_\infty (H \cdot M)_\infty] &= E\Big[\int_0^\infty H_s\,d[M, N]_s\Big] \\ &= E\Big[\sum_s H_s \Delta M_s \Delta N_s\Big]. \end{aligned}$$

Because $N_\infty \in L^2(\Omega)$ is an arbitrary bounded random variable we must have

$$(H \cdot M)_\infty = \int_0^\infty H_s\,dM_s \quad \text{a.s.}$$

Taking right continuous versions of expectations with respect to $\mathscr{F}_t$ we see that $(H \cdot M)_t$ and $\int_0^t H_s\,dM_s$ are indistinguishable. □

Definition 11.14. Suppose $M \in \mathscr{H}^2$. Then $L^2(M)$ denotes the space of optional processes $\{H\}$ such that

$$\|H\|_{[M]} = \Big(E\Big[\int_{[0,\infty[} H_s^2\,d[M, M]_s\Big]\Big)^{1/2} < \infty.$$

From Theorem 11.7(b), note that $L^2(\langle M \rangle)$ is made up of the predictable processes in $L^2(M)$ (taking $N = H \cdot M$).

Definition 11.15. Suppose $H \in L^2(M)$. Then

$$(H \cdot M)_t = \int_{[0,t[} H_s\,dM_s$$

is the unique element $L \in \mathscr{H}^2$ such that for every $N \in \mathscr{H}^2$

$$E[L_\infty N_\infty] = E\left[\int_{[0,\infty[} H_s \, d[M,N]_s\right].$$

$H \cdot M$ is the *optional* stochastic integral of H with respect to M.

Remarks 11.16. Note that from the second inequality of Kunita–Watanabe, Theorem 10.11, $E[\int |H_s| \, |d[M,N]_s|] < \infty$ and so the map $N \to E[\int_{[0,\infty[} H_s \, d[M,N]_s]$ is a continuous linear functional on the Hilbert space $\mathscr{H}^2$. $L \in \mathscr{H}^2$ is, therefore, the unique element giving this linear functional. From Theorem 11.7(b) note that if $H \in L^2(\langle M, M\rangle)$ the stochastic integral defined in Theorem 11.3 coincides with the optional stochastic integral. By considering $N \in \mathscr{H}^2$ stopped at any $T \in \mathscr{T}_0$ we have the following result:

Lemma 11.17. *For any* $N \in \mathscr{H}^2$

$$L_t N_t - \int_{[0,t[} H_s \, d[M,N]_s$$

is a martingale. Furthermore, this process is bounded in $L^1(\Omega)$ *by*

$$L^* N^* + \int_{[0,\infty[} |H_s| \, |d[M,N]_s|.$$

Remarks 11.18. Recall, Corollary 9.36, that any $M \in \mathscr{H}^2$ has an orthogonal decomposition

$$M = M_0 + M^c + M^d,$$

where $M^c \in \mathscr{H}_0^{2,c}$ and $M^d \in \mathscr{H}_0^{2,d}$. Furthermore, there is a sequence of stopping times $\{T_n\}$ which are all positive, with disjoint graphs, such that each T_n is either predictable or totally inaccessible, and such that the $\{T_n\}$ exhaust the jumps of M^d. Then

$$M^d = \sum_n M^n,$$

where

$$M^n = \Delta M_{T_n} I_{\{t \geq T_n\}} - \Pi_p^*(\Delta M_{T_n} I_{\{t \geq T_n\}}).$$

Theorem 11.19. *Suppose* $M \in \mathscr{H}^2$ *has a decomposition as above and* $H \in L^2(M)$. *Then the optional stochastic integral* $L = H \cdot M$ *has an orthogonal decomposition*

$$H \cdot M = H_0 M_0 + H \cdot M^c + \sum_n H \cdot M^n.$$

Also, $E[L_\infty^2] \leq E[\int H_s^2 \, d[M,M]_s]$ *with equality if and only if for every positive predictable stopping time* T

$$E[H_T \Delta M_T | \mathscr{F}_{T-}] = 0.$$

PROOF. The result will be proved by considering the different martingales which occur in the orthogonal decomposition of M.

(i) Consider first a constant martingale $M_t = M_0$. Then if $L_t = H_0M_0$

$$E[L_\infty N_\infty] = E\left[\int_{[0,\infty[} H_s\, d[M, N]_s\right]$$

$$= E[H_0 \Delta M_0 \Delta N_0] = E[H_0 M_0 N_0],$$

for all $N \in \mathcal{H}^2$. Therefore, the constant process H_0M_0 is the required stochastic integral.

(ii) Suppose M is continuous, $M_t = M_t^c \in \mathcal{H}_0^{2,c}$. For the optional integrand $H \in L^2(M)$ there is a predictable process H' such that $\{H \neq H'\}$ is contained in the union of graphs of a countable family of stopping times (see Theorem 6.29). Therefore, because M is continuous,

$$E\left[\int_{[0,\infty[} H_s^2\, d[M^c, M^c]_s\right] = E\left[\int_{[0,\infty[} H_s^2\, d\langle M^c, M^c\rangle_s\right]$$

$$= E\left[\int_{[0,\infty[} (H_s')^2\, d\langle M^c, M^c\rangle_s\right],$$

so $H' \in L^2(\langle M^c\rangle)$. In this case we define $H \cdot M$ to be $H' \cdot M^c = H' \cdot M$, so we are in the situation discussed in Theorem 11.7 and Lemma 11.8 above.

(iii) Suppose now that T is a totally inaccessible stopping time and that $M \in \mathcal{M}(T)$. Write

$$A_t = H_T \Delta M_T I_{\{t \geq T\}}.$$

Because

$$H \in L^2(M), \qquad H_T \Delta M_T \in L^2(\mathcal{F}_T).$$

From Theorem 9.28,

$$L_t = A_t - \Pi_p^* A_t \in \mathcal{M}(T).$$

Furthermore,

$$\Delta L_T = \Delta A_T = H_T \Delta M_T$$

and by Corollary 9.30, for every $N \in \mathcal{H}^2$,

$$E[L_\infty N_\infty] = E[\Delta L_T \Delta N_T]$$

$$= E[H_T \Delta M_T \Delta N_T] = E\left[\int_{[0,\infty[} H_s\, d[M, N]_s\right],$$

because $M \in \mathcal{M}(T)$. Therefore, L is the stochastic integral $H \cdot M$.

(iv) Suppose $T > 0$ is a positive predictable stopping time and that $M \in \mathcal{M}(T)$. Again write $A_t = H_T \Delta M_t I_{\{t \geq T\}}$. Because T is predictable, from Theorem 9.29,

$$\Pi_p^* A_t = E[H_T \Delta M_T | \mathcal{F}_{T-}] I_{\{t \geq T\}}.$$

Then if,

$$L_t = A_t - \Pi_p^* A_t \in \mathcal{M}(T),$$

in this case

$$\Delta L_T = H_T \Delta M_T - E[H_T \Delta M_T | \mathscr{F}_{T-}]$$

and for every $N \in \mathscr{H}^2$, by Corollary 9.30,

$$\begin{aligned} E[L_\infty N_\infty] &= E[\Delta L_T \Delta N_T] \\ &= E[H_T \Delta M_T \Delta N_T] - E[E[H_T \Delta M_T | \mathscr{F}_{T-}] \Delta N_T] \\ &= E[H_T \Delta M_T \Delta N_T] = E\left[\int_{[0,\infty[} H_s \, d[M,N]_s\right]. \end{aligned}$$

Therefore, again L is the stochastic integral $H \cdot M$.

Note, however, an important difference between cases (i), (ii), (iii) and case (iv).

In cases (i), (ii) and (iii)

(a) the processes $\Delta(H \cdot M)_t$ and $H_t \Delta M_t$ are indistinguishable,
(b) $[H \cdot M, N] = H \cdot [M, N]$, for $N \in \mathscr{H}^2$,
(c) $E[(H \cdot M)_\infty^2] = E[\int_{[0,\infty[} H_s^2 \, d[M,M]_s]$.

However, in case (iv), when $M \in \mathcal{M}(T)$ for $T > 0$ predictable,

$$\Delta(H \cdot M)_T = \Delta L_T = H_T \Delta M_T - E[H_T \Delta M_T | \mathscr{F}_{T-}].$$

Therefore, (a) and (b) hold in this case only when $E[H_T \Delta M_T | \mathscr{F}_{T-}] = 0$. In particular, this is so if H is predictable.

Also,

$$\begin{aligned} E[(H \cdot M)_\infty^2] &= E[(\Delta L_T)^2] \\ &= E[(H_T \Delta M_T - E[H_T \Delta M_T | \mathscr{F}_{T-}])^2] \\ &\leq E[H_T^2 \Delta M_T^2] = E\left[\int_{[0,\infty[} H_s^2 \, d[M,M]_s\right], \end{aligned}$$

again with the equality only when

$$E[H_T \Delta M_T | \mathscr{F}_{T-}] = 0.$$

Finally, consider a general $M \in \mathscr{H}^2$ with an orthogonal decomposition

$$M = M_0 + M^c + \sum_n M^n$$

as in Remarks 11.18 above. If $H \in L^2(M)$ the martingales $H_0 M_0 \in \mathcal{M}(0)$, $H \cdot M^c \in \mathscr{H}_0^{2,c}$ and $H \cdot M^n \in \mathcal{M}(T_n)$ are defined in (i)–(iv), and are pairwise orthogonal. Because

$$H \in L^2(M), \qquad E\left[\int_{[0,\infty[} H_s^2 \, d[M,M]_s\right] < \infty$$

so the series

$$H \cdot M_0 + H \cdot M^c + \sum_n H \cdot M^n$$

converges in H^2. Denoting its sum by $L \in \mathscr{H}^2$ we have, by the dominated convergence theorem, that

$$E[L_\infty N_\infty] = E\left[\int_{[0,\infty[} H_s \, d[M, N]_s\right],$$

for every $N \in \mathscr{H}^2$. Therefore

$$L = H \cdot M_0 + H \cdot M^c + \sum_n H \cdot M^n$$

is the stochastic integral $H \cdot M$ and the above sum is its orthogonal decomposition.

Furthermore, because the terms in the sum are orthogonal

$$E[L_\infty^2] \leq E\left[\int H_s^2 \, d[M, M]_s\right],$$

with equality if and only if

$$E[H_{T_n} \Delta M_{T_n} | \mathscr{F}_{T_n-}] = 0$$

for every predictable jump time T_n. Because the equality condition is independent of how the jump times of M are indexed, this condition is the same as stating that

$$E[H_T \Delta M_T | \mathscr{F}_{T-}] = 0 \tag{11.2}$$

for every predictable stopping time $T > 0$. □

Remarks 11.20. From Theorem 9.29 we see that condition (11.2) states that the predictable projection of the process $H_t \Delta M_t$, $\Pi_p(H_t \Delta M_t)$, is evanescent on $]0, \infty[\times \Omega$. For any martingale $L \in \mathscr{H}^2$ we have $E[\Delta L_T | \mathscr{F}_{T-}] = 0$, so condition (11.2) is also equivalent to requiring that the processes $\Delta(H \cdot M)_t$ and $H_t \Delta M_t$ be indistinguishable.

Corollary 11.21. *Finally, with $L = H \cdot M$, if*

$$[L, N] = H \cdot [M, N] \tag{11.3}$$

for every $N \in \mathscr{H}^2$ then, taking $N = L$,

$$\begin{aligned} E[L_\infty^2] &= E[[L, L]_\infty] \\ &= E\left[\int_{[0,\infty[} H_s \, d[M, L]_s\right] \\ &= E\left[\int_{[0,\infty[} H_s^2 \, d[M, M]_s\right]. \end{aligned}$$

Conversely, if we have equality in (11.2), *considering the representation of Remarks* 10.8

$$[L, N] = H \cdot [M, N]$$

for every $N \in \mathcal{H}^2$. *Therefore*, (11.3) *is equivalent to* (11.2).

Remarks 11.22. Recall Definition 5.34, that the filtration $\{\mathcal{F}_t\}$ is left quasi-continuous if $\mathcal{F}_{T-} = \mathcal{F}_T$ for every predictable stopping time T. In this case, for $M \in \mathcal{H}^2$ and $T > 0$ a predictable stopping time, by Corollaries 5.24 and 3.19

$$M_{T-} = E[M_T | \mathcal{F}_{T-}] = M_T,$$

so

$$\Delta M_T = 0.$$

Therefore, condition (11.2) is satisfied and the pathologies in the optional stochastic integral disappear.

Remarks 11.23. So far in this chapter the stochastic integrals have been defined with respect to square integrable martingales in $\mathcal{H}^2$. We now wish to extend the definition to allow integration with respect to local martingales $\mathcal{M}_{\text{loc}}$ (see Definition 9.2).

Recall that $\mathcal{L} = (\mathcal{M}_0)_{\text{loc}}$ and $\mathcal{H}^2_{\text{loc}} \subset \mathcal{M}_{\text{loc}}$ is the space of square integrable local martingales.

Definition 11.24. Suppose M is an adapted process such that $M_0 = 0$ a.s. A stopping time T *reduces* M if M^T is a uniformly integrable martingale. Note from Lemma 9.4 that M^T is then of class D, and $M \in \mathcal{M}_{\text{loc}}$ if there is a sequence $\{T_n\}$ of stopping times such that $\lim T_n = \infty$ and each T_n reduces M.

Lemma 11.25. (i) *If the stopping time T reduces M and S is a stopping time such that $S \leq T$ then, by Doob's optional stopping theorem, S reduces M.*
(ii) *the sum of two local martingales is a local martingale.*

Proof of (ii). Clearly we can consider $M, N \in \mathcal{L}$. Suppose $\{S_n\}$, $S_n \uparrow \infty$ is a sequence of stopping times reducing M and $\{T_n\}$, $T_n \uparrow \infty$ is a sequence of stopping times reducing N. Then, by (i), $S_n \wedge T_n$ is a sequence of stopping times reducing $M + N$. □

Lemma 11.26. (i) *If $M \in \mathcal{L}$ and S, T are stopping times which reduce M, then $S \vee T$ reduces M.*
(ii) *Suppose M is an arbitrary process and there is an increasing sequence of stopping times such that each M^{T_n} is a local martingale. Then M is a local martingale.*

PROOF. (i) The process $M^{S\vee T}=M^S+M^T-M^{S\wedge T}$, and so is a uniformly integrable martingale.

(ii) We can suppose $M_0=0$. Suppose for each n, $R_{nm}\uparrow\infty$ is a sequence of stopping times reducing M^{T_n}. Write $S_{nm}=R_{nm}\wedge T_n$, so $\lim_m S_{nm}=T_n$. Index the stopping times S_{nm} in a single sequence $\{S_k\}$ and write $V_k=S_1\vee S_2\vee\cdots\vee S_k$, so $\lim_k V_k=\infty$. We show the V_k reduce M. Suppose $S_1=S_{n_1m_1},\ldots,S_k=S_{n_km_k}$ and write $r=n_1\vee\cdots\vee n_k$. Now $S_{n_im_i}$ reduces $M^{T_{n_i}}$, but $M^{T_{n_i}}$ and M^{T_r} are the same process up to the stopping time $S_{n_im_i}$. Therefore, $S_{n_im_i}$ reduces M^{T_r}. By part (i), V_k reduces M^{T_r}. However, M and M^{T_r} are the same process up to time V_k. Therefore, V_k reduces M, and $M\in\mathscr{L}$. □

Remarks 11.27. The following results are an extension to continuous parameter martingales of Gundy's decomposition for discrete time martingales. They enable the discussion of local martingales to be reduced to that of square integrable martingales and processes of integrable variation.

Definition 11.28. Suppose $M\in\mathscr{L}$. The stopping time T *strongly reduces* M if T reduces M and the martingale $E[|M_T||\mathscr{F}_t]$ is bounded on $[\![0,T[\![$.

Lemma 11.29. *Suppose S and T are stopping times, $S\le T$ and T strongly reduces $M\in\mathscr{L}$. Then S strongly reduces M.*

PROOF. $M_{T\wedge t}$ is a uniformly integrable martingale, so $|M_{T\wedge t}|$ is a submartingale. Therefore, by Theorem 4.12

$$|M_S|\le E[|M_T||\mathscr{F}_S],$$

and

$$E[|M_S||\mathscr{F}_t]\le E[|M_T||\mathscr{F}_{S\wedge t}]\quad\text{by Corollary 3.33.}$$

Consequently, the martingale $E[|M_S||\mathscr{F}_t]$ is bounded on $[\![0,S[\![$, so S strongly reduces M. □

Lemma 11.30. *Suppose S and T strongly reduce $M\in\mathscr{L}$. Then $S\vee T$ strongly reduces M.*

PROOF. From Lemma 11.26(i), $S\vee T$ reduces M. Now

$$E[|M_{S\vee T}||\mathscr{F}_t]=E[|M_S||\mathscr{F}_t]+E[|M_T||\mathscr{F}_t]-E[|M_{S\wedge T}||\mathscr{F}_t]$$

and the result follows by Lemma 11.29 above. □

Lemma 11.31. *Suppose $\mathscr{M}\in\mathscr{L}$. There is then an increasing sequence $\{T_n\}$ of stopping times such that* $\lim T_n=\infty$ *and each T_n strongly reduces M.*

PROOF. Suppose $\{R_n\}$, with $\lim R_n=\infty$, is the sequence which, by definition, reduces M. Write

$$S_{nm}=R_n\wedge\inf\{t\colon E[|M_{R_n}||\mathscr{F}_t]\ge m\}$$

and index the S_{nm} in a single sequence $\{S_k\}$. Put $T_n = S_1 \vee S_2 \vee \cdots \vee S_n$. From the lemma above it is sufficient to prove that each stopping time S in the sequence S_k strongly reduces M, that is, we must show that $E[|M_S||\mathcal{F}_t]$ is bounded on $[\![0, S[\![$. Suppose $S = S_{nm}$. Then the martingale $Y_t = E[|M_{R_n}||\mathcal{F}_t]$ is bounded by m on $[\![0, S[\![$, and by optional stopping

$$M_S I_{\{t<S\}} = E[M_R I_{\{t<S\}}|\mathcal{F}_S].$$

Therefore,

$$\begin{aligned} E[|M_S|I_{\{t<S\}}|\mathcal{F}_t] &= E[|E|[M_R I_{\{t<S\}}|\mathcal{F}_S]||\mathcal{F}_t] \\ &\leq E[E[|M_R|I_{\{t<S\}}|\mathcal{F}_S]\mathcal{F}_t] \quad \text{by Jensen's inequality} \\ &= E[|M_R|I_{\{t<S\}}|\mathcal{F}_{S\wedge t}] \quad \text{by Corollary 3.33} \\ &= Y_{S\wedge t} I_{\{t<S\}} = Y_t I_{\{t<S\}} \leq m. \end{aligned}$$

□

Theorem 11.32. *Suppose $M \in \mathcal{M}_{\text{loc}}$. Then there is an increasing sequence $\{T_n\}$ of stopping times such that $\lim_T T_n = \infty$ and the stopped martingale M^{T_n} can be written (nonuniquely) as*

$$M^n = M_0 + U^n + V^n.$$

Here, U^n and V^n are both stopped at T_n, $U^n \in \mathcal{H}_0^2$ and V^n is a martingale of integrable variation which is zero at $t = 0$.

PROOF. It is sufficient to prove the result when $M \in \mathcal{L}$. Suppose T strongly reduces M. Then we shall show that M^T has a decomposition $M^T = U + V$ where U and V are of the above form.

As usual, on $\{T = \infty\}$ M_T is defined to be $\lim_{s\to\infty} M_s$. By definition, $|M_T|$ is integrable and the martingale $E[|M_T|\,|\mathcal{F}_t]$ is bounded on $[\![0, T[\![$ by a constant A. For simplicity write $M_t^T = M_t$, $C_t = M_T I_{\{t\geq T\}} = M_t I_{\{t\geq T\}}$, and $X_t = M_t I_{\{t<T\}}$. Then $\{C_t\}$ is a process of integrable variation; write $\tilde{C} = \Pi_p^* C$ for its dual predictable projection.

$V = C - \tilde{C}$ is a martingale of integrable variation which is zero at $t = 0$. The integral of the total variation of V is at most $2E[|M_T|]$.

Define the process U by

$$\begin{aligned} U &= M - V \\ &= X + C - V = X + \tilde{C}. \end{aligned}$$

We must show $U \in \mathcal{H}_0^2$. Write

$$\begin{aligned} M_t^+ &= E[M_T^+|\mathcal{F}_t]; & M_t^- &= E[M_T^-|\mathcal{F}_t], \\ C_t^+ &= M_T^+ I_{\{t\geq T\}}; & C_t^- &= M_T^- I_{\{t\geq T\}}, \\ X_t^+ &= M_t^+ I_{\{t<T\}}; & X_t^- &= M_t^- 1_{\{t<T\}}, \\ U_t^+ &= X_t^+ + (C_t^+)^\sim; & U_t^- &= X_t^- + (C_t^-)^\sim. \end{aligned}$$

Now X^+ is a bounded positive supermartingale. Because

$$
\begin{aligned}
U_t^+ &= X_t^+ + (C_t^+)^\sim \\
&= M_t^+ I_{\{t<T\}} + M_T^+ I_{\{t \geq T\}} - C_t^+ + (C_t^+)^\sim,
\end{aligned}
$$

we see that U^+ is a martingale, and $(C_t^+)^\sim$ must be the integrable increasing predictable process which generates the potential associated with X^+. Therefore, for any stopping time T

$$
X_T^+ = E[\tilde{C}_\infty^+ - \tilde{C}_T^+ | \mathscr{F}_T], \tag{11.4}
$$

and, if T is predictable, by considering a sequence announcing T,

$$
X_{T-}^+ = E[\tilde{C}_\infty^+ - \tilde{C}_{T-}^+ | \mathscr{F}_{T-}].
$$

Furthermore, X_s^+ is bounded by A. Now

$$
\begin{aligned}
(\tilde{C}_\infty^+)^2 &= \int_{[\![0,\infty[\![} (\tilde{C}_s^+ + \tilde{C}_{s-}^+)\, d\tilde{C}_s^+ \\
&= \int_{[\![0,\infty[\![} ((\tilde{C}_\infty^+ - \tilde{C}_s^+) + (\tilde{C}_\infty^+ - \tilde{C}_{s-}^+))\, d\tilde{C}_s^+ \\
&\leq 2 \int_{[\![0,\infty[\![} (\tilde{C}_\infty^+ - \tilde{C}_{s-}^+)\, d\tilde{C}_s^+.
\end{aligned}
$$

Because $\tilde{C}_s^+$ is predictable, when we take the expectation of this expression the process $(\tilde{C}_\infty^+ - \tilde{C}_{s-}^+)$ can be replaced by its predictable projection, which by the above remarks is X_{s-}^+. Therefore,

$$
\begin{aligned}
E[U_\infty^+]^2 = E[(\tilde{C}_\infty^+)^2] &\leq 2E\left[\int_{[\![0,\infty[\![} X_{s-}^+\, d\tilde{C}_s^+\right] \\
&\leq 2AE\left[\int_{[\![0,\infty[\![} d\tilde{C}_s^+\right] \\
&= 2AE[X_0^+] = 2A^2 \quad \text{by (11.4)}.
\end{aligned}
$$

Consequently $U^+ \in \mathscr{H}_0^2$. Similarly $U^- \in \mathscr{H}_0^2$, as is $U = U^+ + U^-$. □

Remark 11.33. In fact it can be shown that U belongs to the space of *bounded mean oscillation martingales*. See [64].

Recall that Definition 9.11 stated that $M, N \in \mathscr{M}_{\text{loc}}$ are *orthogonal* if $MN \in \mathscr{L}$.

The following result is analogous to Corollary 9.36.

Theorem 11.34. *Suppose $M \in \mathscr{L}$.*

Then $\mathscr{M}$ can be written in a unique way as $M = M^c + M^d$, where M^c and M^d are in $\mathscr{L}$, M^c is continuous, and M^d is orthogonal to every continuous local martingale.

PROOF. (a) *Uniqueness*. Suppose

$$M = M_1^c + M_1^d = M_2^c + M_2^d,$$

where all the terms are local martingales, M_1^c and M_2^c are continuous, and M_1^d and M_2^d are orthogonal to very continuous local martingale. Then $M_1^d - M_2^d = M_2^c - M_1^c$ is a continuous local martingale which is orthogonal to itself. $(M_1^d - M_2^d)^2$ is, therefore, a local martingale which is positive and zero at $t = 0$. Consequently, $M_1^d = M_2^d$.

(b) *Existence*. The uniqueness established above allows one to "paste" together martingales obtained on an increasing sequence of stochastic intervals, so it is sufficient to establish the existence of the decomposition up to a stopping time T which strongly reduces M.

If T is such a stopping time then, following the notation of Theorem 11.32,

$$M^T = U + V = U^c + U^d + V.$$

Here $U^c \in \mathscr{H}_0^{2,c}$ and $U^d \in \mathscr{H}_0^{2,d}$ are the unique continuous and totally discontinuous martingales in the decomposition of $U \in \mathscr{H}_0^2$ and V is a martingale of integrable variation.

We must show that $U^d + V$ is orthogonal to every continuous local martingale N. It can be supposed that N is bounded; otherwise consider N stopped at the stopping times $S_n = \inf\{t: |N_t| \geq n\}$. Then $N \in \mathscr{H}^{2,c}$ so N is orthogonal to $U^d \in \mathscr{H}_0^{2,d}$, by definition, and to V by Corollary 9.27, because $\Delta N_t = \Delta N_t^c \equiv 0$. Consequently, $M^c = U^c$ and $M^d = U^d + V$ gives the desired decomposition. □

Lemma 11.35. *Suppose that $M \in \mathscr{L}$. For every finite t the sum $\sum_{s \leq t} \Delta M_s^2(\omega)$ is finite, for almost every ω.*

PROOF. It is enough to show that if T is a stopping time which strongly reduces M then the sum $\sum_{s \leq t} \Delta M_s^2(\omega)$ is a.s. finite. However, $M_t^T = U_t^c + U_t^d + V_t$ in the notation of the above theorem, so

$$\sum_{s \leq t} \Delta M_s^2(\omega) \leq 2\Big(\sum_{s \leq T} (\Delta U_s^d)^2 + \sum_{s \leq T} \Delta V_s^2\Big).$$

By Corollary 9.38

$$\sum_{s \leq T} (\Delta U_s^d)^2 < \infty \quad \text{a.s.}$$

Because V is of integrable variation $\sum_{s \leq T} |\Delta V_s| < \infty$ a.s., so $\sum_{s \leq T} \Delta V_s^2 < \infty$ a.s. and the result follows. □

Definition 11.36. Suppose $M \in \mathscr{L}$ and let $M = M^c + M^d$ be its unique decomposition into a continuous local martingale and a totally discontinuous local martingale. Then the *optional quadratic variation* of M is the increasing process

$$[M, M]_t = \langle M^c, M^c \rangle_t + \sum_{s \le t} \Delta M_s^2 \quad \text{(See Definition 10.14 and Lemma 10.15.)}$$

If $M, N \in \mathscr{L}$

$$[M, N] = \tfrac{1}{2}([M+N, M+N] - [M, M] - [N, N])$$
$$= \langle M^c, N^c \rangle_t + \sum_{s \le t} \Delta M_s \Delta N_s.$$

Remarks 11.37. Corollary 12.24 below shows that $MN - [M, N] \in \mathscr{L}$.

From the uniqueness of the above decompositions and processes, note that if $M \in \mathscr{L}$ and T is a stopping time which strongly reduces M then

$$(M^c)^T = (M^T)^c,$$
$$\langle M^c, M^c \rangle_{t \wedge T} = \langle (M^T)^c, (M^T)^c \rangle_t.$$

If $M \in \mathscr{M}_{\text{loc}}$ and $\bar{M}_t = M_t - M_0 \in \mathscr{L}$ then $[M, M]_t$ is defined to be $[\bar{M}, \bar{M}]_t + M_0^2$.

Lemma 11.38. *Suppose $A \in \mathscr{V}$ is local martingale. Then we also have $A \in \mathscr{A}_{\text{loc}}$.*

Proof. We can suppose $A_0 = 0$. Let $\{T_n\}$ be an increasing sequence of stopping times such that $\lim T_n = \infty$ and each T_n strongly reduces the local martingale A. Write

$$S_n = \inf\left\{t : \int_0^t |dA_s| \ge n\right\},$$

and

$$R_n = T_n \wedge S_n.$$

We must show

$$E\left[\int_{[0, R_n]} |dA_s|\right] < \infty.$$

However,

$$E\left[\int_{[0, R_n[} |dA_s|\right] \le n.$$

From Lemma 11.29 R_n strongly reduces A, so in the notation of Theorem 11.32 A^{T_n} can be expressed as $U + V$, where $U \in \mathscr{H}_0^2$ and $V \in \mathscr{A}_0$. Therefore, $\Delta U_{R_n} \in L^2(\Omega)$ and $\Delta V_{T_n} \in L^1(\Omega)$, so

$$E[|\Delta A_{T_n}|] < \infty$$

and

$$E\left[\int_{[0, R_n]} |dA_s|\right] < \infty. \qquad \square$$

The following result is related to Corollary 7.32.

Lemma 11.39. (i) *Every predictable local martingale M is continuous.*
(ii) *Every predictable local martingale M which is of finite variation is constant.*

Proof. (i) By considering a stopping time which strongly reduces M we can suppose M is uniformly integrable. Because M is predictable its only discontinuities occur at predictable stopping times. Suppose T is a predictable stopping time. Then

$$\Delta M_T = M_T - M_{T-} = M_T - E[M_T|\mathscr{F}_{T-}]$$

by Corollaries 5.24 and 3.19. However, M is predictable so by Theorem 6.30 M_T is $\mathscr{F}_{T-}$ measurable, and $\Delta M_T = 0$.

(ii) If, furthermore, M is of finite variation then $\int_0^t |dM_s|$ is continuous and so, again by stopping the process, we can assume M is of integrable variation. Then by Corollary 7.32, $M_s = M_0$. □

Remark 11.40. If in (ii) above $M_0 = 0$ a.s. then $M_s = 0$ a.s.

Lemma 11.41. *Suppose $A \in \mathscr{A}_{\text{loc}}$. Then there is a unique predictable process $\tilde{A} \in \mathscr{A}_{\text{loc}}$ such that $A - \tilde{A} \in \mathscr{L}$.*

Proof. Again we can suppose $A_0 = 0$. Because A is locally integrable there is an increasing sequence $\{T_n\}$ of stopping times such that $\lim T_n = \infty$ and $E[\int_0^{T_n} |dA_s|] < \infty$. By Corollary 7.31 the dual predictable projection $\Pi_p^* A^{T_n}$ of A^{T_n} is such that $A^{T_n} - \Pi_p^* A^{T_n}$ is a martingale. By uniqueness, if $T_m > T_n$ then $(\Pi_p^* A^{T_m})^{T_n} = \Pi_p^* A^{T_n}$ so we can define a predictable process $\tilde{A}$ by putting $\tilde{A}^{T_n} = \Pi_p^* A^{T_n}$. Clearly $A - \tilde{A}$ is then a local martingale, strongly reduced by the stopping times T_n. □

Definition 11.42. $\tilde{A}$ is the *dual predictable projection* of $A \in \mathscr{A}_{\text{loc}}$ and is written $\Pi_p^*(A)$.

Definition 11.43. An optional process $\{H_t\}$ is *locally bounded* if H_0 is a.s. finite, and if there is an increasing sequence $\{T_n\}$ of stopping times such that $\lim T_n = \infty$ and constants K_n such that

$$|H_t| I_{\{0 < t \leq T_n\}} \leq K_n \quad \text{a.s.}$$

Theorem 11.44. (a) *Suppose $M \in \mathscr{M}_{\text{loc}}$ and H is a predictable locally bounded process. There is then a unique local martingale $H \cdot M$ such that for every bounded martingale N*

$$[H \cdot M, N] = H \cdot [M, N].$$

(Here the right-hand side is just a Stieltjes integral on each sample path.)

(b) $(H \cdot M)_0 = H_0 M_0$, $(H \cdot M)^c = H \cdot M^c$, *and the processes* $\Delta(H \cdot M)_t$ *and* $H_t \Delta M_t$ *are indistinguishable.*

(c) *If the local martingale* M *is also in* $\mathscr{A}_{\text{loc}}$, *then* $H \cdot M$ *can be calculated as the Stieltjes integral along each sample path.*

PROOF. Suppose first that $M_0 = 0$; then we can also assume that $H_0 = 0$. There is an increasing sequence $\{T_n\}$ of stopping times with $\lim T_n = \infty$ such that H^{T_n} is a bounded process and T_n strongly reduces M.

Suppose $M^{T_n} = M^n = U^n + V^n$, where U^n and V^n are as given in Theorem 11.32. The stochastic integral $H \cdot M^n = H \cdot U^n + H \cdot V^n$ is then defined by Theorem 11.3 and Definition 7.10. Furthermore, $H \cdot M^n$ is a uniformly integrable martingale.

If $n < m$ (so that $T_n \leq T_m$ a.s.), because $M^n = (M^m)^{T_n}$ we have

$$H \cdot M^n = (H \cdot M^m)^{T_n}.$$

A process $H \cdot M$ is then defined by putting $(H \cdot M)^{T_n} = H \cdot M^n$. Clearly $H \cdot M$ is local martingale. For $t < T_n$ from

$$\Delta(H \cdot M)_t^{T_n} = H_t \Delta M_t^n,$$

and so by "pasting" together the intervals

$$\Delta(H \cdot M)_t = H_t \Delta M_t.$$

Also, from the definition of M^c and from the proof of Theorem 11.7 we have

$$H \cdot M^c = (H \cdot M)^c.$$

Therefore, for any bounded martingale N

$$[H \cdot M, N]_t = [H \cdot M^c, N^c]_t + \sum_{s \leq t} H_s \Delta M_s \Delta N_s.$$

However, this is equal to

$$(H \cdot [M, N])_t.$$

If $M_0 \neq 0$, $H_0 \neq 0$ then we consider the local martingale $\bar{M} = M - M_0$ and define $H \cdot M = H_0 M_0 + H \cdot \bar{M}$. From the Remarks 11.37 we just add $H_0 M_0 N_0$ to the above identity so we again have that

$$[H \cdot M, N] = H \cdot [M, N].$$

We have, therefore, established (a) and (b).

If $M \in \mathscr{M}_{\text{loc}} \cap \mathscr{A}_{\text{loc}}$ then the stopping times T_n can be chosen so that $M^{T_n} \in \mathscr{A}$. From Theorem 11.13 the stochastic integral $H \cdot M^n$ and the Stieltjes stochastic integral $\int H \, dM^n$ are indistinguishable. This proves (c).

The uniqueness of the stochastic integral $H \cdot M$ follows from the following result, because if $A = H \cdot M$ and $B = \overline{H \cdot M}$ are two processes such that

$$[A, N] = [B, N] = H \cdot [M, N]$$

for every bounded martingale N, then

$$[A-B, N]=0. \qquad \square$$

Theorem 11.45. *Suppose $M \in \mathcal{M}_{\text{loc}}$ and N is any bounded martingale.*

(a) *The process $[M, N]$ is locally integrable, and if T is a stopping time which strongly reduces M then $\int_{]0,T]} |d[M, N]_s|$ is integrable.*
(b) *If there is an increasing sequence of stopping times T_n, with lim $T_n = \infty$, such that $E[[M, N]_{T_n}]=0$, for every bounded martingale N, then $M=0$.*

PROOF. We can suppose $M_0=0$ a.s. Write $M^T = U+V$ as in Theorem 11.32, where $U \in \mathcal{H}_0^2$ and V is a martingale of integrable variation which is 0 at $t=0$. Then by the second inequality of Kunita–Wantabe, Corollary 10.12,

$$E\left[\int |d[U, N]_s|\right] < \infty,$$

and from the definition of $[U, N]$

$$E\left[\int d[U, N]_s\right] = E[U_\infty N_\infty].$$

Also, from Corollary 9.27, $V^c \equiv 0$ so

$$[V, N]_t = \sum_{s \le t} \Delta V_s \Delta N_s.$$

Because V is of integrable variation and N is bounded $\int |d[V, N]_s|$ is integrable. Therefore,

$$\int |d[M^T, N]_s| \le \int |d[U, N]_s| + \int |d[V, N]_s|,$$

so $\int_{]0,T]} |d[M, N]_s|$ is integrable, proving (a) above. Also, $E[\int d[V, N]_s] = E[V_\infty N_\infty]$ by Lemma 9.25.

Now $E[[M, N]_T]=0$ implies that $E[M_T N_\infty]=0$. However, N_∞ is any bounded random variable, so $M_T=0$ a.s. Because M^T is a uniformly integrable martingale $M_0^T=0$ a.s., thus proving (b). If $M_0 \ne 0$ the above results are extended by considering the local martingale $\bar{M} = M - M_0$. $\square$

Definition 11.46. If $M, N \in \mathcal{M}_{\text{loc}}$ are such that $[M, N] \in \mathcal{A}_{\text{loc}}$, then $\langle M, N\rangle$ is defined to be the dual predictable projection of $[M, N]$.

Note that from Theorem 11.45(a) $\langle M, N\rangle$ is defined if $M \in \mathcal{M}_{\text{loc}}$ and N is a local martingale which is locally bounded.

The following two lemmas are used in the sequel.

Lemma 11.47. *Let $\{M_t\}$ be a local martingale, and suppose there is a time $T \le \infty$ such that $E[[M, M]_t] < \infty$ for each $t \in [0, T]$. Then $\{M_t\}$ is a square integral martingale for $t \in [0, T]$.*

PROOF. From Corollary 12.25 below

$$N_t = M_t^2 - [M, M]_t$$

is a local martingale, so there is an increasing sequence $\{T_n\}$ of stopping times such that $\lim T_n = \infty$ a.s. and $N_{t \wedge T_n}$ is a uniformly integrable martingale. For each n, if $0 \le t \le T$,

$$E[N_{t \wedge T_n}] = E[M_{t \wedge T_n}^2 - [M, M]_{t \wedge T_n}] = 0,$$

so

$$E[M_{t \wedge T_n}^2] = E[[M, M]_{t \wedge T_n}]$$
$$\le E[[M, M]_t] < \infty$$

by hypothesis, because $[M, M]$ is an increasing process. Therefore, $M_{t \wedge T_n}^2$ a bounded sequence, and so the sequence $M_{t \wedge T_n}$ is uniformly integrable. Consequently, $\{M_t\}$ is a square integrable martingale on $[0, T]$. □

Lemma 11.48. *Suppose $\{H_t\}$ is a corlol optional (resp. predictable), process. Then $\{H_{t-}\}$ (resp. $\{H_t\}$), is locally bounded.*

PROOF. Write $T_n = \inf\{t: |H_t| \ge n\}$. In the optional case (see Remark 10.16), clearly

$$|H_{t-}| \cdot I_{\{0 \le t \le T_n\}} \le n.$$

In the predictable case each T_n is a predictable stopping time announced by a sequence $\{S_{nm}\}$. Writing

$$S_k = \sup_{\substack{n \le k \\ m \le k}} S_{nm},$$

we have $\lim_k S_k = \infty$ a.s. and $|H_t| I_{\{0 \le t \le S_k\}} \le k$. □

CHAPTER 12

Semimartingales and the Differentiation Rule

Definition 12.1. An adapted process $X = \{X_t\}$, $t \geq 0$, is a *semimartingale* if it has a decomposition of the form

$$X_t = X_0 + M_t + A_t,$$

where $M \in \mathscr{L}$ and $A \in \mathscr{V}_0$.

Remarks 12.2. From our definitions of local martingales and finite variation processes, X is corlol. However, no assumption is made about the integrability of A.

Examples 12.3. Clearly, local martingales and processes of finite variation are semimartingales.

Every supermartingale is a semimartingale, because a supermartingale X has a Doob–Meyer decomposition $X = M - A$ with $M \in \mathscr{M}_{\text{loc}}$ and $A \in \mathscr{A}_{\text{loc}}$.

Suppose the adapted corlol process X has independent increments and values in R^n. Then (see, for example, [50] Theorem 2.78) X is a semimartingale if for every $u \in R^n$ the function $E\ [\exp(i\langle u, X_t\rangle)]$ has finite variation on any compact set.

Remarks 12.4. Because there are local martingales which are processes of finite (indeed, locally integrable), variation the above decomposition is not unique. However, consider two decompositions for a semimartingale X:

$$X_t = X_0 + M_t + A_t = X_0 + \bar{M}_t + \bar{A}_t.$$

Then $N = M - \bar{M}$ is in $\mathscr{L}$, and by Lemma 11.38 is also a process of locally integrable variation. Suppose T is a stopping time which strongly reduces N and which is such that the variation of N^T is integrable.

As in Theorem 11.32 write,

$$N^T = U + V,$$

where V is of integrable variation and $U \in \mathscr{H}_0^2$. Then U is also of integrable variation and so has no continuous part, by Theorem 9.41. Consequently, N^T has no continuous part by Theorem 11.34. Therefore, $M^c = \bar{M}^c$, and the continuous part of the martingale in the decomposition of X is independent of the decomposition.

Definition 12.5. Write

$$M^c = X^c.$$

Then X^c is the *continuous martingale part* of X.

Definition 12.6. Suppose X is a semimartingale. Then the *optional quadratic variation* of X is the process

$$[X, X]_t = \langle X^c, X^c \rangle_t + \sum_{s \leq t} \Delta X_s^2.$$

$A \in \mathscr{V}_0$ and $\sum_{s \leq t} \Delta A_s^2 \leq K \sum_{s \leq t} |\Delta A_s| < \infty$ for some K. Also, from Lemma 11.35, $\sum_{s \leq t} \Delta M_s^2 < \infty$. Therefore, this process is almost surely finite because $\Delta X_s^2 \leq 2(\Delta M_s^2 + \Delta A_s^2)$. As before, if Y is a second semimartingale define $[X, Y] = \frac{1}{2}([X+Y, X+Y] - [X, X] - [Y, Y])$. If $[X, Y] \in \mathscr{A}_{\text{loc}}$ then the *predictable quadratic variation* of X and Y is the process

$$\langle X, Y \rangle = \Pi_p^*[X, Y].$$

Theorem 12.7. *Suppose X and Y are two semimartingales, and H and K are measurable processes, then*

$$\int_0^\infty |H_s||K_s||d[X, Y]_s| \leq \left(\int_0^\infty H_s^2 \, d[X, X]_s\right)^{1/2} \left(\int_0^\infty K_s^2 \, d[Y, Y]_s\right)^{1/2}.$$

PROOF. The process $[X, Y]$ is bilinear in X and Y, and $[X, X]$ is positive, so the proof is exactly the same as that of Theorem 10.11. □

Theorem 12.8. *Suppose $\{M_t\} \in \mathscr{L}$ and $\{A_t\} \in \mathscr{V}$. If $\{A_t\}$ is predictable then $[M, A]_t \in \mathscr{L}$.*

PROOF. By definition

$$[M, A]_t = \sum_{s \leq t} \Delta M_s \Delta A_s.$$

However, the local martingale M has a unique decomposition as the sum of a continuous local martingale M^c and a discontinuous martingale M^d. Furthermore, the discontinuous local martingale can be decomposed into the sum of a discontinuous local martingale M^{dp} which has only accessible jump times, and a discontinuous local martingale M^{dq} which has only totally inaccessible jump times.

Because A is predictable the processes $\{A_t\}$ and $\{A_{t-}\}$ are locally bounded by Lemma 11.48. Therefore,

$$\sum_{s\leq t} \Delta M_s \Delta A_s = \sum_{s\leq t} \Delta A_s \Delta M_s^{dp}$$

$$= \int_0^t \Delta A_s \, dM_s^{dp},$$

and so $[M, A]_t$ is a local martingale. □

Theorem 12.9. *Suppose $X = X_0 + M + A$ is a semimartingale. Here $M \in \mathcal{L}$ and $A \in \mathcal{V}_0$. As in Definition* 11.42, *suppose that H is a predictable locally bounded process.*

Then the process

$$H \cdot X = H_0 X_0 + H \cdot M + H \cdot A$$

is a semimartingale. $H \cdot X$ is independent of the decomposition X, and the processes $H \cdot X^c$ and $(H \cdot X)^c$ and $H_t \Delta X_t$ and $\Delta(H \cdot X)_t$ are indistinguishable.

Furthermore, if T is any stopping time the processes $H \cdot X^T$ and $(H \cdot X)^T$ are indistinguishable.

PROOF. Suppose $X = X_0 + \bar{M} + \bar{A}, \bar{M} \in \mathcal{L}, \bar{A} \in \mathcal{V}_0$, is a second decomposition of X. Then $M - \bar{M} = \bar{A} - A \in \mathcal{L} \cap \mathcal{V}_0$, and so $M - \bar{M}$ is a local martingale which is locally of integrable variation. Therefore, by Theorem 11.44(c) the stochastic integral $H \cdot (M - \bar{M})$ is equal to the Stieltjes integral $H \cdot (A - \bar{A})$, and so

$$H_0 X_0 + H \cdot M + H \cdot A = H_0 X_0 + H \cdot \bar{M} + H \cdot \bar{A}.$$

Because $X^c = M^c$ the processes $H \cdot X^c$ and $(H \cdot X)^c$ are indistinguishable by Theorem 11.44(b). Similarly

$$H_t \Delta X_t = H_t(\Delta M_t + \Delta A_t)$$

$$= \Delta(H \cdot M)_t + \Delta(H \cdot A)_t = \Delta(H \cdot X)_t.$$

Also, $X^T = I_{[\![0,T]\!]} X$ so clearly

$$H \cdot X^T = (H \cdot X)^T.$$ □

Remarks 12.10. The main result of this chapter, the differentiation rule, will now be proved in several stages. It is a generalization of the Ito differentiation rule established in [48], for stochastic integrals with respect to Brownian motion. The rule is first established for a continuous, bounded, real semimartingale.

Theorem 12.11. *Suppose the real semimartingale $X = X_0 + M + A$ is such that $|X_0| \leq K$ a.s., $M \in \mathcal{H}_0^{2,c}$ is bounded by K, and $A \in \mathcal{V}_0^c$ is such that*

$$\int_0^\infty |dA_s| \leq K \quad a.s.$$

Let F be a real valued function on $\mathbf{R}$ which is twice continuously differentiable. Then

$$F(X_t) = F(X_0) + \int_0^t F'(X_s)\, dM_s + \int_0^t F'(X_s)\, dA_s$$
$$+ \tfrac{1}{2} \int_0^t F''(X_s)\, d\langle M, M\rangle_s.$$

That is, the processes on the left- and right-hand side are indistinguishable.

PROOF. Write

$$I_1 = \int_0^t F'(X_s)\, dM_s, \qquad I_2 = \int_0^t F\,(X_s)\, dA_s, \qquad I_3 = \int_0^t F'(X_s)\, d\langle M, M\rangle_s.$$

Now

$$|X| \leq 3K,$$

so write C for a constant greater than

$$\sup\{|F'(x)| + |F''(x)|: -3K \leq x \leq 3K\}.$$

If $a, b \in [-3K, 3K]$,

$$F(b) - F(a) = (b-a)F'(a) + \tfrac{1}{2}(b-a)^2 F''(a) + r(a,b),$$

where, because F'' is uniformly continuous on $[-3K, 3K]$, $|r(a,b)| \leq \varepsilon(|b-a|)(b-a)^2$. Here $\varepsilon(s)$ is an increasing function of s such that $\lim_{s\to 0} \varepsilon(s) = 0$.

A *stochastic* subdivision of $[0, t]$ is now defined by putting $t_0 = 0$,

$$t_{i+1} = t \wedge (t_i + a) \wedge \inf\{s > t_i : |M_s - M_{t_i}| > a \text{ or } |A_s - A_{t_i}| > a\},$$

where a is any positive real number.

Then as $a \to 0$ the steps of the subdivision, $\sup(t_{i+1} - t_i)$, converge uniformly to 0, and the random variables $\sup|M_{t_{i+1}} - M_{t_i}| \leq a$, $\sup|A_{t_{i+1}} - A_{t_i}| \leq a$, tend uniformly to 0. Therefore, the variation of X on the interval $[t_i, t_{i+1}]$ is bounded by 4a. Now

$$F(X_t) - F(X_0) = \sum_i F'(X_{t_i})(X_{t_{i+1}} - X_{t_i}) + \tfrac{1}{2} \sum_i F''(X_{t_i})(X_{t_{i+1}} - X_{t_i})^2$$
$$+ \sum_i r(X_{t_i}, X_{t_{i+1}})$$
$$= S_1 + \tfrac{1}{2} S_2 + R, \quad \text{say}.$$

We shall show that as $a \to 0$, S_1 converges in measure to $I_1 + I_2$, S_2 converges in measure to I_3, and R converges in measure to 0.

Write

$$S_1 = U_1 + U_2,$$

where

$$U_1 = \sum_i F'(X_{t_i})(M_{t_{i+1}} - M_{t_i}), \qquad U_2 = \sum_i F'(X_{t_i})(A_{t_{i+1}} - A_{t_i}).$$

Step 1. We show that U_1 converges to I_1 in $L^2(\Omega)$.

Write

$$I_1 = \sum_i \int_{t_{i+1}}^{t_i} F'(X_s)\, dM_s.$$

The martingale property implies different terms in the sum are mutually orthogonal, so

$$\begin{aligned}
\|U_1 - I_1\|_2^2 &= \sum_i \left\| \int_{t_i}^{t_{i+1}} (F'(X_s) - F'(X_{t_i}))\, dM_s \right\|_2^2 \\
&= E\left[\sum_i \int_{t_i}^{t_{i+1}} (F'(X_s) - F'(X_{t_i}))^2\, d\langle M, M\rangle_s \right] \\
&\leq E[\{\sup_i \sup_{t_i \leq s \leq t_{i+1}} (F'(X_s) - F'(X_{t_i}))^2\} \langle M, M\rangle_t].
\end{aligned}$$

By uniform continuity, the supremum tends uniformly to zero. $\langle M, M\rangle_t$ is integrable, so the result follows by the monotone convergence theorem.

Step 2. We show that U_2 converges to I_2 in $L^1(\Omega)$.

$$\begin{aligned}
|U_2 - I_2| &\leq \sum_i \int_{t_i}^{t_{i+1}} |F'(X_s) - F(X_{t_i})|\, |dA_s| \\
&\leq \{\sup_i \sup_{t_i \leq s \leq t_{i+1}} |F'(X_s) - F'(X_{t_i})|\} \int_0^t |dA_s|.
\end{aligned}$$

Again, by uniform continuity of F' and the monotone convergence theorem, $\|U_2 - I_2\|_1$ converges to 0.

Step 3. Writing

$$S_2 = V_1 + V_2 + V_3,$$

where

$$\begin{aligned}
V_1 &= \sum_i F''(X_{t_i})(A_{t_{i+1}} - A_{t_i})^2, \\
V_2 &= 2 \sum_i F''(X_{t_i})(A_{t_{i+1}} - A_{t_i})(M_{t_{i+1}} - M_{t_i}), \\
V_3 &= \sum_i F''(X_{t_i})(M_{t_{i+1}} - M_{t_i})^2
\end{aligned}$$

we first show that V_1 and V_2 converge to 0 both a.s. and in $L^1(\Omega)$. However,

$$|V_1| \leq C \sup_i |A_{t_{i+1}} - A_{t_i}| \int_0^t |dA_s|$$
$$\leq aCK,$$

so as $a \to 0$, $V_1 \to 0$, and the result follows. Similarly,

$$|V_2| \leq 2C \sup_i |M_{t_{i+1}} - M_{t_i}| \int_0^t |dA_s|$$
$$\leq 2aCK.$$

Step 4. We show that V_3 converges to I_3 in measure.

First recall that M is bounded by K so

$$E[\langle M, M\rangle_\infty - \langle M, M\rangle_t | \mathscr{F}_t] = E[M_\infty^2 | \mathscr{F}_t] - M_t^2 \leq K^2.$$

Therefore,

$$E[\langle M, M\rangle_\infty^2] = 2E\left[\int_0^\infty (\langle M, M\rangle_\infty - \langle M, M\rangle_t)\, d\langle M, M\rangle_t\right]$$
$$= 2E\left[\int_0^\infty (E[M_\infty^2 | \mathscr{F}_t] - M_t^2)\, d\langle M, M\rangle_t\right]$$
$$\leq 2K^2 E[\langle M, M\rangle_\infty] \leq 2K^4.$$

Consequently $\langle M, M\rangle_\infty \in L^2(\Omega)$ and the martingale $M^2 - \langle M, M\rangle$ is actually in $\mathscr{H}_0^2$. Write

$$J_3 = \sum_i F''(X_{t_i})(\langle M, M\rangle_{t_{i+1}} - \langle M, M\rangle_{t_i}).$$

Then the same argument as Step 2 shows that J_3 converges to

$$I_3 = \int_0^t F''(X_s)\, d\langle M, M\rangle_s$$

in $L^1(\Omega)$. Therefore, certainly J_3 converges to I_3 in measure. We shall show that V_3 converges to J_3 in $L^2(\Omega)$.

Because $M^2 - \langle M, M\rangle$ is a martingale

$$E[(M_{t_{i+1}} - M_{t_i})^2 - \langle M, M\rangle_{t_{i+1}} + \langle M, M\rangle_{t_i} | \mathscr{F}_{t_i}] = 0.$$

Therefore, distinct terms in the sum defining $V_3 - J_3$ are orthogonal and

$$\|V_3 - J_3\|_2^2 = \sum_i E[F''(X_{t_i})^2((M_{t_{i+1}} - M_{t_i})^2 - \langle M, M\rangle_{t_{i+1}} + \langle M, M\rangle_{t_i})^2].$$

However, $F''(X_{t_i})^2 \leq C^2$ and $(\alpha - \beta)^2 \leq 2(\alpha^2 + \beta^2)$ so

$$\|V_3 - J_3\|_2^2 \leq 2C^2 \sum_i E(M_{t_{i+1}} - M_{t_i})^4 + 2C^2 \sum_i E(\langle M, M\rangle_{t_{i+1}} - \langle M, M\rangle_{t_i})^2.$$

The second sum here is treated similarly to V_1 in Step 3: because $\langle M, M\rangle$ is uniformly continuous on $[0, t]$, $\sup_i (\langle M, M\rangle_{t_{i+1}} - \langle M, M\rangle_{t_i})$ converges almost surely to zero as $a \to 0$ and is bounded by $\langle M, M\rangle_t$. Therefore,

$$2C^2 \sum_i E(\langle M, M\rangle_{t_{i+1}} - \langle M, M\rangle_{t_i})^2$$
$$\leq 2C^2 E(\sup_i(\langle M, M\rangle_{t_{i+1}} - \langle M, M\rangle_{t_i})\langle M, M\rangle_t).$$

Now $\langle M, M\rangle_t \in L^2(\Omega)$, so the second sum converges to zero by Lebesgue's dominated convergence theorem.

For the first sum,

$$2C^2 \sum_i E(M_{t_{i+1}} - M_{t_i})^4 \leq 2C^2 E(\sup_i (M_{t_{i+1}} - M_{t_i})^2 \sum_i (M_{t_{i+1}} - M_{t_i})^2)$$
$$\leq 2C^2 a^2 E\Big(\sum_i (M_{t_{i+1}} - M_{t_i})^2\Big) = 2C^2 a^2 E[M_t^2],$$

which again converges to zero as $a \to 0$. (Note that it is only here, where we use the fact that $|M_{t_{i+1}} - M_{t_i}| \leq a$, that the random character of the partition $\{t_i\}$ is used.)

We have, thus, shown that $V_3 - J_3$ converges to zero in $L^2(\Omega)$, and so V_3 converges to J_3 in measure. However, J_3 converges to I_3 in measure so V_3 converges to I_3 in measure.

Step 5. Finally, we show that the remainder term R converges to 0 as $a \to 0$.

We have observed that the remainder term r in the Taylor expansion is such that

$$|r(a, b)| \leq \varepsilon(|b - a|)(b - a)^2,$$

where ε is an increasing function and $\lim_{s\to 0} \varepsilon(s) = 0$. Therefore,

$$|R| \leq \sum_i (X_{t_{i+1}} - X_{t_i})^2 \varepsilon(|X_{t_{i+1}} - X_{t_i}|)$$
$$\leq 2\varepsilon(2a) \sum_i ((A_{t_{i+1}} - A_{t_i})^2 + (M_{t_{i+1}} - M_{t_i})^2).$$

Now

$$E\Big[\sum_i (M_{t_{i+1}} - M_{t_i})^2\Big] = E[M_t^2]$$

is independent of the partition, and

$$E\Big[\sum_i (A_{t_{i+1}} - A_{t_i})^2\Big] \leq aE\Big[\sum_i |A_{t_{i+1}} - A_{t_i}|\Big] \leq Ka.$$

Because $\lim_{a\to 0} \varepsilon(2a) = 0$

$$\lim_{a\to 0} |E(R)| \leq \lim_{a\to 0} E(|R|) = 0.$$

For a fixed t, therefore,

$$F(X_t) = F(X_0) + \int_0^t F'(X_s)\,dM_s + \int_0^t F'(X_s)\,dA_s + \tfrac{1}{2}\int_0^t F''(X_s)\,d\langle M, M\rangle_s,$$

almost surely. Because all the processes are corlol the two sides are indistinguishable. □

Remark 12.12. The differentiation rule will next be proved for a function F which is twice continuously differentiable, and which has bounded first and second derivatives, and a semimartingale X of the form

$$X_t = X_0 + M_t + A_t,$$

where $X_0 \in L^1(\Omega)$ $M \in \mathscr{H}_0^2$ and $A \in \mathscr{A}_0$. That is the following result will be proved after the lemmas and remarks below:

Theorem 12.13. *Suppose the semimartingale X is of the form*

$$X_t = X_0 + M_t + A_t,$$

$$X_0 \in L^1(\Omega, \mathscr{F}_0), \qquad M \in \mathscr{H}_0^2, \quad A \in \mathscr{A}_0,$$

and F is twice continuously differentiable with bounded first and second derivatives. Then the following two processes, the right- and left-hand sides below, are indistinguishable:

$$F(X_t) = F(X_0) + \int_0^t F'(X_{s-})\,dX_s + \tfrac{1}{2}\int_0^t F''(X_{s-})\,d\langle X^c, X^c\rangle_s + \sum_{0<s\le t} (F(X_s) - F(X_{s-}) - F'(X_{s-})\Delta X_s).$$

Remarks 12.14. Note that the series is absolutely convergent because, if C is a bound for $|F''|$, then by Taylor's theorem

$$\sum_s |F(X_s) - F(X_{s-}) - F'(X_{s-})\Delta X_s| \le \frac{C}{2} \sum_s \Delta X_s^2,$$

and the right-hand side is finite as in Definition 12.6.

Also, because $\langle X^c, X^c\rangle$ is a continuous process,

$$\int F''(X_{s-})\,d\langle X^c, X^c\rangle_s = \int F''(X_s)\,d\langle X^c, X^c\rangle_s.$$

However, the final two terms can be written as

$$\tfrac{1}{2}\int_0^t F''(X_{s-})\,d[X, X]_s + \sum_{0<s\le t} (F(X_s) - F(X_{s-}) - F'(X_{s-})\Delta X_s - \tfrac{1}{2}F''(X_{s-})\Delta X_s^2).$$

This representation is of interest because whilst the predictable quadratic variation process $\langle X^c, X^c \rangle$ depends on the underlying measure the optional quadratic variation process $[X, X]$ does not.

Note that Theorem 12.13 is already established when M and A are continuous and bounded.

Lemma 12.15. *Suppose the differentiation rule of Theorem* 12.13 *is true for all semimartingales of the form*

$$Y_t = Y_0 + N_t + B_t,$$

where Y_0 belongs to some dense set in $L^1(\Omega)$, N belongs to some dense set in $\mathcal{H}_0^2$, and B belongs to some dense set in $\mathcal{A}_0$.

Then Theorem 12.13 *is true for general semimartingales of the stated form.*

PROOF. Write C for the uniform bound for F' and F''. Consider a general semimartingale of the form

$$X_t = X_0 + M_t + A_t,$$

where $X_0 \in L^1(\Omega)$, $M \in \mathcal{H}_0^2$, $A \in \mathcal{A}_0$.

Suppose $\{Y^n\}$, $n = 1, 2, \ldots,$ is a sequence of semimartingales in the dense subspace for which the differentiation rule is true, and which are such that, if

$$Y_t^n = Y_0^n + N_t^n + B_t^n, \qquad n = 1, 2, \ldots,$$

then

$$\sum_n \|Y_0^n - X_0\|_1 < \infty, \tag{12.1}$$

$$\sum_n \|N_\infty^n - M_\infty\|_2 < \infty, \tag{12.2}$$

and

$$\sum_n \|B^n - A\|_v < \infty. \tag{12.3}$$

(Here, for a process $A \in \mathcal{A}_0$, $\|A\|_v = E[\int_0^\infty |dA_s|]$.)

Step 1. Inequality (12.1) implies that Y_0^n converges to X_0 both almost surely and in $L^1(\Omega)$. From Doob's inequality, Theorem 4.2(5), (12.2) implies that $(N^n - M)_\infty^*$ converges to zero almost surely. Therefore, the trajectories of N^n converge to those of M uniformly, almost surely, as do their left limits. Furthermore, $(N^n)^* \leq M^* + \sum_k (M - N^k)^*$ and so $(N^n)_\infty^*$ is uniformly bounded in $L^2(\Omega)$.

Also, the trajectories of B^n converge uniformly, almost surely, to those of A.

Therefore we certainly have that $F(Y_t^n)$ converge to $F(X_t)$ and $F(Y_0^n)$ converge to $F(X_0)$, almost surely, and so in measure.

The remaining terms will be considered separately:

Step 2. Write

$$I_1 = \int_0^t F'(Y_{s-}^n)\, dN_s^n - \int_0^t F'(X_{s-})\, dM_s,$$

and decompose I_1 as $I_2 + I_3$, where

$$I_2 = \int_0^t (F'(Y_{s-}^n) - F'(X_{s-}))\, dM_s,$$

$$I_3 = \int_0^t F'(Y_{s-}^n)(dN_s^n - dM_s).$$

Then

$$\|I_2\|_2^2 = E\left[\int_0^t (F'(Y_{s-}^n) - F'(X_{s-}))^2\, d\langle M, M\rangle_s\right]$$

and, because F' is bounded, this converges to zero as $n \to \infty$ by the dominated convergence theorem. Also,

$$\|I_3\|_2^2 \le C^2 E[(N^n - M)_\infty^2],$$

so I_3 converges to zero as $n \to \infty$. That is, I_1 converges to zero in L^2, and so certainly in measure, as $n \to \infty$.

Step 3. Write

$$I_4 = \int_0^t F'(Y_{s-}^n)\, dB_s^n - \int_0^t F'(X_{s-})\, dA_s$$
$$= I_5 + I_6,$$

where

$$I_5 = \int_0^t (F'(Y_{s-}^n) - F'(X_{s-}))\, dA_s,$$

and

$$I_6 = \int_0^t F'(Y_{s-}^n)(dB_s^n - dA_s)_t.$$

Again, because $|I_5| \le 2C \int_0^t |dA_s|$, the dominated convergence theorem applies, and as

$$\|I_5\|_1 \le E\left[\int_0^t |F'(Y_{s-}^n) - F'(X_{s-})|\, |dA_s|\right],$$

we see that I_5 converges to zero in $L^1(\Omega)$. Also,

$$\|I_6\|_1 \le CE\left[\int_0^t |dB_s^n - dA_s|\right]$$
$$\le C\|B^n - A\|_v.$$

Therefore, I_6 converges to zero in $L^1(\Omega)$. Consequently, I_4 converges to zero in $L^1(\Omega)$, and so in measure.

Step 4. Note first that

$$\langle Y^{nc}, Y^{nc}\rangle_s - \langle X^c, X^c\rangle_s = \langle Y^{nc} + X^c, Y^{nc} - X^c\rangle_s.$$

By the Kunita–Watanabe inequality, Corollary 10.12,

$$E\left[\int_0^\infty |d\langle Y^{nc} + X^c, Y^{nc} - X^c\rangle_s|\right] \leq \|Y^{nc} + X^c\|_{\mathcal{H}^2}\|Y^{nc} - X^c\|_{\mathcal{H}^2}.$$

However, by definition $Y^{nc} = N^{nc}$ and $X^c = M^c$. Also N^n converges to M in $\mathcal{H}^2$. Because the projection onto $\mathcal{H}^{2c}$ is continuous, N^{nc} converges to M^c in $\mathcal{H}^2$. Therefore,

$$E\left[\int_0^\infty |d\langle Y^{nc}, Y^{nc}\rangle_s - d\langle X^c, X^c\rangle_s|\right]$$

converges to zero. Writing

$$I_7 = \int_0^t F''(Y^n_{s-})\, d\langle Y^{nc}, Y^{nc}\rangle_s - \int_0^t F''(X_{s-})\, d\langle X^c, X^c\rangle_s$$

the same arguments as in Step 3 now show that I_7 converges to zero in $L^1(\Omega)$, and so in measure.

Step 5. Write

$$\begin{aligned} I_8 = \sum_{0<s\leq t} &((F(Y^n_s) - F(Y^n_{s-}) - F'(Y^n_{s-})\Delta Y^n_s) \\ &- (F(X_s) - F(X_{s-}) - F'(X_{s-})\Delta X_s)). \end{aligned}$$

We shall show that I_8 converges to zero almost surely, and so in measure. From remarks in Step 1 the trajectories of Y^n converge uniformly, almost surely, to those of X. There is, therefore, a constant C' such that

$$\begin{aligned} |F(Y^n_s) - F(Y^n_{s-}) - F'(Y^n_{s-})\Delta Y^n_s| &\leq C(\Delta Y^n_s)^2 \\ &\leq C'(\Delta X_s)^2. \end{aligned}$$

Now $\sum_{0<s\leq t} (\Delta X_s)^2$ is almost surely convergent, so the series

$$\sum_{0<s\leq t} (F(Y^n_s) - F(Y^n_{s-}) - F'(Y^n_{s-})\Delta Y^n_s)$$

is absolutely convergent, uniformly in n. Its sum, therefore, converges almost surely to

$$\sum_{0<s\leq t} (F(X_s) - F(X_{s-}) - F'(X_{s-})\Delta X_s).$$

That is, I_8 converges to zero almost surely, and so in measure.

Consequently, letting $n \to \infty$ the differentiation rule of Theorem 12.13 is true for the semimartingale X. □

Remarks 12.16. We now consider the sequence of semimartingales $\{Y^n\}$ by which we approximate the semimartingale $X_t = X_0 + M_t + A_t$ of Theorem 12.13.

From Theorems 6.46, 6.27 and 5.16 there is a sequence of stopping times $\{T_n\}$, with disjoint graphs, which exhaust the jumps of M and A, and such that each stopping time T_n is >0, and either predictable or totally inaccessible.

Write

$$L_t^n = \Delta M_{T_n} I_{\{t \geq T_n\}} - \Pi_p^*(\Delta M_{T_n} I_{\{t \geq T_n\}}) \quad \text{(see Theorems 9.28 and 9.29)}$$

and

$$C_t^n = \Delta A_{T_n} I_{\{t \geq T_n\}}.$$

Then

$$M = M^c + \sum_n L^n \quad \text{in } \mathcal{H}^2$$

and

$$A = A^c + \sum_n C^n \quad \text{in } \mathcal{A}.$$

Define

$$Y^N = M^c + \sum_{n \leq N} L^n + A^c + \sum_{n \leq N} C^n,$$

so that Y^N approaches the semimartingale X, in the sense that $M^c + \sum_{n \leq N} L^n$ approaches M in $\mathcal{H}^2$ and $A^c + \sum_{n \leq N} C^n$ approaches A in $\mathcal{A}$. Theorem 12.13 will, therefore, be proved if we can establish it for semimartingales Y^N which have only finitely many jumps at the disjoint stopping times $T_1, \ldots, T_N$. However, in the above decomposition of Y^N the martingales L^n are also processes of integrable variation; let us forget these processes are martingales (because the decomposition of a semimartingale is not unique), and include the processes L^n in the integrable variation part of the semimartingale. That is, we have now reduced the semimartingale we need consider in Theorem 12.13 to a semimartingale $X_t = X_0 + M_t + A_t$, where $M \in \mathcal{H}_0^{2,c}$ (i.e. M is continuous) and $A \in \mathcal{A}_0$ has at most N jumps. (12.4)

Suppose the N jumps of such a semimartingale X are indexed in increasing order as $S_1, \ldots, S_N$, where $0 < S_1 \leq S_2 \leq \cdots \leq S_N \leq \infty$.

Lemma 12.17. *The semimartingale we need consider can be further restricted so that, if $X_t = X_0 + M_t + A_t$, then $M \in \mathcal{H}_0^{2,c}$ and M is bounded, $A \in \mathcal{A}$ has at most N jumps and $\int_0^\infty |dA_s^c|$, is bounded, and X_0 is bounded.*

PROOF. If X has the properties (12.4) above and $k>0$ write

$$R=\inf\left\{t: |M_t|\geq k \text{ or } \int_0^t |dA_s|\geq k\right\}.$$

Then R is a stopping time.

Put

$$\bar{M}_t = M_{t\wedge R},$$

$$\bar{A}_t = A_{t\wedge R},$$

and

$$X_0 = X_0 I_{\{|X_0|\leq k\}}.$$

As $k\to\infty$, $\bar{M}$ approaches M in $\mathcal{H}^2$, $\bar{A}$ approaches A in $\mathcal{A}$, and $\bar{X}_0$ approaches X_0 in $L^1(\Omega)$. Certainly $\int_0^\infty |d\bar{A}_s^c|<\infty$, so the result will follow from Lemma 12.15. □

PROOF OF THEOREM 12.13. From Lemma 12.15, Remarks 12.16, and Lemma 12.17, the result of Theorem 12.13 will follow if it can be proved for a semimartingale

$$X_t = X_0 + M_t + A_t,$$

where $M\in\mathcal{H}_0^{2,c}$ and M is bounded, $A\in\mathcal{A}_0$ has at most N jumps and $\int_0^\infty |dA_s^c|<\infty$, and $X_0\in L^1(\Omega,\mathcal{F}_0)$ is bounded.

However, note that the two sides of the differentiation formula have the same jump at time t: because $\langle X^c, X^c\rangle$ is continuous the jump of the right-hand side is

$$F'(X_{t-})\Delta X_t + (F(X_t)-F(X_{t-})-F'(X_{t-})\Delta X_t) = F(X_t)-F(X_{t-}),$$

which is the jump of the left-hand side at t. Consider the continuous semimartingale

$$\bar{X}_t = X_0 + M_t + A_t^c.$$

Then from Theorem 12.11 the differentiation rule is true for $\bar{X}$. Furthermore, if the jumps of X are indexed in increasing order as above, as $0<S_1\leq\cdots\leq S_N\leq\infty$, then $X_t=\bar{X}_t$ on $[\![0,S_1[\![$. Therefore, $X_{t-}=\bar{X}_{t-}$ on $[\![0,S_1]\!]$ so

$$\int_0^t F'(X_{s-})\,dM_s = \int_0^t F'(\bar{X}_{s-})\,dM_s$$

on $[\![0,S_1]\!]$. Also,

$$\int_0^t F'(X_{s-})\,dA_s = \int_0^t F'(\bar{X}_{s-})\,dA_s^c \quad \text{on } [\![0,S_1[\![.$$

Because $X^c = \bar{X}^c = M$ and the formula is true for $\bar{X}$ (on $[\![0, \infty[\![$), the differentiation formula is true for X on $[\![0, S_1[\![$. However, the two sides of the formula have equal jumps at S_1, so the formula is true on $[\![0, S_1]\!]$. The same reasoning establishes the formula on $[\![S_1, S_2]\!]$, and so on, up to $[\![S_N, \infty[\![$. Theorem 12.13 is, therefore, proved.

Remarks 12.18. The differentiation rule will now be extended to the situation when X is a general semimartingale and F is a twice continuously differentiable function (not necessarily having bounded derivatives).

Theorem 12.19. *Suppose X is a semimartingale and F a twice continuously differentiable function. Then $F(X)$ is a semimartingale, and, with equality denoting indistinguishability:*

$$F(X_t) = F(X_0) + \int_0^t F'(X_{s-})\, dX_s + \tfrac{1}{2}\int_0^t F''(X_s)\, d\langle X^c, X^c\rangle_s$$
$$+ \sum_{0<s\le t} (F(X_s) - F(X_{s-}) - F'(X_{s-})\Delta X_s).$$

PROOF. First note that $F'(X_s)$ is corlol, and so $F'(X_{s-})$ is predictable. Also, by considering stopping times such as

$$S = \inf\{t: |F'(X_s)| \ge n\},$$

we see that $F'(X_{s-})$ is locally bounded. Similarly remarks apply to $F''(X_{s-})$. Therefore, the two integrals are well defined.

For any $\omega \in \Omega$ the trajectory $X_s(\omega)$, $0 \le s \le t$, remains in a compact interval $[-C(t, \omega), C(t, \omega)]$. On such a compact interval the second derivative of F is bounded by some constant $K(t, \omega)$. Therefore, if $s \le t$

$$|F(X_s(\omega)) - F(X_{s-}(\omega)) - F'(X_{s-}(\omega))\Delta X_s(\omega)| \le \tfrac{1}{2} K(t, \omega)\Delta X_s^2(\omega).$$

As in Definition 12.6, we know that $\sum_{s\le t} \Delta X_s^2$ is finite almost surely. Therefore, for any t the sum occurring on the right-hand side of the differentiation rule is almost surely absolutely convergent.

Suppose the general semimartingale of the theorem has a decomposition

$$X_t = X_0 + M_t + A_t,$$

where $M \in \mathscr{L}$ and $A \in \mathscr{V}_0$, and suppose there is a stopping time T which has the following properties:

(1) X_0 is bounded on $\{T > 0\}$,
(2) T strongly reduces M (so that, in particular, by Jensen's inequality M is bounded on $[\![0, T[\![$),
(3) $\int_{]0,T[} |dA_s|$ is bounded, and
(4) M and A are stopped at T.

Write K for a constant such that

$$|X| \le K \quad \text{on } [\![0, T[\![.$$

Define a process Y by

$$Y_t = X_t - \Delta X_T I_{\{t \geq T\}}.$$

Note that

$$Y_0 = 0 \quad \text{on } \{T = 0\}.$$

Because T strongly reduces M, as in Theorem 11.32, M can be written as $U + V$, where $U \in \mathscr{H}_0^2$ and V is a martingale of integrable variation. Also, U and V are stopped at time T. Then Y can be expressed as

$$Y_t = X_0 I_{\{T>0\}} + U_t + (V_t - \Delta V_T I_{\{t \geq T\}} + A_t - \Delta A_T I_{\{t \geq T\}} - \Delta U_T I_{\{t \geq T\}}),$$

so that Y is a semimartingale of the kind discussed in Theorem 12.13 above, because $U \in \mathscr{H}_0^2$ and the term in the brackets is of integrable variation.

Write G for a twice continuously differentiable function, which has bounded first and second derivatives and which is equal to F on $[-K, K]$. If $B_t = A_t - \Delta X_T I_{\{t \geq T > 0\}}$ we have $Y_t = Y_0 + M_t + B_t$. Because F and G are equal on the interval $[-K, K]$, in which Y takes values, applying Theorem 12.13 to Y and G (and writing F for G), we have

$$F(Y_t) = F(Y_0) + \int_0^t F'(Y_{s-})\, dM_s + \int_0^t F'(Y_{s-})\, dB_s$$

$$+ \tfrac{1}{2} \int_0^t F''(Y_{s-})\, d\langle M^c, M^c\rangle_s + \sum_{0<s\leq t} ((F(Y_s) - F(Y_{s-}) - F'(Y_{s-})\Delta Y_s).$$

However, X and Y are the same process on the interval $[\![0, T[\![$, so X_{s-} and Y_{s-} are the same process on the interval $[\![0, T]\!]$ (as long as $T > 0$). Because M is stopped at T and zero at $t = 0$

$$\int_0^t F'(Y_{s-})\, dM_s = \int_0^t F'(X_{s-})\, dM_s,$$

and also

$$\int_0^t F''(Y_{s-})\, d\langle M^c, M^c\rangle_s = \int_0^t F''(X_{s-})\, d\langle M^c, M^c\rangle_s.$$

Also, A and B are the same process on the interval $[\![0, T[\![$, so

$$\int_0^t F'(Y_{s-})\, dB_s = \int_0^t F'(X_{s-})\, dA_s \quad \text{for } t < T.$$

Therefore, if $t < T$

$$F(X_t) = F(X_0) + \int_0^t F'(X_{s-})\, dX_s + \tfrac{1}{2} \int_0^t F''(X_{s-})\, d\langle X^c, X^c\rangle_s$$

$$+ \sum_{0<s\leq t} ((F(X_s) - F(X_{s-}) - F\,(X_{s-})\Delta X_s).$$

On the set $\{T=0\}$ the two sides are trivially equal. However, we have seen in Theorem 12.13 that the two sides above have the same jumps so they are equal on the interval $[\![0, T]\!]$. Because M and A were supposed stopped at T, X is stopped at T and the identity is established for all t.

We must finally show that for a general semimartingale $X_t = X_0 + M_t + A_t$, where $M \in \mathscr{L}$ and $A \in \mathscr{A}_0$, there is a sequence of stopping times $\{T_n\}$ which increase to ∞ and for which the stopped semimartingale X^{T_n} satisfies the hypotheses above. Because $M \in \mathscr{L}$ there is an increasing sequence of stopping times $\{R_n\}$ which strongly reduce M. Write

$$V_n = \inf\left\{t: \int_0^t |dA_s| \geq n\right\},$$

$$W_n = 0 \quad \text{if } |X_0| > n \quad \text{and} \quad W_n = \infty \text{ otherwise},$$

$$T_n = R_n \wedge V_n \wedge W_n.$$

Then the stopping times T_n have the required properties, so that the differentiation rule is true for X^{T_n}. That is, it is true for X_t if $t \leq T_n$. Letting $n \to \infty$ the differentiation rule is true for a general (scalar) semimartingale. □

Remark 12.20. The differentiation rule for a vector (R^n) valued semimartingale is established by the same method. However, the notation becomes very involved so the vector form of the theorem is now stated without proof.

Theorem 12.21. *Suppose X is a process with values in R^n, each of whose components X^i is a semimartingale. Suppose F is a real valued twice continuously differentiable function on R^n. Then $F(X_t)$ is a semimartingale and, with equality denoting indistinguishability:*

$$F(X_t) = F(X_0) + \sum_{i=1}^n \int_{]0,t]} \frac{\partial}{\partial x_i} F(X_{s-})\, dX_s^i$$

$$+ \frac{1}{2} \sum_{i,j=1}^n \int_0^t \frac{\partial^2}{\partial x_i\, \partial x_j} F(X_{s-})\, d\langle X^{ic}, X^{jc}\rangle_s$$

$$+ \sum_{0<s\leq t} \left(F(X_s) - F(X_{s-}) - \sum_{i=1}^n \frac{\partial}{\partial x_i} F(X_{s-})\Delta X_s^i\right).$$

The following corollaries are immediate applications of the differentiation rule.

Corollary 12.22. *If X and Y are semimartingales then the product XY is a semimartingale and*

$$X_t Y_t = \int_{]0,t]} X_{s-}\, dY_s + \int_{]0,t]} Y_{s-}\, dX_s + [X, Y]_t.$$

That is, in differential form

$$d(XY)_t = X_{t-}\, dY_t + Y_{t-}\, dX_t + d[X, Y]_t.$$

Corollary 12.23. *If X is a process of finite variation then*

$$d(XY) = X_{t-}\, dY_t + Y_t\, dX_t.$$

PROOF. Suppose a decomposition of X is $X_t = X_0 + M_t + A_t$, $M \in \mathscr{L}$, $A \in \mathscr{V}_0$. Then M is a local martingale of finite variation, so $M \in \mathscr{A}_{\text{loc}}$ by Lemma 11.38. Let T be a stopping time which strongly reduces M and such that the variation of M^T is integrable. As in Theorem 11.32 write $M^T = U + V$, where $U \in \mathscr{H}_0^2$ and $V \in \mathscr{A}_0$. Then $U \in \mathscr{H}_0^2 \cap \mathscr{A}$, so by Theorem 9.41, U has no continuous part. Therefore, $M^c = X^c = 0$, and $\langle X^c, Y^c \rangle = 0$, so

$$[X, Y]_t = \sum_{s \le t} \Delta X_s \Delta Y_s,$$

and

$$d[X, Y]_t = \Delta Y_t\, dX_t.$$

Consequently,

$$Y_{t-}\, dX_t + d[X, Y]_t = Y_t\, dX_t. \qquad \square$$

Corollary 12.24. *If M and N are local martingales $MN - [M, N] \in \mathscr{L}$ and*

$$M_t N_t - [M, N]_t = \int_0^t M_{s-}\, dN_s + \int_0^t N_{s-}\, dM_s.$$

Corollary 12.25. *If M is a local martingale the process*

$$M_t^2 - [M, M]_t = 2 \int_{]0,t]} M_{s-}\, dM_s$$

is a local martingale which is zero at $t = 0$.

Remarks 12.26. The differentiation rule will now be used to prove Lévy's characterization of Brownian motion, and a similar result for the Poisson process. Related martingale representation results will also be obtained.

Definition 12.27. Suppose $\{B_t\}$, $t \ge 0$, is an adapted stochastic process on a filtered probability space $(\Omega, \mathscr{F}, P, \mathscr{F}_t)$. Then $\{B_t\}$ is a *Brownian motion* if for every pair $s, t \in [0, \infty)$ with $s \le t$, $B_t - B_s$ is a Gaussian random variable which is independent of $\mathscr{F}_s$ and of variance $t - s$.

Theorem 12.28. *Suppose $\{B_t\}$ is a continuous martingale on the filtered probability space $(\Omega, \mathscr{F}, P, \mathscr{F}_t)$ such that $B_t^2 - t$ is a martingale. Then $\{B_t\}$ is a Brownian motion.*

PROOF. Consider any random variable $X \in L^2(\mathscr{F}_\infty)$ and write X_t for the martingale $E[X|\mathscr{F}_t]$. Suppose that the product X_tB_t is a martingale; then we shall first show that for every real number u and every pair s, t with $s<t$

$$E[e^{iu(B_t-B_s)}X_t|\mathscr{F}_s]=X_s e^{-u^2(t-s)/2}. \tag{12.5}$$

Notice that for $v \geq 0$ we could define $\mathscr{F}'_v = \mathscr{F}_{s+v}, X'_v = X_{s+v}, B'_v = B_{s+v} - B_s$, $t'=t-s$, so that, by such a time displacement, it is sufficient to establish (12.5) when $s=0$ and $B_0=0$. That is, we must prove that

$$E[e^{iuB_t}X_t|\mathscr{F}_0]=X_0 e^{-u^2t/2}.$$

However, applying the differentiation rule to the function $F(x)=e^{iux}$:

$$e^{iuB_t}=1+iu\int_0^t e^{iuB_s}\,dB_s-\frac{u^2}{2}\int_0^t e^{iuB_s}\,ds, \tag{12.6}$$

because $\langle B, B\rangle_s = s$ by hypothesis. Consider $A \in \mathscr{F}_0$ and write

$$f(s)=\int_A e^{iuB_s}X_s\,dP.$$

Multiply (12.6) above by X_t and integrate over A. Then on the left we obtain $f(t)$.

On the right the following terms are obtained:

(1) $\int_A X_t\,dP=\int_A X_0\,dP$, because X_t is a martingale.

(2) Because X_tB_t is a martingale, X is orthogonal to B, and so $X_t\int_0^t e^{iuB_s}\,dB_s$ is a martingale. therefore,

$$\int_A\left(X_t\int_0^t e^{iuB_s}\,dB_s\right)dP=0.$$

(3) By the differentiation rule

$$X_t\int_0^t e^{iuB_s}\,ds=\int_0^t X_s e^{iuB_s}\,ds+\int_0^t\left(\int_0^s e^{iuB_v}\,dv\right)dX_s,$$

so

$$\int_A\left(X_t\int_0^t e^{iuB_s}\,ds\right)dP=\int_0^t f(s)\,ds.$$

The function $f(s)$, therefore, satisfies the equation

$$f(t)=\int_A X_0\,dP-\frac{u^2}{2}\int_0^t f(s)\,ds,$$

which has the solution

$$f(t)=\left(\int_A X_0\,dP\right)e^{-u^2t/2}.$$

That is,

$$E[e^{iuB_t}X_t|\mathscr{F}_0]=X_0\,e^{-u^2t/2}.$$

Re-scaling the time and taking $X=1$ we have

$$E[e^{iu(B_t-B_s)}|\mathscr{F}_s]=e^{-u^2(t-s)/2}.$$

Therefore, the conditional law of B_t-B_s, given $\mathscr{F}_s$, is a fixed centred, Gaussian law, of variance $(t-s)$, and so $\{B_t\}$ is a Brownian motion. □

The following extension can be proved the same way:

Corollary 12.29. *Suppose $\{B_t\}=\{B_t^1,\ldots,B_t^n\}$ is a continuous vector martingale with values in R^n such that*

$$\langle B^i,B^j\rangle_t=\delta^{ij}t.$$

Then $\{B_t\}$ is an n-dimensional Brownian motion (that is, a vector of independent Brownian motions).

Definition 12.30. Suppose $\{P_t\}$ is a purely discontinuous, increasing, adapted process on the filtered probability space $(\Omega,\mathscr{F},P,\mathscr{F}_t)$, $t\geq 0$, all of whose jumps equal $+1$.

Denote the sequence of jump times by $S_1,S_2,\ldots$. Then P_t is a *Poisson process* if the random variables $S_1,S_2-S_1,\ldots,S_n-S_{n-1},\ldots$ are exponentially distributed with parameter 1, and are, respectively, independent of $\mathscr{F}_0,\mathscr{F}_{S_1},\ldots,\mathscr{F}_{S_{n-1}}\ldots.$

Theorem 12.31. *Suppose $\{Q_t\}$ is a purely discontinuous martingale on the filtered probability space $(\Omega,\mathscr{F},P,\mathscr{F}_t)$, $t\geq 0$, all of whose jumps equal $+1$. If Q_t^2-t is a martingale then $P_t=Q_t+t$ is a Poisson process.*

Proof. We can suppose $Q_0=P_0=0$. Because Q_t is purely discontinuous

$$E[[Q,Q]_t]=E[Q_t^2]<\infty,$$

but because $\Delta Q_s=+1$,

$$[Q,Q]_t=\sum_{s\leq t}\Delta Q_s^2=\sum_{s\leq t}\Delta Q_s.$$

Write

$$P_t=\sum_{s\leq t}\Delta Q_s.$$

Then P_t is integrable, because $[Q,Q]_t$ is. Furthermore, Q is a compensated sum of jumps, by Theorem 9.24, so

$$Q_t=P_t-\Pi_p^*P_t.$$

However,

$$\begin{aligned}P_t - t &= [Q, Q]_t - t \\ &= (Q_t^2 - t) + ([Q, Q]_t - Q_t^2),\end{aligned}$$

and is therefore, a martingale. Consequently,

$$\Pi_p^* P_t = t.$$

That is

$$Q_t = P_t - t.$$

Applying the differentiation rule to the martingale Q_t and the function $F(x) = e^{iux}$, $u \in R$, from time 0 to the first jump time $S = S_1$, we have

$$e^{iuQ_S} = 1 + iu \int_0^S e^{iuQ_{v-}} dQ_v + (e^{iuQ_S} - e^{iuQ_{S-}} - iu\, e^{iuQ_{S-}} \Delta Q_S).$$

Now $Q_{S-} = -S$ and $\Delta Q_S = 1$ so

$$(1 + iu)\, e^{-iuS} = 1 + iu \int_0^S e^{iuQ_{v-}} dQ_v.$$

Taking the conditional expectation with respect to $\mathscr{F}_0$ we have

$$E[u^{-iuS} | \mathscr{F}_0] = (1 + iu)^{-1}.$$

Therefore, S is independent of $\mathscr{F}_0$ and is exponentially distributed with parameter 1.

A time translation and a similar argument shows that $S_n - S_{n-1}$ is independent of $\mathscr{F}_{n-1}$, and is exponentially distributed with parameter 1. Therefore, P_t is a Poisson process. □

Remark 12.32. The following proofs are adapted from those of Davis [13].

Theorem 12.33. *Suppose $\{B_t\}$, $t \geq 0$, is a Brownian motion on the filtered probability space $(\Omega, \mathscr{F}, P, \mathscr{F}_t)$. Write*

$$\mathscr{G}_t^0 = \sigma\{B_s : s \leq t\}$$

and $\{\mathscr{G}_t\}$ for the completion of $\{\mathscr{G}_t^0\}$, so that the filtration $\{\mathscr{G}_t\}$ is certainly right continuous.

Then every random variable $X \in L^2(\Omega, \mathscr{G}_\infty)$ can be represented as a stochastic integral

$$X = E[X | \mathscr{G}_0] + \int_0^\infty H_s\, dB_s,$$

where $\{H_s\}$ is a $\mathscr{G}_t$-predictable process and $E[\int_0^\infty H_s^2\, ds] < \infty$. Furthermore,

$$\begin{aligned}E[X | \mathscr{G}_t] &= E[X | \mathscr{F}_t] \\ &= E[X | \mathscr{G}_0] + \int_0^t H_s\, dB_s.\end{aligned}$$

PROOF. We can restrict our attention to the filtration $\mathscr{G}_t$ and by subtraction we can suppose that $X_0 = E[X|\mathscr{G}_0] = 0$. Consider any $n \in N$ and write Y_t^n for the projection of the square integrable martingale $X_t = [X|\mathscr{G}_t]$ onto the subspace of $\mathscr{H}^2$ generated by $\{B_{t\wedge n}\}$. From Theorem 11.11 Y_t^n is a stochastic integral of the form $H^n \,.\, B^n$. Clearly for $s \leq n$, $H_s^n = H_s^{n+1}$, so by pasting together we obtain a $\mathscr{G}_t$-predictable integrand H such that $E[\int_0^\infty H_s^2\, ds] < \infty$, and a process $Y_t = \int_0^t H_s\, dB_s$. By construction we then have that $\langle X - Y, B\rangle = 0$. We can, therefore, suppose at the beginning of the proof that X_t is orthogonal to B_t, and it remains to be shown that this then implies that $X = X_\infty = 0$ a.s.

We next establish that all $\mathscr{G}_t$ stopping times are predictable. Suppose T is any $\mathscr{G}_t$ stopping time and write

$$p_t = I_{t \geq T}.$$

Then p_t is a submartingale, and it has a Doob–Meyer decomposition (Theorem 8.15),

$$p_t = r_t + q_t,$$

where q_t is a martingale and r_t is a predictable increasing process with jumps bounded by 1. Write

$$\tau_n = \inf\{t : r_t \geq n\},$$

$$q_t^n = q_{t\wedge\tau_n},$$

and

$$L = 1 - \tfrac{1}{2} q_\infty^n.$$

Then $L \geq \frac{1}{2}$ a.s. and $EL = 1$, so a new probability measure Q can be defined on $(\Omega, \mathscr{G}_\infty)$ by putting

$$\frac{dQ}{dP} = L,$$

so that Q is mutually absolutely continuous with respect to P. Now q_t^n is a square integrable martingale, and, as a compensated jump process it is orthogonal to every continuous martingale. Therefore, in particular it is orthogonal to B_t and $B_t^2 - t$. That is, $q_t^n B_t$ and $q_t^n(B_t^2 - t)$ are martingales, so B_t and $B_t^2 - t$ are martingales on $(\Omega, \mathscr{G}_t, Q)$. Therefore, B_t is a Brownian motion under measure Q as well as measure P, so P and Q coincide on $\mathscr{G}_\infty$. Consequently, $q_\infty^n = 0$ a.s., so $p_t = r_t$ a.s. That is, p_t is a predictable process and T is a predictable stopping time. In particular, every accessible stopping time is predictable, and so by Theorem 5.36 the filtration $\mathscr{G}_t$ is quasi-left continuous, $\mathscr{G}_{T-} = \mathscr{G}_T$ for every predictable stopping time. However, every stopping time T is predictable, so for every $M \in \mathscr{H}^2(\mathscr{G}_t)$

$$M_{T-} = E[M_T|\mathscr{G}_{T-}] = E[M_T|\mathscr{G}_T] = M_T.$$

Therefore, every M has continuous paths. For the martingale X_t of the first paragraph write

$$\sigma_n = \inf\{t: |X_t| \geq n\},$$

and

$$X_t^n = \frac{1}{2n} X_{t\wedge\sigma_n}.$$

Then $|X_t^n| \leq \frac{1}{2}$ and is also orthogonal to both B_t and $B_t^2 - t = \int_0^t B_s\, dB_s$. Writing

$$\Lambda = 1 + X_\infty^n$$

the same argument as above shows that $X_\infty^n = 0$ a.s. Therefore, letting $n \to \infty$: $X = X_\infty = 0$ a.s.

Finally, suppose $\{Z_t\}$ is a right continuous version of the martingale $E[X|\mathscr{G}_t]$. Then, by the above result, Z_t has a representation as a stochastic integral with respect to B_t. However, B_t is an $\mathscr{F}_t$ martingale so Z_t is also a square integrable $\mathscr{F}_t$ martingale. Therefore, the limit $Z_\infty = \lim_{t\to\infty} Z_t$ exists both almost surely and in L^2, and $Z_t = E[Z_\infty|\mathscr{F}_t]$. However, X is $\mathscr{G}_\infty$ measurable, so applying the convergence theorem with respect to the filtration $\mathscr{G}_t$ we see that $\lim_{t\to\infty} Z_t = X = Z_\infty$. Therefore,

$$E[X|\mathscr{G}_t] = E[X|\mathscr{F}_t] \quad \text{a.s.}$$

for all $t \geq 0$. □

Remark 12.34. Suppose $\{B_t\} = \{B_t^1, \ldots, B_t^n\}$ is an n-dimensional Brownian motion and that

$$\mathscr{G}_t = \sigma\{B_s: s \leq t\}.$$

The same kind of argument shows that if $\{X_t\}$ is a square integrable $\mathscr{G}_t$ martingale then there are $\mathscr{G}_t$-predictable processes

$$H_t^i, \quad i = 1, \ldots, n, \qquad E\left[\int_0^\infty (H_s^i)^2\, ds\right] < \infty, \quad \text{such that}$$

$$X_t = E[X_\infty|\mathscr{G}_0] + \sum_{i=1}^n \int_0^t H_s^i\, dB_s^i.$$

Theorem 12.35. *Suppose $\{P_t\}$, $t \geq 0$, is a Poisson process on the filtered probability space $(\Omega, \mathscr{F}, P, \mathscr{F}_t)$. Write $\mathscr{G}_t^0 = \sigma\{P_s: s \leq t\}$ and $\{\mathscr{G}_t\}$ for the completion of $\{\mathscr{G}_t^0\}$ so that the filtration $\{\mathscr{G}_t\}$ is certainly right continuous.*

Then every random variable $X \in L^2(\Omega, \mathscr{G}_\infty)$ can be represented as a stochastic integral

$$X = E[X|\mathscr{G}_0] + \int_0^\infty H_s\, dQ_s,$$

where

$$Q_t = P_t - t, \quad \{H_s\} \text{ is } \mathcal{G}_t\text{-predictable} \quad \text{and} \quad E\left[\int_0^\infty H_s^2\,ds\right] < \infty.$$

Furthermore,

$$E[X|\mathcal{G}_t] = E[X|\mathcal{F}_t] \quad \text{a.s. for all } t.$$

PROOF. Similar remarks to those at the beginning of Theorem 12.33 show that we must establish that, if X_t is a square integrable $\mathcal{G}_t$ martingale which is orthogonal to Q_t, then $X_\infty = 0$ a.s.

However, suppose $\{T_i\}$, $i = 1, 2, \ldots$, are the jump times of the Poisson process, and $[\![T]\!]$ is the graph of any accessible $\mathcal{G}_t$-stopping time T. Then, because each jump time T_i is totally inaccessible

$$[\![T]\!] \cap \left(\bigcup_i [\![T_i]\!]\right)$$

is evanescent. Therefore, the process $p_t = I_{t \geq T}$ has no discontinuities in common with P_t, and neither has its compensating process r_t because this is predictable.

Again write

$$p_t = r_t + q_t,$$

$$\tau_n = \inf\{t : r_t \geq n\},$$

$$q_t^n = q_{t \wedge \tau_n},$$

and

$$L = 1 - \tfrac{1}{2} q_\infty^n.$$

Then, because $L \geq \frac{1}{2}$ a.s. and $EL = 1$, a new probability measure P^* can be defined on $(\Omega, \mathcal{G}_\infty)$ by putting

$$\frac{dP^*}{dP} = L.$$

As noted above, the martingale q_t^n has no discontinuities in common with the Poisson process P_t, so the martingale q_t^n is orthogonal to Q_t and $Q_t^2 - t$. Therefore, Q_t and $Q_t^2 - t$ are martingales under measure P^*, so P and P^* coincide on $\mathcal{G}_\infty$, and $q_\infty^n = 0$ a.s. Again we can deduce that the accessible stopping time T is predictable, so the filtration $\mathcal{G}_t$ is quasi-left continuous.

Now suppose, as in the first paragraph of this proof, that X_t is a square integrable martingale which is orthogonal to Q_t. Then X_t has no discontinuities in common with P_t; if T is a $\mathcal{G}_t$ stopping time such that $P\{X_T \neq X_{T-}\} > 0$ then $[\![T]\!] \cap (\bigcup_i [\![T_i]\!])$ is evanescent, and the same argument as above shows that T must be predictable. However, the filtration is quasi-left continuous so

$$X_{T-} = E[X_T|\mathcal{G}_{T-}] = E[X_T|\mathcal{G}_T] = X_T.$$

Therefore, X_t is sample continuous and the proof is concluded by the same arguments as those of Theorem 12.33. □

Remarks 12.36. We have seen in the proof of Theorem 11.32 that if T is a stopping time which strongly reduces the local martingale $M \in \mathscr{L}$ then M^T can be expressed as $U+V$, where $U \in \mathscr{H}^2$ and V is of integrable variation. Then, from Definition 11.36 and Remarks 11.37:

$$([M,M]_T)^{1/2} = ([M^T, M^T]_\infty)^{1/2}$$
$$\leq [U,U]_\infty^{1/2} + [V,V]_\infty^{1/2}.$$

Because $U \in \mathscr{H}^2$, $[U,U]_\infty^{1/2} \in L^2(\Omega) \subset L^1(\Omega)$. Furthermore, V is a martingale of integrable variation, and so without any continuous part, and $[V,V]_\infty^{1/2} = (\sum_s \Delta V_s^2)^{1/2} \leq k.\sum_s |\Delta V_s| \in L^1(\Omega)$. Therefore, for every local martingale M the increasing process $[M,M]_t^{1/2}$ is locally integrable.

This observation motivates the following definition:

Definition 12.37. A semimartingale X is called a *special semimartingale* if there is a decomposition

$$X_t = X_0 + M_t + A_t, \qquad M \in \mathscr{L}, \qquad A \in \mathscr{V}_0$$

in which A is locally of integrable variation.

Theorem 12.38. *The following conditions are equivalent*:

(a) *X is a special semimartingale,*
(b) *the increasing process $(\sum_{0<s\leq t} \Delta X_s^2)$ is locally integrable,*
(c) *for every decomposition of X as above the process A is locally of integrable variation,*
(d) *there is a decomposition of the above form in which A is predictable.*

Furthermore, if X is a special semimartingale the decomposition, as in (d), *in which A is predictable is unique.*

Definition 12.39. the decomposition of a special semimartingale as in (d) is called the *canonical decomposition*.

PROOF. We shall show that (d)⇒(a)⇒(b)⇒(c)⇒(d).

(d)⇒(a). Here we are assuming that $X_t = X_0 + M_t + A_t$, where $A_0 = 0$ a.s. and A is predictable. Therefore, the process $\int_0^t |dA_s|$ is predictable (see Lemma 7.5), and the stopping time

$$S_k = \inf\left\{t: \int_0^t |dA_s| \geq k\right\} > 0$$

is predictable. Suppose $\{S_m^k\}$, $m = 1, 2, \ldots$, is a sequence of stopping times which announce S_k. Then the variation of A on $[\![0, S_m^k]\!]$ is less than k and so integrable. Writing $T_n = \sup_{k \le n, m \le n} S_m^k$ we see that $\lim T_n = \infty$ and $A^{T_n} \in \mathscr{A}$, so that A is locally integrable.

(a)$\Rightarrow$(b). Write

$$X_t = X_0 + M_t + A_t.$$

Then

$$\Big(\sum_{s \le t} \Delta X^2\Big)^{1/2} \le \Big(\sum_{s \le t} \Delta M_s^2\Big)^{1/2} + \Big(\sum_{s \le t} \Delta A_s^2\Big)^{1/2}$$

$$\le [M, M]_t^{1/2} + k \sum_{s \le t} |\Delta A_s|$$

$$\le [M, M]_t^{1/2} + k \int_0^t |dA_s|,$$

and this is locally integrable because A is locally integrable.

(b)$\Rightarrow$(c).

$$\Big(\sum_{s \le t} \Delta A_s^2\Big)^{1/2} \le \Big(\sum_{s \le t} \Delta X_s^2\Big)^{1/2} + [M, M]_t^{1/2},$$

so if (b) holds the left-hand side is an increasing locally integrable process. Suppose $\{S_n\}$ is an increasing sequence of stopping times, $\lim S_n = \infty$, such that

$$E\Big[\Big(\sum_{s \le S_n} \Delta A_s^2\Big)^{1/2}\Big] < \infty,$$

and write

$$R_n = \inf\Big\{t : \int_0^t |dA_s| \ge n\Big\}.$$

If $T_n = R_n \wedge S_n$ then certainly

$$E[|A_{T_n}|] < \infty,$$

so

$$E\Big[\int_0^{T_n} |dA_s|\Big] \le n + E[|\Delta A_{T_n}|] < \infty.$$

(c)$\Rightarrow$(d). Suppose $X_t = X_0 + M_t + A_t$ where A is locally integrable. By Lemma 11.41 there is then a unique predictable process $\tilde{A}$ such that $A - \tilde{A}$ is a local martingale. Writing

$$X_t = X_0 + (M_t + A_t - \tilde{A}_t) + \tilde{A}_t,$$

the term in brackets is a local martingale and $\tilde{A} \in \mathscr{A}_{\text{loc}}$ is predictable.

Finally, suppose

$$X_t = X_0 + M_t + A_t = X_0 + \bar{M}_t + \bar{A}_t,$$

where M and $\bar{M}$ are in $\mathscr{L}$ and A and $\bar{A}$ are both predictable. Then $\bar{M}-M=A-\bar{A}$ is a predictable local martingale which is of finite variation and locally of integrable variation. Therefore, localizing we can apply Corollary 7.32 to see it is the zero process. Therefore, the cannonical decomposition of a special semimartingale is unique. □

Corollary 12.40. *The following conditions are equivalent*:

(a) *X is a special semimartingale,*
(b) *the increasing process $X_t^* = \sup_{s\le t} |X_s|$ is locally integrable,*
(c) *the increasing process $D_t = \sup_{s\le t} |\Delta X_s|$ is integrable.*

PROOF. (a)⇒(b). If X is a special semimartingale there is a decomposition $X_t = X_0 + M_t + A_t$ in which $A \in \mathscr{A}_{\text{loc}}$.

If M is a local martingale there is, from Lemma 11.31, an increasing sequence $\{T_n\}$ of stopping times which strongly reduce M. Thus $M_{T_n} \in L^1(\Omega)$ and $|M_t|$ is bounded on $[\![0, T_n[\![$ so certainly $M_t^* \in \mathscr{A}_{\text{loc}}$. Now

$$X_t^* \le |X_0| + M_t^* + \int_0^t |dA_s|$$

so

$$X_t^* \in \mathscr{A}_{\text{loc}}.$$

(b)⇒(c). This follows because $|\Delta X_t| = |X_t - X_{t-}| \le 2X_t^*$.

(c)⇒(a). Suppose $X_t = X_0 + M_t + A_t$ is a decomposition of X. Now

$$|\Delta A_t| \le |\Delta M_t| + |\Delta X_t| \le 2M_t^* + D_t,$$

so the process $|\Delta A_t| \in \mathscr{A}_{\text{loc}}$. Consider an increasing sequence of stopping times $\{R_n\}$, such that $\lim R_n = \infty$ and

$$E[\sup_{s\le R_n} |\Delta A_s|] \le E[2M_{R_n}^* + D_{R_n}] < \infty$$

for all n. Write

$$S_n = \inf\left\{t : \int_0^t |dA_s| \ge n\right\} \quad \text{and} \quad T_n = S_n \wedge R_n.$$

Then

$$\int_0^{T_n} |dA_s| = \int_0^{T_n^-} |dA_s| + |\Delta A_{T_n}|$$

$$\le n + \sup_{s\le R_n} |\Delta A_s| \in L^1(\Omega).$$

Therefore, $A \in \mathscr{A}_{\text{loc}}$ and the result is proved. □

Theorem 12.41. *Suppose X is a process and $\{T_n\}$ a sequence of stopping times such that* $\lim T_n = \infty$. *Then X is a (special) semimartingale if, and only if, X^{T_n} is a (special) semimartingale for each n.*

PROOF. Clearly, if X is a (special) semimartingale then X^{T_n} is a (special) semimartingale for each n.

If S and T are stopping times and X^S and X^T are (special) semimartingales then the same is true of $X^{S\vee T} = X^S + X^T - X^{S\wedge T}$. Therefore, we can assume that $\{T_n\}$ is an increasing sequence of stopping times with the stated properties.

If X^{T_n} is a special semimartingale for each n it has a unique canonical decomposition

$$X_t^{T_n} = X_0 + M_t^n + A_t^n.$$

However,

$$(X^{T_{n+1}})^{T_n} = X^{T_n}, \quad \text{so } (M^{n+1})^{T_n} = M^n \quad \text{and} \quad (A^{n+1})^{T_n} = A^n.$$

The processes $\{M^n\}$ and $\{A^n\}$ can, therefore, be "pasted" together to give a local martingale M and a predictable process $A \in \mathscr{A}_{\text{loc}}$, so the process X in this case is a special semimartingale.

In the general case we just know that X^{T_n} is a semimartingale for each n. However, X is certainly a corlol process, so the process

$$V_t = \sum_{0<s\leq t} \Delta X_s I_{\{|\Delta X_s| \geq 1\}}$$

is of finite variation, as is

$$Y = X - V - X_0.$$

For each n, $Y^{T_n} = X_t^{T_n} - V_t^{T_n} - X_0$ is a semimartingale whose jumps are all bounded by 1. Therefore, by Corollary 12.40(c), Y^{T_n} is a special semimartingale. By the first part of this proof Y is then a special semimartingale and $X_t = X_0 + Y_t + V_t$ is a semimartingale. □

Remark 12.42. We now define and state, without proofs, some of the properties of the *Stratonovich* integral. This type of integral was first described by Stratonovich [74]. It is further investigated by Meyer [64], and is particularly suitable for geometric, and some filtering applications ([16, 65]).

Definition 12.43. Suppose H and X are two (real) continuous semimartingales. The *Stratonovich* integral H with respect to X is the process

$$(H \circ X)_t = (S) \int_0^t H_s \, dX_s = \int_0^t H_s \, dX_s + \tfrac{1}{2}\langle H^c, X^c \rangle_t,$$

where the integral on the right is the usual stochastic integral defined in

Chapter 11 above, and H^c, X^c denote the continuous martingale parts of H and X, respectively.

The usual stochastic integral is the limit in probability of sums

$$\sum_i H_{t_i}(X_{t_{i+1}} - X_{t_i}),$$

where $0 = t_0 < t_1 < \cdots < t_n = t$ is a subdivision of $[0, t]$. However, the Stratonovich integral is the limit in probability of sums of the form

$$\sum_i H_{s_i}(X_{t_{i+1}} - X_{t_i}), \quad \text{where } s_i = \tfrac{1}{2}(t_i + t_{i+1}).$$

Alternatively, the Stratonovich integral is the limit of Stieltjes integrals of the form

$$\int_0^t H_s^n \, dX_s^n,$$

where H^n, X^n are the polygonal functions obtained by linear interpolation from H and X between the times $\{t_i^n\}$ of the nth dyadic subdivision of $[0, t]$.

Lemma 12.44. *The Stratonovich integral satisfies the usual differentiation rule. That is, suppose X is a continuous semimartingale with a decomposition*

$$X_t = X_0 + M_t + A_t, \qquad t \geq 0,$$

(so $M = X^c$), and let f be a three times continuously differentiable function. Then $f'(X_t)$ is a continuous semimartingale and

$$(S)\int_0^t f'(X_s)\, dX_s = f(X_t) - f(X_0).$$

PROOF. Write $H_t = f'(X_t)$ so that

$$H_t = H_0 + \int_0^t f''(X_s)\, dM_s + \int_0^t f''(X_s)\, dA_s + \tfrac{1}{2}\int_0^t f'''(X_s)\, d\langle M, M\rangle_s.$$

Then the continuous martingale part of H is

$$H_t^c = \int_0^t f''(X_s)\, dM_s,$$

so that

$$\langle H^c, X^c\rangle_t = \int_0^t f''(X_s)\, d\langle M, M\rangle_s.$$

By definition

$$(S)\int_0^t H_s\, dX_s = \int_0^t f'(X_s)\, dX_s + \tfrac{1}{2}\int_0^t f''(X_s)\, d\langle M, M\rangle_s$$

and by the Ito differentiation rule this is equal to $f(X_t) - f(X_0)$. □

Definition 12.45. Suppose $\{X_t\}$ is a real semimartingale and f a continuously differentiable function.

Write Y for the process

$$Y_t = f(X_t),$$

and $\mathscr{D}$ for the set of all processes which can be expressed in this form.

Remarks 12.46. The Stratonovich integral can be defined for integrands $H \in \mathscr{D}$ and general (noncontinuous) semimartingale integrators X by putting:

$$(S)\int_0^t H_{s-}\, dX_s = \int_0^t H_{u-}\, dX_u + \tfrac{1}{2}[H, X^c]_t.$$

If f is a twice continuously differentiable function it can then be shown, as above, that

$$(S)\int_0^t f'(X_{s-})\, dX_s + \sum_{s \leq t} \{f(X_s) - f(X_{s-}) - f'(X_{s-})\Delta X_s\} = f(X_t) - f(X_0).$$

That is, the Stratonovich integral satisfies the same formula as when X is a process of bounded variation. Further details can be found in Meyer ([64] p. 360). Another natural definition is given by Marcus [82].

CHAPTER 13

The Exponential Formula and Girsanov's Theorem

Convention 13.1. As in previous chapters we shall consider real valued processes X_t and functions $f(t)$ defined for $t \geq 0$. In Lemma 13.2 below we shall, however, suppose that $f(0-) = g(0-) = 0$, so that $\Delta f(0) = f(0)$ and $\Delta g(0) = g(0)$. In Lemma 13.4 $x(0-) \neq 0$ in general.

Lemma 13.2. *Suppose $f(t)$, $g(t)$ are corlol functions of bounded variation. Then*

$$f(t)g(t) = \int_{[0,t]} f(s-)\, dg(s) + \int_{[0,t]} g(s)\, df(s).$$

If F is once continuously differentiable

$$\begin{aligned} F(f(t)) &= F(0) + \int_{[0,t]} F'(f(s-))\, df(s) \\ &\quad + \sum_{0 \leq s \leq t} \{F(f(s)) - F(f(s-)) - F'(f(s-))\Delta f(s)\} \\ &= \int_0^t F'(f(s))\, df^c(s) + \sum_{0 \leq s \leq t} (F(f(s)) - F(f(s-)). \end{aligned}$$

Here, f^c is the continuous part of the function of bounded variation f.

PROOF. These formulae are just special cases of Theorem 12.19 and Corollary 12.23 obtained by taking care with the jump at 0, and combining (or subtracting) the remaining jump terms from the integrals. Because we are considering a function of bounded variation we only require F to be once continuously differentiable. □

Corollary 13.3. *If $f(t)$ is increasing and corlol, then applying the second identity above to $F(x)=x^n$ we see that*

$$d(f(t)^n)\geq nf(t)^{n-1}\,df(t).$$

Lemma 13.4 *Suppose $f(t)$ is a corlol function bounded variation. Then there is at most one locally bounded solution of the equation*

$$x(t)-x(0-)=\int_{[0,t]} x(s-)\,df(s) \tag{13.1}$$

on $[0,\infty[$.

PROOF. Consider two such solutions $z_1(t), z_2(t)$ of (13.1). Then their difference is a solution $x(t)$ of (13.1) such that $x(0-)=0$. We shall show that the only solution of (13.1) such that $x(0-)=0$ is the function $x(t)=0$ for all $t\geq 0$.

If

$$x(0-)=0, \qquad \int_{[0,t]} x(s-)\,df(s)=\int_{]0,t]} x(s-)\,df(s)$$

and the mass of f at 0 does not enter the calculation so we can assume that $f(0-)=0$.

Write

$$g(t)=\int_0^t |df(s)|$$

for the total variation of f on $[0,t]$ and

$$B=\sup_{s\leq t}|x(s)|.$$

Then for $0\leq r\leq t$

$$\begin{aligned}|x(r)|&=\left|\int_0^r x(s-)\,df(s)\right|\leq\int_0^r |x(s-)|\,dg(s)\\&\leq Bg(r).\end{aligned}$$

Re-substituting this upper bound

$$\begin{aligned}|x(r)|&\leq\int_0^r |x(s-)|\,dg(s)\leq B\int_0^r g(s-)\,dg(s)\\&\leq Bg^2(r)/2 \quad \text{by Corollary 13.3.}\end{aligned}$$

Again:

$$\begin{aligned}|x(r)|&\leq\int_0^r |x(s-)|\,dg(s)\leq (B/2)\int_0^r g^2(s-)\,dg(s)\\&\leq Bg^3(r)/3!\end{aligned}$$

Therefore, for all $n = 1, 2, \ldots$ and $0 \leq r \leq t$

$$|x(r)| \leq Bg^n(r)/n!$$

That is, $x(r) = 0$ for all r. □

Theorem 13.5. *Suppose $\{X_t\}$, $t \geq 0$, is a semimartingale and suppose $X_{0-} = 0$ a.s. Then there is a unique semimartingale $\{Z_t\}$, $t \geq 0$, such that*

$$Z_t = Z_{0-} + \int_{[0,t]} Z_{s-}\, dX_s. \tag{13.2}$$

Here Z_{0-} is a given initial value. Furthermore, Z_t is given by the expression

$$Z_t = Z_{0-} \exp(X_t - \tfrac{1}{2}\langle X^c, X^c\rangle_t) \prod_{0 \leq s \leq t} (1 + \Delta X_s)\, e^{-\Delta X_s}, \tag{13.3}$$

for $t \geq 0$, where the infinite product is absolutely convergent almost surely.

Remarks 13.6. The notation $Z_t = Z_{0-}\mathscr{E}(X)_t$ is sometimes used.

This result was first obtained by Doléans–Dade [25].

The most common situation encountered is when $X_0 = 0$ and $Z_{0-} = Z_0$ so that the integral equation defining Z_t can be written

$$Z_t = Z_0 + \int_0^t Z_{s-}\, dX_s.$$

The proof will proceed by first showing that the process defined by (13.3) exists. Then, that it is a semimartingale, and finally that it satisfies (13.2).

Clearly we may suppose that $Z_{0-} = 1$. The only difficulty with the existence of (13.3) concerns the infinite product.

Lemma 13.7. *The infinite product*

$$V_t = \prod_{0 \leq s \leq t} (1 + \Delta X_s)\, e^{-\Delta X_s}$$

(taking $V_{0-} = 1$) is absolutely convergent, and defines a purely discontinuous process of finite variation.

PROOF. By the formula for integration by parts, the product of a finite number of purely discontinuous functions of finite variation is a function of the same type.

On any finite interval the number of jumps such that $|\Delta X_s| \geq \frac{1}{2}$ is finite. Therefore, by the above remark, the product

$$V'_t = \prod_{s \leq t} (1 + \Delta X_s I\{|\Delta X_s| \geq \tfrac{1}{2}\}) \exp(-\Delta X_s I\{|\Delta X_s| \geq \tfrac{1}{2}\})$$

is a purely discontinuous function of finite variation.

Consider the remaining terms, in the product

$$V_t'' = \prod_{s\le t} (1+\Delta X_s I\{|\Delta X_s|<\tfrac{1}{2}\}) \exp(-\Delta X_s I\{|\Delta X_s|<\tfrac{1}{2}\}).$$

Because, from Lemma 11.35 and Definition 12.6, the series $\sum_{s\le t} \Delta X_s^2$ converges, the series

$$\sum_{s\le t} (\log(1+\Delta X_s I\{|\Delta X_s|<\tfrac{1}{2}\}) - \Delta X_s I\{|\Delta X_s|<\tfrac{1}{2}\})$$

is absolutely convergent.

Therefore, $\log V_t''$ is a purely discontinuous function of finite variation. Consequently, by the differentiation rule,

$$V_t'' = \exp(\log V_t'')$$

is a function of the same type, as is

$$V_t = V_t' V_t''. \qquad \square$$

PROOF OF THEOREM 13.5. Write

$$W_t = X_t - \tfrac{1}{2}\langle X^c, X^c\rangle_t$$

and

$$F(x, y) = e^x y.$$

Then (with $Z_{0-} = 1$),

$$Z_t = F(W_t, V_t),$$

and the existence of Z_t and absolute convergence of the infinite product are consequences of Lemma 13.7 above. By the dfferentiation rule

$$Z_t - Z_{0-} = I_1 + I_2 + I_3 + I_4,$$

where

$$I_1 = \int_{[0,t]} Z_{s-}\, dW_s, \qquad I_2 = \int_{[0,t]} e^{W_{s-}}\, dV_s,$$

$$I_3 = \tfrac{1}{2}\int_{[0,t]} Z_{s-}\, d\langle W^c, W^c\rangle_s,$$

and

$$I_4 = \sum_{0\le s\le t} (Z_s - Z_{s-} - Z_{s-}\Delta W_s - e^{W_{s-}} \Delta V_s).$$

Now

$$dW_s = dX_s - \tfrac{1}{2}\, d\langle X^c, X^c\rangle_s$$

and

$$W^c = X^c, \qquad d\langle W^c, W^c\rangle_s = d\langle X^c, X^c\rangle_s.$$

Therefore

$$I_1+I_3=\int_{[0,t]} Z_{s-}\, dX_s.$$

Because V is purely discontinuous and of finite variation

$$I_2=\sum_{0\le s\le t} e^{W_{s-}}\Delta V_s.$$

Also,

$$Z_s=Z_{s-}(1+\Delta X_s)$$

and

$$Z_{s-}\Delta W_s=Z_{s-}\Delta X_s,$$

so

$$I_2+I_4=0.$$

That is, Z_t satisfies

$$Z_t=Z_{0-}+\int_{[0,t]} Z_{s-}\, dX_s. \tag{13.4}$$

Uniqueness. Suppose $\bar{Z}_t$ is a solution of (13.4), with $Z_{0-}=1$. With

$$W_t=X_t-\tfrac{1}{2}\langle X^c, X^c\rangle_t \quad \text{and } F \text{ as above,}$$

write

$$\bar{V}_t=F(-W_t, \bar{Z}_t)=e^{-W_t}\bar{Z}_t \quad \text{so } \bar{V}_{0-}=1.$$

Then by the differentiation rule

$$\bar{V}_t-\bar{V}_{0-}=J_1+J_2+J_3+J_4+J_5,$$

where

$$J_1=\int_{[0,t]} -\bar{V}_{s-}\, dW_s, \qquad J_2=\int_{[0,t]} e^{-W_{s-}}\, d\bar{Z}_s,$$

$$J_3=\tfrac{1}{2}\int_{[0,t]} \bar{V}_{s-}\, d\langle W^c, W^c\rangle_s,$$

$$J_4=-\int_{[0,t]} e^{-W_s}\, d\langle W^c, \bar{Z}^c\rangle_s,$$

$$J_5=\sum_{0\le s\le t}(\bar{V}_s-\bar{V}_{s-}+\bar{V}_{s-}\Delta W_s-e^{-W_{s-}}\Delta\bar{Z}_s).$$

From (13.4)

$$d\bar{Z}_s=\bar{Z}_{s-}\, dX_s.$$

Also

$$dW_s = dX_s - \tfrac{1}{2}\, d\langle X^c, X^c\rangle_s$$

and

$$W^c = X^c.$$

Therefore

$$\begin{aligned} d\langle W^c, \bar{Z}^c\rangle_s &= d\langle X^c, \bar{Z}^c\rangle_s \\ &= \bar{Z}_{s-}\, d\langle X^c, X^c\rangle_s. \end{aligned}$$

Now

$$\begin{aligned} J_1 + J_2 &= \int_{[0,t]} -\bar{V}_{s-}\, dW_s + \int_{[0,t]} e^{-W_{s-}} \bar{Z}_{s-}\, dX_s. \\ &= \tfrac{1}{2} \int_{[0,t]} \bar{V}_s\, d\langle X^c, X^c\rangle_s, \end{aligned}$$

so

$$J_1 + J_2 + J_3 = \int_{[0,t]} \bar{V}_s\, d\langle X^c, X^c\rangle_s.$$

However,

$$J_4 = -\int_{[0,t]} e^{-W_{s-}} \bar{Z}_{s-}\, d\langle X^c, X^c\rangle_s,$$

so

$$J_1 + J_2 + J_3 + J_4 = 0.$$

Again,

$$\Delta \bar{Z}_s = \bar{Z}_{s-} \Delta X_s$$

and

$$\Delta W_s = \Delta X_s,$$

so

$$J_5 = \sum_{0 \le s \le t} \Delta \bar{V}_s.$$

This series is absolutely convergent, and the above calculations show that

$$\bar{V}_t - \bar{V}_{0-} = J_5 = \sum_{0 \le s \le t} \Delta \bar{V}_s.$$

Therefore, V_t is a purely discontinuous process of finite variation.

$$\begin{aligned} \bar{V}_s = e^{-W_s} \bar{Z}_s &= e^{-W_{s-}}\, e^{-\Delta X_s} \bar{Z}_{s-}(1 + \Delta X_s) \\ &= \bar{V}_{s-}\, e^{-\Delta X_s}(1 + \Delta X_s), \end{aligned}$$

and

$$\Delta \bar{V}_s = \bar{V}_{s-}((1+\Delta X_s)\, e^{-\Delta X_s} - 1).$$

Consider the function

$$A_t = \sum_{0 \le s \le t} (e^{-\Delta X_s}(1+\Delta X_s) - 1).$$

Except for a finite number on any finite interval, each term of this series is bounded by $k\Delta X_s^2$ so the series is absolutely convergent. A_t is a purely discontinuous process of finite variation, and

$$\bar{V}_t = \bar{V}_{0-} + \int_{[0,t]} \bar{V}_{s-}\, dA_s, \qquad \bar{V}_{0-} = 1.$$

However, from Lemma 13.4 this integral equation has a unique solution $\bar{V}$. Therefore $\bar{Z}_t = e^{W_t}\bar{V}_t$ is also unique. □

Corollary 13.8. *If X and Y are semimartingales then*

$$\mathscr{E}(X)\mathscr{E}(Y) = \mathscr{E}(X + Y + [X, Y]).$$

Proof. This follows from the absolute convergence of the infinite product, because $(1+\Delta X_s)(1+\Delta Y_s) = (1+\Delta X_s + \Delta Y_s + \Delta X_s \Delta Y_s)$ and the $\Delta X_s \Delta Y_s$ terms cancel in the exponential on the right. Alternatively, writing $U = \mathscr{E}(X)$ and $V = \mathscr{E}(Y)$, consider

$$(UV)_t = X_{0-}Y_{0-} + \int_0^t U_{s-}\, dV_s + \int_0^t V_{s-}\, dU_s + [U, V]_t.$$

However,

$$U_t = X_{0-} + \int_0^t U_{s-}\, dX_s$$

and

$$V_t = Y_{0-} + \int_0^t V_{s-}\, dY_s,$$

so

$$(UV)_t = (XY)_{0-} + \int_0^t (UV)_{s-}\, d(X_s + Y_s + [X, Y]_s).$$

That is,

$$(UV)_t = \mathscr{E}(X + Y + [X, Y]).$$ □

Remarks 13.9. Suppose $(\Omega, \mathscr{F}, P, \mathscr{F}_t, t \ge 0)$ is a filtered probability space, with the filtration satisfying the usual conditions. If Q is a second probability measure on $(\Omega, \mathscr{F}, P)$ which is equivalent to P (that is $Q \ll P$ and $P \ll Q$), then the filtration also satisfies the usual conditions under Q. Furthermore, the evanescent sets, and the optional and predictable σ-fields are the same for Q and P.

Write

$$M_\infty = \frac{dQ}{dP},$$

and $\{M_t\}$ for the corlol version of the uniformly integrable (P) martingale $\{E[M_\infty|\mathscr{F}_t]\}$, $t \geq 0$.

By the optional stopping theorem, for every stopping time T, M_T is a density for the restriction of Q to $\mathscr{F}_T$ with respect to P.

Now $\{M_t\}$ is a positive martingale, and because Q and P are equivalent M_∞ is a.s. strictly positive. Therefore, from Theorem 4.16 the stopping time

$$T = \inf\{t\colon M_t = 0, \text{ or } t > 0 \text{ and } M_{t-} = 0\}$$

must be ∞ a.s. That is, for almost every ω the function $M_s(\omega)$ is bounded below on $[0, \infty]$ by a strictly positive number.

Lemma 13.10. *A process $\{X_t\}$, $t \geq 0$, is a local martingale under measure Q if and only if the process $\{X_t M_t\}$ is a local martingale under measure P.*

PROOF. Because Q and P are equivalent, an increasing sequence of stopping times $\{T_n\}$ which reduces X also reduces XM. It is, therefore, sufficient to prove the result for martingales. Suppose $s \leq t$ and $A \in \mathscr{F}_s$. Then

$$\int_A X_t \, dQ = \int_A X_t M_t \, dP = \int_A X_s M_s \, dP$$

$$= \int_A X_s \, dQ. \qquad \square$$

Corollary 13.11. *The process $\{M_t^{-1}\}$, $t \geq 0$, is a martingale under measure Q.*

Theorem 13.12. *A process $\{X_t\}$, $t \geq 0$, is a semimartingale under measure Q if and only if it is a semimartingale under measure P.*

PROOF. By definition, a semimartingale is the sum of a local martingale and a process of finite variation. We need only prove the theorem in one direction and we can suppose $X_0 = 0$.

If $\{X_t\}$ is a semimartingale under P, then by the product rule $\{X_t M_t\}$ is also a semimartingale under P, which has a decomposition

$$X_t M_t = N_t + A_t, \qquad N \in \mathscr{L}, \qquad A \in \mathscr{V}.$$

Therefore

$$X_t = N_t/M_t + A_t/M_t.$$

By the lemma, the first term is a local martingale under Q, and the second term is the product of the Q-semimartingale A of finite variation and the Q-martingale M_t^{-1}. □

Remarks 13.13. Suppose $\{X_t\}$, $t \geq 0$, is a local martingale under measure P. Then the above result states that X_t is certainly a semimartingale under measure Q. The following result, which is a generalization due to Yoeurp [80], of the work of Girsanov [43], gives an explicit decomposition of this Q-martingale.

Theorem 13.14. *Suppose $\{X_t\}$ is a P-local martingale, which is zero at $t = 0$.*

(a) *$\{X_t\}$ is a special semimartingale under Q if the process $\langle X, M\rangle_t$ exists, and then the canonical decomposition of $\{X_t\}$ under Q is:*

$$X_t = \left(X_t - \int_0^t M_{s-}^{-1}\, d\langle X, M\rangle_s\right) + \int_0^t M_{s-}^{-1}\, d\langle X, M\rangle_s.$$

Here the first term is a local martingale under Q, and the second is a predictable process of finite variation.

(b) *In general, the process*

$$X_t - \int_0^t M_s^{-1}\, d[X, M]_s$$

is a local martingale under Q.

PROOF. (a) Suppose $\langle X, M\rangle$ exists. Then

$$(XM)_t - \langle X, M\rangle_t$$

is a local martingale, and if $A \in \mathscr{V}_0$ is a predictable process, by Corollary 12.22,

$$(AM)_t - \int_0^t M_{s-}\, dA_s = \int_0^t A_{s-}\, dM_s + [M, A]_t. \tag{13.5}$$

From Theorem 12.8 $[M, A]_t \in \mathscr{L}$, so the right-hand side is a local martingale.

Therefore $M_t(X - A)_t$ is a local martingale under P if and only if $Y_t = \int_0^t M_{s-}\, dA_s - \langle X, M\rangle_t$ is a local martingale under P. The simplest situation when this is the case is when $Y_t = 0$, and then because M_{s-} is a.s. never zero,

$$A_t = \int_0^t M_{s-}^{-1}\, d\langle X, M\rangle_s.$$

Consequently, $A_t \in \mathscr{V}_0$ is a predictable process such that $X_t - A_t$ is a local martingale under Q, and

$$X_t = (X_t - A_t) + A_t$$

is a special semimartingale under Q.

Conversely, if X_t is a special semimartingale under Q there is a predictable process A_t of finite variation such that $X_t - A_t$ is a local martingale under Q. Then $M_t(X_t - A_t)$ is a local martingale under P, and from (13.5) $M_tX_t - \int_0^t M_{s-}\, dA_s$ is a local martingale under P. However, $\int_0^t M_{s-}\, dA_s$ is a predictable process of finite variation and so in $\mathscr{A}_{\text{loc}}$ by Lemma 7.39. Therefore, M_tX_t is a special semimartingale under P. This implies the existence of $\langle X, M\rangle_t$ by Definition 12.6.

(b) Because M_s is a.s. never zero we can consider the optional process in $\mathscr{V}_0$

$$B_t = \int_0^t M_s^{-1}\, d[X, M]_s.$$

Because B_t is of finite variation we have from Corollary 12.23

$$M_tB_t = \int_0^t M_s\, dB_s + \int_0^t B_{s-}\, dM_s,$$

so

$$M_tB_t - \int_0^t M_s\, dB_s = M_tB_t - [X, M]_t$$

is a local martingale under P. From the definition of $[X, M]_t$, $X_tM_t - [X, M]_t$ is a local martingale under P. Therefore, $M_t(X_t - B_t)$ is a local martingale under P, and so $X_t - B_t$ is a local martingale under Q. □

Theorem 13.15. *Again let P and Q be equivalent probability measures. Suppose H_t is a predictable locally bounded process, and suppose X_t is a semimartingale, (under both P and Q by the above result). Then the stochastic integrals: $(H_P \cdot X)_t$ under measure P and $(H_Q \cdot X)_t$ under measure Q are indistinguishable processes.*

PROOF. Clearly integrals with respect to processes of finite variation are the same under either measure, so it is sufficient to consider the case when X_t is local martingale under measure P. Write

$$Y_t = (H_P \cdot X)_t.$$

Then by the above Theorem 13.14(b)

$$\tilde{Y}_t = Y_t - \int_0^t M_s^{-1}\, d[Y, M]_s$$

is a local martingale under measure Q. Now

$$[Y, M]_t = (H \cdot [X, M])_t$$

and for any P-semimartingale U_t:

$$[\tilde{Y}, U]_t = \left[Y - \left(\frac{H}{M}\right) \cdot [X, M], U \right]_t$$

$$= [Y, U]_t - \sum_{0 \le s \le t} \frac{H_s}{M_s} \Delta X_s \Delta M_s \Delta U_s$$

$$= \left(H \cdot \left[X - \frac{1}{M}[X, M], U \right] \right)_t$$

$$= (H \cdot [\tilde{X}, U])_t,$$

where $\tilde{X}$ is the Q local martingale

$$X_t - (M^{-1} \cdot [X, M])_t.$$

If U is an Q-local martingale this identity is the relation which characterizes the stochastic integral $H_Q \cdot \tilde{X}$ (see Theorem 11.44).

Therefore,

$$\tilde{Y}_t = (H_Q \cdot \tilde{X})_t.$$

However, because

$$X_t = \tilde{X}_t + (M^{-1} \cdot [X, M])_t,$$

$$(H_Q \cdot X)_t = (H_Q \cdot \tilde{X})_t + \left(\frac{H}{M} \cdot [X, M] \right)_t$$

$$= \tilde{Y}_t + \left(\frac{H}{M} \cdot [X, M] \right)_t$$

$$= (H_P \cdot X)_t.$$

□

Remarks 13.16. We now discuss the relationship between the exponential formula, the change of probability measure and the form of Girsanov's theorem given above.

From Theorem 13.5 if X is a semimartingale the unique solution of the equation

$$Z_t = 1 + \int_0^t Z_{s-} \, dX_s$$

is

$$Z_t = \exp(X_t - \tfrac{1}{2}\langle X^c, X^c \rangle_t) \prod_{0 \le s \le t} (1 + \Delta X_s) \, e^{-\Delta X_s}.$$

Clearly, if X is a local martingale then Z is a local martingale. Furthermore Z_t is strictly positive if and only if $\Delta X_s > -1$ a.s. for all s.

Lemma 13.17. *If X is a local martingale with $\Delta X_s > -1$ a.s. for all s then Z is a supermartingale.*

Z is a martingale if and only if $E[Z_t] = 1$ for each $t \geq 0$.

PROOF. Because

$$Z_t = 1 + \int_0^t Z_{s-}\, dX_s$$

Z is clearly a local martingale. Suppose $\{T_n\}$ is an increasing sequence of stopping times which strongly reduce Z. Writing $Z_t^n = Z_{T_n \wedge t}$ we have for $s \leq t$,

$$E[Z_t^n | \mathscr{F}_s] = Z_s^n \quad \text{a.s.}$$

Because $Z_s^n \geq 0$ and $\lim Z_s^n = Z_s$ a.s., by Fatou's lemma $Z_s \geq E[Z_t | \mathscr{F}_s]$ a.s., so Z is a supermartingale. The second assertion follows because $E[Z_0] = 1$. □

Remarks 13.18. If Z_t is a uniformly integrable positive martingale then $Z_\infty = \lim_{t \to \infty} Z_t$ exists, and

$$E[Z_\infty | \mathscr{F}_t] = Z_t \quad \text{a.s.}$$

Consequently,

$$E[Z_\infty] = E[Z_0] = 1.$$

Furthermore, in this situation a new probability measure Q can be defined on $(\Omega, \mathscr{F})$ by putting

$$\frac{dQ}{dP} = Z_\infty.$$

Q is equivalent to P if and only if $Z_\infty > 0$ a.s.

General conditions on the semimartingale X_t under which Z_t is uniformly integrable are given by Memin and Shiryayev [83].

The following form of Girsanov's theorem was proved by Van Schuppen and Wong [72].

Theorem 13.19. *Suppose the exponential $Z_t = \mathscr{E}(X)_t$ is, as above, a uniformly integrable positive martingale and that a new probability measure Q is defined on $(\Omega, \mathscr{F})$ by putting*

$$\frac{dQ}{dP} = Z_\infty.$$

If N_t is a local martingale under measure P, and the process $\langle N, X \rangle_t$ exists under P, then

$$\tilde{N}_t = N_t - \langle N, X \rangle_t$$

is a local martingale under Q.

PROOF. Because $dQ/dP = Z_\infty$, Z_t plays the role of M_t in part (a) of Theorem 13.14. However,

$$Z_t = 1 + \int_0^t Z_{s-}\, dX_s,$$

so

$$\langle N, Z\rangle_t = \int_0^t Z_{s-}\, d\langle N, X\rangle_s$$

and

$$\int_0^t Z_{s-}^{-1}\, d\langle N, Z\rangle_s = \langle N, X\rangle_t.$$

That is, from part (a) of Theorem 13.14,

$$\tilde{N}_t = N_t - \langle N, X\rangle_t$$

is a local martingale under Q. □

Remark 13.20. We now investigate the quadratic variation process of N_t under Q following Neveu [67].

Lemma 13.21. *Suppose for $\theta \in R$*

$$Y_t^\theta = \exp[\theta N_t - \tfrac{1}{2}\theta^2 A_t]$$

is a martingale, and suppose there is an open neighbourhood I of $\theta = 0$ such that for all $\theta \in I$ and all t (P a.s.),

(a) $|Y_t^\theta| \le a$,
(b) $|(d/d\theta) Y_t^\theta| \le b$, *and*
(c) $|(d^2/d\theta^2) Y_t^\theta| \le c$.

Here, a, b, c are nonrandom constants which depend on I, but not on t.
Then the processes $\{N_t\}$ and $\{N_t^2 - A_t\}$ are martingales.

PROOF. Take $s \le t$ and $A \in \mathscr{F}_s$. Then using (a) and (b):

$$\begin{aligned}\int_A E\left[\left(\frac{d}{d\theta} Y_t^\theta\right)_{\theta=0} \middle| \mathscr{F}_s\right] dP &= \int_A \left(\frac{d}{d\theta} Y_t^\theta\right)_{\theta=0} dP \\ &= \left(\frac{d}{d\theta}\int_A Y_t^\theta\, dP\right)_{\theta=0} \\ &= \left(\frac{d}{d\theta}\int_A Y_s^\theta\, dP\right)_{\theta=0} \\ &= \int_A \left(\frac{d}{d\theta} Y_s^\theta\right)_{\theta=0} dP.\end{aligned}$$

That is,

$$E\left[\left(\frac{d}{d\theta}Y_t^\theta\right)_{\theta=0}\middle|\mathscr{F}_s\right]=\left(\frac{d}{d\theta}Y_s^\theta\right)_{\theta=0}\quad P \text{ a.s.}$$

so

$$E[N_t|\mathscr{F}_s]=N_s\quad\text{a.s.}$$

The second assertion follows similarly using (c), because $((d^2/d\theta^2)Y_t^\theta)_{\theta=0}=N_t^2-A_t$. □

Theorem 13.22. *Suppose* $\{N_t\}$ *and* $\{A_t\}$, $t\geq 0$, *are continuous stochastic processes where* $N_0=0$ *a.s. Then the following statements are equivalent*:

(a) $\{N_t\}$ *is a continuous local martingale, with* $\langle N, N\rangle_t=A_t$,
(b) *for all* $\theta\in R, \{Y_t^\theta\}$ *is a continuous local martingale, where* $Y_t=\exp[\theta N_t-\frac{1}{2}\theta^2A_t]$.

PROOF. (a)⇒(b). By the differentiation rule

$$dY_t^\theta=Y_t^\theta\,d(\theta N_t-\tfrac{1}{2}\theta^2A_t)+\tfrac{1}{2}Y_t^\theta\theta^2\,d\langle N, N\rangle_t$$

so

$$Y_t^\theta=1+\int_0^t\theta Y_s^\theta\,dN_t,$$

and Y_t^θ is a continuous local martingale of the stated form.

(b)⇒(a). Consider the stopping times

$$T_n=\inf\{t\colon t\leq T, |N_t|\geq n \text{ or } |A_t|\geq n\}.$$

Then

$$Y_{t\wedge T_n}^\theta=\exp[\theta N_{t\wedge T_n}-\tfrac{1}{2}\theta^2A_{t\wedge T_n}],$$

and for θ in some bounded neighbourhood I of 0,

$$|Y_{t\wedge T_n}^\theta|\leq\exp[n|\theta|+\tfrac{1}{2}\theta^2n]\leq a_n(I),$$

$$\left|\frac{d}{d\theta}Y_{t\wedge T_n}^\theta\right|\leq|Y_{t\wedge T_n}^\theta|\,|N_{t\wedge T_n}-\theta A_{t\wedge T_n}|\leq b_n(I),$$

$$\left|\frac{d^2}{d\theta^2}Y_{t\wedge T_n}^\theta\right|\leq|Y_{t\wedge T_n}^\theta|((N_{t\wedge T_n}-\theta A_{t\wedge T_n})^2+A_{t\wedge T_n})\leq c_n(I).$$

Therefore, by Lemma 13.21 $\{N_{t\wedge T_n}\}$ and $\{N_{t\wedge T_n}^2-A_{t\wedge T_n}\}$ are continuous martingales, and the result follows. □

Corollary 13.23. *Let* $\{N_t\}=\{N_t^1,\ldots,N_t^m\}$, $t\geq 0$, *be a continuous stochastic process with values in* R^m *and* $N_0=0$ *a.s., and let* $\{A_t\}=\{A_t^{ij}\}$, $t\geq 0, 1\leq i$,

$j \leq m$, be a continuous matrix valued process. Then the following statements are equivalent:

(a) *N_t is a continuous R^m-valued local martingale, and $\langle N^i, N^j \rangle = A_t^{ij}$,*
(b) *for all $\theta \in R^m$, $\{Y_t^\theta\}$ is a continuous local martingale, where $Y_t^\theta = \exp[(\theta, N_t) - \frac{1}{2}(\theta, A_t \theta)]$.*

PROOF. (a)$\Rightarrow$(b). This is proved as before by the differentiation rule.
(b)$\Rightarrow$(a). Write

$$X_t = \sum_{i=1}^{m} \theta^i N_t^i,$$

$$B_t = (\theta, A_t \theta) = \sum_{i,j} A_t^{ij} \theta^i \theta^j.$$

Then for every $\alpha \in R$ and $\theta \in R^m$ the hypothesis (b) implies that

$$Y_t^{\alpha\theta} = \exp[\alpha X_t - \tfrac{1}{2}\alpha^2 B_t]$$

is a continuous local martingale. Therefore, $\{X_t\}$ is a continuous local martingale and $\langle X, X \rangle_t = B_t$. The result follows. $\square$

Remark. The following result gives an alternative proof of Theorem 13.19 in the continuous case, and also describes the quadratic variation under the new measure Q.

Theorem 13.24. *Suppose for a continuous local martingale $\{X_t\}$ the exponential $Z_t = \mathscr{E}(X_t)$ is a uniformly integrable positive martingale, and that a new measure Q is defined on $(\Omega, \mathscr{F})$ by putting $dQ/dP = Z_\infty$.*

Let $\{N_t\} = \{N_t^1, \ldots, N_t^m\}$ be an m-dimensional continuous local martingale, with values in R^m, under measure P. Then $\{\tilde{N}_t\} = \{\tilde{N}_t^1, \ldots, \tilde{N}_t^m\}$ is a continuous local martingale under Q, where $\tilde{N}_t^i = N_t^i - \langle N^i, X \rangle_t$, and the quadratic variation under Q of $\tilde{N}$ is equal to the quadratic variation under P of N.

That is $\langle N^i, N^j \rangle_t^Q = \langle N^i, N^j \rangle_t^P$.

PROOF. Because X_t and each N_t^i are continuous local martingales the process $\langle N^i, X \rangle$ exists and $\tilde{N}_t^i = N_t^i - \langle N^i, X \rangle_t$ is a local martingale under Q by Theorem 13.19.

However, we shall prove the result using Corollary 13.23 above. Write $A_t^{ij} = \langle N^i, N^j \rangle_t$, and A_t for the $m \times m$ matrix process with entries A_t^{ij}. Then, by Corollary 13.23, the result will follow if we show that for every $\theta \in R^m$ the process

$$\tilde{Y}_t^\theta = \exp[(\theta, \tilde{N}_t) - \tfrac{1}{2}(\theta, A_t \theta)]$$

is a continuous martingale under Q. This is so if $Z_t \tilde{Y}_t^\theta$ is a continuous local martingale under P. However, for any $\theta \in R^m$ if

$$H_t = (\theta, N_t) = \sum_{i=1}^{m} \theta^i N_t^i,$$

then

$$\begin{aligned}\langle H, H\rangle_t &= \sum_{i,j} \theta^i \theta^j \langle N^i, N^j\rangle_t \\ &= (\theta, A_t \theta)\end{aligned}$$

and

$$\langle H, X\rangle_t = \sum_{i=1}^m \theta^i \langle N^i, X\rangle_t.$$

Therefore,

$$\begin{aligned}(\theta, \tilde{N}_t) - \tfrac{1}{2}(\theta, A_t\theta) + X_t - \tfrac{1}{2}\langle X, X\rangle_t &= (\theta, N_t) - \sum_{i=1}^m \theta^i \langle N^i, X\rangle_t \\ &\quad - \tfrac{1}{2}\langle H, H\rangle_t + X_t - \tfrac{1}{2}\langle X, X\rangle_t \\ &= H_t + X_t - \tfrac{1}{2}\langle H + X, H + X\rangle_t,\end{aligned}$$

so

$$\begin{aligned}Z_t Y_t^\theta &= \exp(X_t - \tfrac{1}{2}\langle X, X\rangle_t)\exp((\theta, \tilde{N}_t) - \tfrac{1}{2}(\theta, A_t\theta)) \\ &= \exp(H_t + X_t - \tfrac{1}{2}\langle H + X, H + X\rangle_t) \\ &= \mathscr{E}(H + X)_t.\end{aligned}$$

Because H and X are continuous local martingales under P, $Z_t Y_t^\theta$ is also. □

Corollary 13.25 (Girsanov). *Suppose* $\{B_t\} = \{B_t^1, \ldots, B_t^m\}$, $t \in [0, T]$, *is an m-dimensional Brownian motion on a filtered probability space* $(\Omega, \mathscr{F}, \mathscr{F}_t, P)$. *Let* $f: \Omega \times [0, T] \to R^m$ *be a predictable process such that*

$$\int_0^T |f_t|^2\, dt < \infty \quad a.s.$$

Write

$$\xi_0^t(f) = \exp\left[\sum_{i=1}^m \int_0^t f_s^i\, dB_s^i - \tfrac{1}{2}\int_0^t |f_s|^2\, ds\right]$$

and suppose

$$E[\xi_0^T(f)] = 1.$$

If Q is the probability measure on $(\Omega, \mathscr{F})$ *defined by* $dQ/dP = \xi_0^T(f)$, *then* $\{\tilde{B}_t\}$ *is an n-dimensional Brownian motion on* $(\Omega, \mathscr{F}, \mathscr{F}_t, Q)$, *where*

$$\tilde{B}_t^i = B_t^i - \int_0^t f_s^i\, ds.$$

PROOF. We can apply Theorem 13.24 above with $N_t^i = B_t^i$ and

$$X_t = \sum_{i=1}^{m} \int_0^t f_s^i \, dB_s^i$$

so

$$\begin{aligned}\tilde{N}_t^i &= N_t^i - \langle N^i, X \rangle_t \\ &= B_t^i - \int_0^t f_s^i \, ds = \tilde{B}_t^i.\end{aligned}$$

Theorem 13.24 then states that, for $t \in [0, T]$, $\{\tilde{B}_t^i\}$ is a continuous local martingale under measure Q and

$$\langle \tilde{B}^i, \tilde{B}^j \rangle_t^Q = \langle B^i, B^i \rangle_t^P = \delta_{ij} t.$$

By Lemma 11.46 $\{\tilde{B}_t\}$ is a continuous martingale for $t \in [0, T]$ and, by Corollary 12.29, an m-dimensional Brownian motion. □

Definition 13.26. Under the above conditions we have shown that the original Brownian motion process $\{B_t\}$ is a *weak solution* of the stochastic differential equation

$$dx = f(t, \omega) \, dt + dw_t, \qquad x_0 = 0.$$

(Here $\{w_t\}$ is an m-dimensional Brownian motion.)

That is, we have constructed a probability measure Q on $(\Omega, \mathscr{F})$ and a new Brownian motion $\{\tilde{B}_t\}$ such that

$$dB = f(t, \omega) \, dt + d\tilde{B}_t.$$

It is clearly of importance to known when the condition $E[\xi_0^T(f)] = 1$ is satisfied. The following result, due to Novikov [68], gives the strongest possible sufficient condition. We shall first establish the result for one dimensional processes, $m = 1$.

Theorem 13.27. *Let $\{B_t\}$ be a Brownian motion as in Definition* 12.27. *Suppose $f \colon \Omega \times [0, T] \to R$ is a predictable process such that*

$$\int_0^T |f_s|^2 \, ds < \infty \quad a.s.$$

Then if

$$E\left[\exp\left(\tfrac{1}{2} \int_0^T |f_s|^2 \, ds\right)\right] < \infty,$$

the supermartingale $\{\xi_0^t(f)\}$, $0 \le t \le T$, is a martingale, and

$$E[\xi_0^t(f)] = 1 \quad \textit{for all } t \in [0, T].$$

PROOF.

$$\xi_0^t(f) = \mathscr{E}\left(\int_0^t f_s \, dB_s\right),$$

and so ξ is a supermartingale by Lemma 13.17.

Consider $a > 0$ and write

$$\tau_a = \begin{cases} \inf\left\{t \leq T : \int_0^t f_s \, dB_s - \int_0^t f_s^2 \, ds = -a\right\}, \\ T, \quad \text{if} \inf_{t \in [0,T]} \left\{\int_0^t f_s \, dB_s - \int_0^t f_s^2 \, ds\right\} > -a. \end{cases}$$

We first show that for $\lambda \leq 0$

$$E[\xi_0^{\tau_a}(\lambda f)] = 1.$$

By the differentiation rule,

$$\xi_0^{\tau_a}(\lambda f) = 1 + \lambda \int_0^{\tau_a} \xi_0^s(\lambda f) f_s \, dB_s.$$

From Theorem 11.3, the stochastic integral here will have expectation zero if

$$E\left[\int_0^{\tau_a} \xi_0^s(\lambda f)^2 f_s^2 \, ds\right] < \infty. \tag{13.6}$$

For $x \geqq 0$, $x \leq 2 \exp(x/2)$, so

$$\begin{aligned} E\left[\int_0^{\tau_a} f_s^2 \, ds\right] &\leq 2E\left[\exp\left(\tfrac{1}{2} \int_0^{\tau_a} f_s^2 \, ds\right)\right] \\ &\leq 2E\left[\exp\left(\tfrac{1}{2} \int_0^T f_s^2 \, ds\right)\right] < \infty \end{aligned} \tag{13.7}$$

by hypothesis. Now for $0 \leq s \leq \tau_a$, because $\lambda \leq 0$:

$$\begin{aligned} \xi_0^s(\lambda f) &= \exp\left(\lambda \int_0^s f_u \, dB_u - \frac{\lambda^2}{2} \int_0^s f_u^2 \, du\right) \\ &= \exp\left\{\lambda\left(\int_0^s f_u \, dB_u - \int_0^s f_u^2 \, du\right)\right\} \exp\left\{\left(\lambda - \frac{\lambda^2}{2}\right) \int_0^s f_u^2 \, du\right\} \\ &\leq \exp\left\{\lambda\left(\int_0^s f_u \, dB_u - \int_0^s f_u^2 \, du\right)\right\} \leq \exp(-a\lambda). \end{aligned}$$

Therefore, for $0 \leq s \leq \tau_a$

$$\xi_0^s(\lambda f)^2 \leq \exp(2a|\lambda|)$$

and so (13.6) follows, using (13.7).

We now show that for $\lambda \leq 1$

$$E[\xi_0^{\tau_a}(\lambda f)] = 1.$$

Write

$$\rho_{\tau_a}(\lambda f) = e^{\lambda a} \xi_0^{\tau_a}(\lambda f),$$

So that if $\lambda \leq 0$ we have already proved that

$$E[\rho_{\tau_a}(\lambda f)] = e^{\lambda a}. \tag{13.8}$$

Furthermore, write

$$A(\omega) = \int_0^{\tau_a} f_s^2 \, ds$$

$$B(\omega) = \int_0^{\tau_a} f_s \, dB_s - \int_0^{\tau_a} f_s^2 \, ds + a \geq 0.$$

Let

$$z = 1 - (1 - \lambda)^2 = 2\lambda - \lambda^2$$

so that

$$\lambda = 1 - \sqrt{1 - z}.$$

Note that $0 \leq z \leq 1$ implies $0 \leq \lambda \leq 1$.

Write

$$\begin{aligned} u(z) &= \rho_{\tau_a}(\lambda f) \\ &= \exp\left\{\frac{z}{2} A(\omega) + (1 - \sqrt{1 - z}) B(\omega)\right\}. \end{aligned}$$

For $z < 1$ we see that $u(z)$ is a.s. analytic, and so it can be expressed a.s. as a power series

$$u(z) = \sum_{k=0}^{\infty} p_k(\omega) z^k.$$

Because $A(\omega)$ and $B(\omega)$ are nonnegative, and $1 - \sqrt{1 - z}$ has an expansion with positive coefficients, we see that $p_k(\omega) \geq 0$ a.s., $k = 0, 1, \ldots$. Now

$$\xi_0^t(\lambda f), \lambda \leq 1,$$

is certainly a supermartingale by Lemma 13.17, so for $z \leq 1$

$$E[u(z)] \leq e^{\lambda a} = \exp\{a(1 - \sqrt{1 - z})\}.$$

Certainly for any particular $z' \in [0, 1]$

$$E[u(z')] < \infty.$$

Therefore, if $|z|<z'$

$$E\left[\sum_{k=0}^{\infty} |z|^k p_k(\omega)\right] \le E\left[\sum_{k=0}^{\infty} (z')^k p_k(\omega)\right]$$
$$= E[u(z')] < \infty,$$

and so by Fubini's theorem, if $|z|<1$

$$E[u(z)] = E\left[\sum_{k=0}^{\infty} z^k p_k(\omega)\right]$$
$$= \sum_{k=0}^{\infty} z^k E[p_k(\omega)].$$

Now for $|z|<1$

$$\exp\{a(1-\sqrt{1-z})\} = \sum_{k=0}^{\infty} c_k z^k,$$

with coefficients $c_k \ge 0$, $k=0, 1, \ldots$. However, if $-1<z\le 0$ it follows from (13.8) that

$$E[u(z)] = \exp\{a(1-\sqrt{1-z})\},$$

that is

$$\sum_{k=0}^{\infty} z^k E[p_k(\omega)] = \sum_{k=0}^{\infty} z^k c_k.$$

Consequently, $E[p_k(\omega)] = c_k$, $k=0, 1, \ldots$, and for $0 \le z < 1$

$$E[u(z)] = \sum_{k=0}^{\infty} z^k c_k = \exp\{a(1-\sqrt{1-z})\}.$$

Therefore,

$$E[\rho_{\tau_a}(\lambda f)] = e^{\lambda a}, \quad \text{for all } \lambda < 1.$$

Because $A(\omega) \ge 0$ a.s. and $B(\omega) \ge 0$ a.s. we have that

$$\rho_{\tau_a}(\lambda f) = \exp\left\{\left(\lambda - \frac{\lambda^2}{2}\right) A(\omega) + \lambda B(\omega)\right\}$$

increases to $\rho_{\tau_a}(f)$ as λ approaches 1 from below. By Lebesgue's monotone convergence theorem

$$\lim_{\lambda \uparrow 1} E[\rho_{\tau_a}(\lambda f)] = E[\rho_{\tau_a}(f)]$$
$$= \lim_{\lambda \uparrow 1} e^{\lambda a} = e^a.$$

Therefore,

$$E[\xi_0^{\tau_a}(f)] = 1.$$

Consequently,

$$E[\xi_0^{\tau_a}(f)I_{T>\tau_a}]+E[\xi_0^{\tau_a}(f)I_{T=\tau_a}]=1=E[\xi_0^{\tau_a}(f)I_{T>\tau_a}]+E[\xi_0^T(f)I_{T=\tau_a}]$$

so

$$E[\xi_0^T(f)]=E[\xi_0^T(f)I_{T>\tau_a}]+1-E[\xi_0^{\tau_a}(f)I_{T>\tau_a}]. \tag{13.9}$$

However, $\lim_{a\to\infty} I_{T>\tau_a}=0$ a.s. and

$$E[\xi_0^T(f)]\le 1$$

so

$$\lim_{a\to\infty} E[\xi_0^T(f)I_{T>\tau_a}]=0.$$

On the set $\{T>\tau_a\}$

$$\begin{aligned}\xi_0^{\tau_a}(f)&=\exp\left\{-a+\tfrac{1}{2}\int_0^{\tau_a} f_s^2\,ds\right\}\\ &\le\exp\left\{-a+\tfrac{1}{2}\int_0^{T} f_s^2\,ds\right\},\end{aligned}$$

and therefore

$$E\left[I_{T>\tau_a}\xi_0^{\tau_a}(f)\right]\le e^{-a}E\left[\exp\left(\tfrac{1}{2}\int_0^T f_s^2\,ds\right)\right],$$

which converges to zero as $a\to\infty$. Letting $a\to\infty$ in (13.9) we see that

$$E[\xi_0^T(f)]=1.$$

As in Lemma 13.17, because $\{\xi_0^t(f)\}$ is a supermartingale for $0\le t\le T$ this implies that $\{\xi_0^t(f)\}$ is a martingale. □

Remarks 13.28. The above theorem remains true if T is replaced by any $\mathscr{F}_t$ stopping time. Furthermore, the theorem is true if $T\equiv\infty$.

Corollary 13.29. *Suppose $\{B_t\}$ is a Brownian motion on the filtered probability space $(\Omega,\mathscr{F},P,\mathscr{F}_t)$, and let τ be an $\mathscr{F}_t$ stopping time such that $Ee^{\tau/2}<\infty$. Then $E[e^{B_\tau-(\tau/2)}]=1$.*

Proof. Take $f_s\equiv 1$ in the above theorem. □

Example 13.30. If $|f_s|\le A<\infty$ a.s. for $s\le T$ then $E[\xi_0^T(f)]=1$, because

$$E\left[\exp\left(\tfrac{1}{2}\int_0^T f_s^2\,ds\right)\right]\le\exp\left(\frac{TA^2}{2}\right)<\infty.$$

Example 13.31. Write

$$T_n = \inf\begin{cases} t \le T: \int_0^t f_s^2\,ds = n, \\ T, \quad \text{if} \int_0^t f_s^2\,ds < n. \end{cases}$$

Then

$$E\left[\exp\left(\tfrac{1}{2}\int_0^{T_n} f_s^2\,ds\right)\right] \le e^{n/2} < \infty,$$

so

$$E[\xi_0^{T_n}(f)] = 1.$$

Example 13.32. Suppose

$$\sup_{s \le T} E[\exp \delta f_s^2] < \infty$$

for some $\delta > 0$. Then by Jensen's inequality:

$$\begin{aligned} \exp\left(\tfrac{1}{2}\int_0^T f_s^2\,ds\right) &= \exp\left(\frac{1}{T}\int_0^T \frac{Tf_s^2}{2}\,ds\right) \\ &\le \frac{1}{T}\int_0^T \exp\left(\frac{Tf_s^2}{2}\right) ds. \end{aligned}$$

If $T \le 2\delta$

$$E \exp\left(\tfrac{1}{2}\int_0^T f_s^2\,ds\right) \le \sup_{0 \le s \le T} E[\exp \delta f_s^2] < \infty.$$

Therefore, by Theorem 13.27,

$$E[\xi_0^T(f)] = 1.$$

Suppose $T > 2\delta$. In this case write

$$\xi_0^T(f) = \xi_0^{t_1}(f) \times \xi_{t_1}^{t_2}(f) \times \cdots \times \xi_{t_{n-1}}^{t_n}(f),$$

where $0 < t_1 < t_2 < \cdots < t_n = T$ and $\max_i (t_{i+1} - t_i) \le 2\delta$. Then

$$E[\xi_{t_i}^{t_{i+1}}(f) | \mathscr{F}_{t_i}] = 1 \quad \text{a.s.}$$

and

$$\begin{aligned} E[\xi_0^T(f)] &= E[E[\xi_0^{t_{n-1}}(f)\xi_{t_{n-1}}^{t_n}(f) | \mathscr{F}_{t_{n-1}}]] \\ &= E[\xi_0^{t_{n-1}}(f)] = \cdots = E[\xi_0^{t_1}(f)] = 1, \end{aligned}$$

by the above.

Example 13.33. Suppose $\{B_t\}$, $t \in [0, T]$, is one-dimensional Brownian motion and $f(t, B(\omega))$ is a predictable function such that

$$|f(t, B(\omega))| \le K(1 + B_t^*(\omega)),$$

where

$$B_t^*(\omega) = \sup_{s \le t} |B_s(\omega)|.$$

Then for some constants K_1, K_2

$$\exp(\delta f_s^2) \le K_1 \exp(\delta K_2 B_s^*(\omega)^2).$$

By Doob [27]

$$E[\exp(\delta K_2 B_T^*(\omega)^2)] = K_3 \int_{-\infty}^{\infty} \exp((\delta K_2 - T^{-1})x^2)\, dx < \infty$$

for small enough $\delta > 0$. Therefore,

$$\sup_{s \le T} E[\exp \delta f_s^2] < \infty$$

for this $\delta > 0$. By Example 13.32 $(\xi_0^t(f), \mathscr{F}_t)$ is then a martingale.

Example 13.34. Finally we show that the condition $E[\exp(\frac{1}{2}\int_0^T f_s^2\, ds)] < \infty$ cannot be weakened, by giving an example on $[0, \infty]$, i.e. $T = \infty$, where for an arbitrary $\varepsilon > 0$, $0 < \varepsilon < \frac{1}{2}$,

$$E\left[\exp(\tfrac{1}{2} - \varepsilon) \int_0^\infty f_s^2\, ds\right] < \infty$$

does not imply

$$E[\xi_0^\infty(f)] = 1.$$

Again let $\{B_t\}$ be a one-dimensional Brownian, $0 < \varepsilon < \frac{1}{2}$, and $a > 0$. Write

$$T_\varepsilon = \inf(t\colon B_t - (1 - \varepsilon)t = -a),$$
$$T_\varepsilon^n = \inf(t\colon n \le B_t \le -a + (1 - \varepsilon)t).$$

We first show that

$$E[\exp(\tfrac{1}{2} - \varepsilon)T_\varepsilon^n] = V_n(0), \tag{13.10}$$

where

$$V_n(x) = e^x(e^{-2\varepsilon n} - e^{-(1-2\varepsilon)a - n})(e^{-(a+2\varepsilon n)} - e^{-(1-2\varepsilon)a})^{-1}$$
$$+ e^{(1-2\varepsilon)x}(e^{-(a+n)} - 1)(e^{-(a+2\varepsilon n)} - e^{-(1-2\varepsilon)a})^{-1}$$

is the solution of the equation

$$V_n''(x) - 2(1 - \varepsilon)V_n'(x) + (1 - 2\varepsilon)V_n(x) = 0, \tag{13.11}$$

satisfying

$$V_n(-a) = V_n(n) = 1.$$

To establish (13.10) consider the function

$$W_n(x) = V_n(x)\, e^{(1/2-\varepsilon)t}.$$

By the Ito differentiation rule, if $x_t = B_t - (1-\varepsilon)t$, for any integer $N \geq 1$

$$V_n(x_{T_\varepsilon^n \wedge N}) \exp[(\tfrac{1}{2}-\varepsilon)T_\varepsilon^n \wedge N] = V_n(0) + \int_0^{T_\varepsilon^n \wedge N} V_n'(x_s) \exp[(\tfrac{1}{2}-\varepsilon)s]\, dW_s$$

using (13.11). Now $V_n'(x)$ is bounded if $-a \leq x \leq n$ so

$$E\left[\int_0^{T_\varepsilon^n \wedge N} V_n'(x_s) \exp[(\tfrac{1}{2}-\varepsilon)s]\, dW_s\right] = 0.$$

Therefore,

$$E[V_n(X_{T_\varepsilon^n \wedge N}) \exp[(\tfrac{1}{2}-\varepsilon)T_\varepsilon^n \wedge N]] = V_n(0). \tag{13.12}$$

From the explicit form for $V_n(x)$ we see that

$$0 < \inf_{-a \leq x \leq n} V_n(x) < \sup_{-a \leq x \leq n} V_n(x) < \infty,$$

and so

$$E[\exp(\tfrac{1}{2}-\varepsilon)T_\varepsilon^n \wedge N] \leq \frac{V_n(0)}{\inf V_n(x)} < \infty.$$

Consequently, letting $N \to \infty$ we have

$$E[\exp(\tfrac{1}{2}-\varepsilon)T_\varepsilon^n] \leq \frac{V_n(0)}{\inf_{-a \leq x \leq n} V_n(x)} < \infty.$$

Furthermore, because

$$V_n(X_{T_\varepsilon^n \wedge N}) \exp[(\tfrac{1}{2}-\varepsilon)T_\varepsilon^n \wedge N] \leq \exp[(\tfrac{1}{2}-\varepsilon)T_\varepsilon^n] \sup_{-a \leq x \leq n} V_n(x),$$

we can let $N \to \infty$ in (13.12) to obtain

$$E[V_n(X_{T_\varepsilon^n}) \exp[(\tfrac{1}{2}-\varepsilon)T_\varepsilon^n]] = V_n(0).$$

However,

$$V_n(X_{T_\varepsilon^n}) = 1 \quad \text{a.s.}$$

so we have

$$E[\exp[(\tfrac{1}{2}-\varepsilon)T_\varepsilon^n]] = V_n(0).$$

Letting $n \to \infty$ we see that

$$E[\exp[(\tfrac{1}{2}-\varepsilon)T_\varepsilon]] = \exp((1-2\varepsilon)a) < \infty.$$

However, if

$$f_t^\varepsilon = I_{t \le T_\varepsilon},$$

then

$$\begin{aligned}\xi_0^{T_\varepsilon}(f^\varepsilon) &= \exp(B_T - \tfrac{1}{2}T_\varepsilon)\\ &= \exp(B_{T_\varepsilon} - (1-\varepsilon)T_\varepsilon)\exp((\tfrac{1}{2}-\varepsilon)T_\varepsilon)\\ &= e^{-a}\exp[(\tfrac{1}{2}-\varepsilon)T_\varepsilon].\end{aligned}$$

Therefore

$$E[\xi_0^{T_\varepsilon}(f^\varepsilon)] = e^{-2\varepsilon a} < 1.$$

Remarks 13.35. The vector form of Theorem 13.27 is established by similar methods and will now be stated without proof.

Theorem 13.36. *Suppose* $\{B_t\} = \{B_t^1, \dots, B_t^m\}, t \in [0, T]$ *is an* m-*dimensional Brownian motion on a filtered probability space* $(\Omega, \mathcal{F}, \mathcal{F}_t, P)$. Let $f: \Omega \times [0, T] \to R^m$ *be a predictable process such that*

$$E\left[\exp \tfrac{1}{2}\left(\int_0^T |f_s|^2\, ds\right)\right] < \infty.$$

Then

$$E[\xi_0^T(f)] = 1,$$

where

$$\xi_0^t(f) = \exp\left[\sum_{i=1}^m \int_0^t f_s^i\, dB_s^i - \tfrac{1}{2}\int_0^t |f_s|^2\, ds\right].$$

Because of its importance in applications we now give an alternative proof of the vector form of Example 13.33 based on the original techniques of Girsanov [43] and Beneš [4].

Lemma 13.37. *Suppose* $\{B_t\}$, $t \in [0, T]$, *is an* m-*dimensional Brownian motion on a filtered probability space* $(\Omega, \mathcal{F}, \mathcal{F}_t, P)$ *and* $f(t, B(\omega))$ *is a predictable function such that*

$$|f(t, B(\omega))| \le K(1 + B_t^*(\omega))$$

for some $K < \infty$. (*Here, as in Definition* 6.37, *if* x_t *is a continuous function from* $[0, T]$ *with values in* R^m, $x_t^* = \sup_{0 \le s \le t} |x_s|$.) *Then* $\{\xi_0^t(f)\}$, $t \in [0, T]$, *is an* $\mathcal{F}_t$ *martingale.*

PROOF. Consider the space $\mathscr{C}$ of m-dimensional continuous functions on $[0, T]$. For each positive integer N consider the set

$$C_N(t) = \{x \in \mathscr{C}: 1 + 2e^{2K^2}(K^2T + x_t^{*2}) < N^2K^{-2}\}.$$

Then $C_N(t)$ is a monotonic increasing (in N) family of open sets such that if $s \le t$ then

$$C_N(t) \subset C_N(s).$$

Because $\bigcup_N C_N(1)$ is the whole space of continuous functions, for any $\varepsilon > 0$ there is an $N(\varepsilon)$ such that

$$P\{\omega : B(\cdot, \omega) \in C_N(1)\} > 1 - \varepsilon, \tag{13.13}$$

if $N > N(\varepsilon)$. Also, if $s < t$ and $x(\cdot) \in C_N(s)$ but $x(\cdot) \notin C_N(t)$ then there is a τ, $s < \tau < t$, such that

$$x(\cdot) \in C_N(u) \quad \text{for } u < \tau \quad \text{but } x(\cdot) \notin C_N(\tau). \tag{13.14}$$

Let $\{y_t\}$ be the process defined by

$$y(t, \omega) = B(t, \omega) - \int_0^t f(s, B(\cdot, \omega))\, ds.$$

Write

$$D_N(t) = \{\omega : y(\cdot, \omega) \in C_N(t)\}.$$

Then certainly $D_N(t) \in \mathscr{F}_t$. The growth condition on f implies

$$|B_t|^2 \le 2|y_t|^2 + 2K^2\left(T + \int_0^t B_s^{*2}\, ds\right)$$

so

$$B_t^{*2} \le 2y_t^{*2} + 2K^2\left(T + \int_0^t B_s^{*2}\, ds\right).$$

Therefore, by Gronwall's inequality (Lemma 14.20),

$$B_t^{*2} \le 2e^{2K^2}(K^2T + y_t^{*2}).$$

If $y(\cdot, \omega) \in C_N(t)$ then

$$K^2(1 + B_t^{*2}) \le N^2.$$

That is, if $\omega \in D_N(t)$

$$|f(t, \omega)| \le 2N.$$

Define

$$f_N(t, \omega) = \begin{cases} f(t, \omega)), & \text{if } \omega \in D_N(t), \\ 0, & \text{if } \omega \notin D_N(t). \end{cases}$$

Then $|f_N(t, \omega)| \le 2N$ for all ω. Consider the process $\{y_N(t, \omega)\}$ (under measure P), defined by

$$y_N(t, \omega) = B(t, \omega) - \int_0^t f_N(s, B(\cdot, \omega))\, ds.$$

By property (13.14)

$$D_N(t)=\{\omega: y(\cdot,\omega)\in C_N(t)\}=\{\omega: y_N(\cdot,\omega)\in C_N(t)\}.$$

Indeed, if $y(t,\omega)\in C_N(t)$ then

$$y_N(s,\omega)=y(s,\omega) \quad \text{for all } s\leq t.$$

Now $y(0,\omega)=y_N(0,\omega)=0\in R^m$, so if $y_N(\cdot,\omega)\notin C_N(t)$ then, by property (13.14), there is a time τ, $0<\tau<t$ such that $y_N(\ ,\omega)\in C_N(u)$ for all $u<\tau$ but $y_N(\cdot,\omega)\notin C_N(\tau)$. However, by construction,

$$y_N(u,\omega)=y(u,\omega) \quad \text{for } u<\tau.$$

Therefore, $y(\cdot,\omega)\notin C_N(\tau)$ and so $y(\cdot,\omega)\notin C_N(t)$. By Example 13.30 $E[\xi_0^T(f_N)]=1$ so we can define a new probability measure $\tilde{P}_N$ on $(\Omega,\ )$ by putting

$$\frac{d\tilde{P}_N}{dP}=\xi_0^T(f_N).$$

From Corollary 13.25 $\{y_N(t)\}$ is an m-dimensional Brownian motion under $\tilde{P}_N$. Therefore,

$$\begin{aligned} P\{\omega: B(\cdot,\omega)\in C_N(1)\} &= \tilde{P}_N\{\omega: y_N(\cdot,\omega)\in C_N(1)\} \\ &= \int_\Omega I_{D_N(1)}\, d\tilde{P}_N = \int_{D_N(1)} \xi_0^T(f_N)\, dP \\ &= \int_{D_N(1)} \xi_0^T(f)\, dP. \end{aligned}$$

However, from (13.13) this tends to 1 as $N\to\infty$. That is

$$E[\xi_0^T(f)]=1$$

and by Lemma 13.17 $\{\xi_0^t(f)\}$ is then a martingale. □

Theorem 13.38. *Suppose* $\{B_t\}=\{B_t^1,\ldots,B_t^m\}$, $t\in[0,T]$, *is an m-dimensional Brownian motion on a filtered probability space* $(\Omega,\mathcal{F},\mathcal{F}_t,P)$. *Let*

$$f(t,B(\omega))=(f^1(t,B(\omega)),\ldots,f^m(t,B(\omega)))$$

be a predictable function such that

$$|f(t,B(\omega))|\leq K(1+B_t^*(\omega)), \tag{13.15}$$

where

$$B_t^*(\omega)=\sup_{s\leq t}|B_s(\omega)|.$$

Then P^* *is a probability measure on* $(\Omega,\mathcal{F})$ *if* $dP^*/dP=\xi_0^T(f)$. *Furthermore,* $\{W_t\}=\{W_t^i,\ldots,W_t^m\}$ *is an m-dimensional Brownian motion on* $(\Omega,\mathcal{F},\mathcal{F}_t,P^*)$ *if* $W_t^i=B_t^i-\int_0^t f_s^i\,ds$, *that is* $dW_t=dB_t-f_t\,dt$.

PROOF. From Lemma 13.37 we deduce that the linear growth condition implies

$$E[\xi_0^t(f)] = 1.$$

Therefore, P^* is a probability measure, and the final statement follows from Girsanov's result, Corollary 13.25. □

Remark 13.39. Again, under the above conditions, this result states that the process $\{B_t\}$ is a weak solution of the stochastic differential equation $dx_t = f_t\,dt + dW_t$, $x_0 = 0 \in R^m$, under the measure P^*. That is, to solve (weakly) such an equation, where the drift coefficient satisfies a linear growth condition (13.15), one considers a Brownian motion $\{B_t\}$ and constructs the new probability measure P^*. Under P^* the process $\{B_t\}$ is a weak solution, in the sense that

$$B_t - \int_0^t f(s, B(\omega))\,ds$$

is a P^*-Brownian motion.

CHAPTER 14

Strong Solutions of Stochastic Differential Equations

Suppose $(\Omega, \mathscr{F}, P, \mathscr{F}_t)$, $t \geq 0$, is a filtered probability space satisfying the usual conditions of Convention 5.1, and let Z be an R^n-valued semimartingale, with $Z_0 = 0$.

Definition 14.1. Write $\mathscr{D}$ for the space of corlol functions defined on R^+ with values in R^d. The *coefficient* of the stochastic differential equation is a map

$$f: R^+ \times \Omega \times \mathscr{D} \to L(R^n, R^d)$$

(the set of $n \times d$ matrices), such that, if X is any corlol adapted process, then the map

$$f \circ X: R^+ \times \Omega \to L(R^n, R^d)$$

defined by

$$(f \circ X)(t, \omega) = f(t, \omega, X.(\omega))$$

is locally bounded and predictable.

Definition 14.2. Suppose X is a corlol adapted process such that

$$X_t = H_t + \int_0^t (f \circ X)_s \, dZ_s \quad \text{a.s.} \tag{14.1}$$

for $t \geq 0$, where H is a given corlol adapted process. Then X is said to be a *strong solution* of (14.1).

Example 14.3. If $B_t = (B_t^1, \ldots, B_t^n)$ is an n-dimensional Brownian motion,

$$Z_t = (B_t, t), \qquad f(t, \omega, X.(\omega)) = f(t, X_{t-}(\omega)) \quad \text{and} \quad H_t = x_0 \in R^d$$

then the stochastic differential equation (14.1) has the form

$$X_t^i = x_0 + \sum_{j=1}^{n} \int_0^t \sigma_{ij}(s, X_{s-})\, dB_s^j + \int_0^t g_i(s, X_{s-})\, ds, \qquad i = 1, \ldots, d.$$

This is usually abbreviated as

$$X_t = x_0 + \int_0^t \sigma(s, X_{s-})\, dB_s + \int_0^t g(s, X_{s-})\, ds.$$

Example 14.4. If $f: R \to R$ is continuous and bounded and we take $X_t = x_0$ for $t \leq 0$ then, for any $\varepsilon > 0$, the "delayed" stochastic differential equation

$$X_t = x_0 + \int_0^t f(X_{s-\varepsilon})\, dZ_s$$

always has a unique solution, obtained by integrating over successive intervals $[0, \varepsilon]$, $]\varepsilon, 2\varepsilon[$, ..., etc.

Remark 14.5. For simplicity of exposition we shall establish the existence of solutions of (14.1), under suitable hypotheses, in the scalar case when $n = d = 1$. The proof in the general case is similar. The method is a small generalization of that given by Doléans-Dade and Meyer [26], in that our coefficient functions depend on the past of the trajectory.

Theorem 14.6. *Suppose Z is a (scalar) semimartingale such that $Z_0 = 0$ a.s. and H is an adapted corlol process. Then the stochastic differential equation*

$$X_t(\omega) = H_t(\omega) + \int_0^t f(s, \omega, X.(\omega))\, dZ_s(\omega) \tag{14.2}$$

has a unique strong solution X if f satisfies the following conditions:

(a) *f satisfies the conditions of Definition* 14.1 (*with $n = d = 1$*),
(b) *if X and Y are two corlol adapted processes then*

$$|f(s, \omega, X.(\omega)) - f(s, \omega, Y.(\omega))| \leq K(|X(\omega) - Y(\omega)|_s^*),$$

where, as usual,

$$|X(\omega)|_s^* = \sup_{0 \leq u \leq s} |X_u(\omega)|.$$

Remarks 14.7. "Unique" here means that any two solutions are indistinguishable processes.

The theorem will be proved in several steps, the first showing that we can suppose all the jumps of Z are bounded.

Lemma 14.8. *Suppose that for each Lipschitz constant K there is a number $\alpha > 0$, which may depend on K, such that the existence and uniqueness of* *a solution of* (14.2) *is known under the following hypothesis*: (J^1) *All jumps of the semimartingale* Z *are of modulus* $\leq \alpha$.

Then the existence and uniqueness of a solution of (14.2) *is known without this restriction.*

PROOF. Suppose $T_1, \ldots, T_n$ are the times at which Z has jumps of modulus greater than α. Write

$$Z_t^1 = Z_t I_{\{t<T_1\}} + Z_{T_1-} I_{\{t\geq T_1\}},$$

so the jumps of Z^1 are of modulus $\leq \alpha$. Write

$$H_t^1 = H_t I_{\{t<T_1\}} + H_{T_1-} I_{\{t\geq T_1\}}.$$

Then the process X is a solution of (14.2) on $[\![0, T_1[\![$ if and only if the process

$$X_t^1 = X_t I_{\{t<T_1\}} + X_{T_1-} I_{\{t\geq T_1\}}$$

is a solution of

$$X_t^1 = H_t^1 + \int_0^t f(s, \cdot, X^1.(\cdot))\, dZ_s^1 \tag{14.3}$$

on $[\![0, \infty[\![$. However, (14.3) has a unique solution by hypothesis. Therefore, (14.2) has a unique solution on $[\![0, T_1[\![$.

If we put

$$\Delta X_{T_1} = \Delta H_{T_1} + f(T_1, \cdot, X.(\cdot))\Delta Z_{T_1},$$

we obtain a (unique) solution on $[\![0, T_1]\!]$. By transferring the origin to T_1 we can establish the existence and uniqueness of a solution on $[\![T_1, T_2]\!]$ in the same way. Thus the existence and uniqueness without hypothesis (J^1) is established. □

Lemma 14.9. *Suppose the existence and uniqueness of a solution of* (14.2) *is known under the following hypotheses*:

Z is a semimartingale of the form $Z = N + A$, where: (J^2) *N is a square integrable martingale, zero at* 0, *and* $[N, N]_\infty \leq b$, *where* $b \leq 1$ *is a constant which depends on the Lipschitz constant K of f*; *and* (J^3) *A is a predictable process of finite variation, zero at* 0, *and*

$$\int_0^\infty |dA_s| \leq b.$$

Then the existence and uniqueness of a solution is known without these hypotheses.

PROOF. The existence and uniqueness will be established for a semimartingale Z, which is zero at 0, whose jumps are smaller in modulus than $a = b/4$. Then, by Lemma 14.8, the existence and uniqueness will hold in general.

If Z is a semimartingale, zero at 0, whose jumps are smaller in modulus than $b/4$, then by Corollary 12.40(c), Z is a special semimartingale. Consequently Z has a canonical decomposition

$$Z = N + A,$$

where $N \in \mathscr{L}$ and $A \in (\mathscr{A}_0)_{\text{loc}}$ is predictable.

By stopping if necessary, we can suppose Z is bounded and N is uniformly integrable. Then, if T is a totally inaccessible stopping time

$$\Delta A_T = 0$$

and

$$|\Delta Z_T| = |\Delta N_T| \leq b/4.$$

If T is a predictable stopping time

$$|\Delta N_T| = |\Delta Z_T - E[\Delta Z_T|\mathscr{F}_{T-}]| \leq b/4$$

and

$$|\Delta A_T| = |E[\Delta Z_T|\mathscr{F}_{T-}]| \leq b/4.$$

Define a sequence of stopping times by:

$$T_0 = 0,$$

$$T_n = \inf\left\{t > T_{n-1}: [N, N]_t - [N, N]_{T_{n-1}} \geq b/2 \text{ or } \int_{T_{n-1}}^t |dA_s| \geq b/2\right\}.$$

The jumps of $[N, N]$ are always less than $(b/2)^2 \leq b/2$. The jumps of A are always less that $b/4$.

Therefore,

$$[N, N]_{T_n} - [N, N]_{T_{n-1}} \leq b, \qquad \int_{T_{n-1}}^{T_n} |dA_s| \leq b.$$

Now $\{X_t\}$ is a solution of (14.2) on $[\![0, T_1]\!]$ if and only if on $[\![0, \infty[\![$

$$X_t^1 = H_t^1 + \int_0^t f(s, \cdot, X^1.(\cdot))\, dZ_s^1,$$

where $X_t^1 = X_t^{T_1}$, $H_t^1 = H_t^{T_1}$, $Z_t^1 = Z_t^{T_1}$. However, Z^1 is a semimartingale satisfying hypotheses (J^2) and (J^3) so we know there exists a unique solution on $[\![0, T_1]\!]$. Transferring the origin to T_1, the existence and uniqueness of a solution is established on $[\![T_1, T_2]\!]$, and so on. □

Remarks 14.10. The number b will be chosen below, so that it is less than $1 \wedge (3K)^{-2}$. We now show we can suppose $H = 0$.

Lemma 14.11. *Suppose the existence and uniqueness of a solution of* (14.2) *is known when $H = 0$. Then the existence and uniqueness is known in the general case.*

PROOF. The process $\{X_t\}$ is a solution of

$$X_t(\omega)=H_t(\omega)+\int_0^t f(s,\omega,X.(\omega))\,dZ_s(\omega),$$

if and only if the process $\{\bar{X}_t\}$, where $\bar{X}_t=X_t-H_t$, is a solution of

$$\bar{X}_t(\omega)=\int_0^t \bar{f}(s,\omega,\bar{X}.(\omega))\,dZ_s(\omega)). \tag{14.4}$$

Here, for any corlol adapted process $\{Y_t\}$,

$$\bar{f}(s,\omega,Y.(\omega))=f(s,\omega,(Y+H).(\omega)).$$

The semimartingale Z is the same in both equations and f and $\bar{f}$ have the same Lipschitz constant. Therefore, a solution of (14.2) exists under the same conditions that ensure a solution of (14.4). □

Lemma 14.12. *Suppose the existence and uniqueness of a solution of* (14.4) *is known under the following hypothesis:*

(J^4) $$|f(s,\omega,0)|\le C,$$

in addition to hypotheses (J^1), (J^2) *and* (J^3). *Then the existence and uniqueness is known without these hypotheses. Here* $\{0\}$ *denotes the identically zero process.*

PROOF. Consider the following sequence of predictable stopping times:

$$T_0=0$$
$$T_n=\inf\{t\colon |f(s,\omega,0)|\geqq n\}.$$

Write, for any corlol adapted process $\{X_t\}$,

$$f^n(s,\omega,X.(\omega))=f(s,\omega,X.(\omega))I_{\{0<s\le T_n(\omega)\}},$$
$$X_t^n(\omega)=X_t^{T_n}(\omega),\qquad Z_t^n(\omega)=Z_t^{T_n}(\omega).$$

Then $\{X_t\}$ is a solution of (14.4) is and only if for every n,

$$X_t^n=\int_0^t f^n(s,\cdot,X^n.)\,dZ_s^n. \tag{14.5}$$

Because $|f^n(s,\cdot,0)|\le n$ each of the equations (14.5) has a unique solution. Therefore (14.4) has a unique solution. □

Lemma 14.13. *Suppose f satisfies hypotheses* (a) *and* (b) *of Theorem* 14.6, *and that*

(J^4) $$|f(s,\omega,0)|\le C$$

for some constant $C<\infty$.

Suppose further that Z is a special semimartingale with a canonical decomposition $Z = N + A$ such that

$$(\mathrm{J}^2) \qquad N \in \mathscr{H}_0^2 \quad \text{and} \quad [N, N]_\infty \le b,$$

$$(\mathrm{J}^3) \qquad A \in \mathscr{A}_0, A \text{ is predictable} \quad \text{and} \quad \int_0^\infty |dA_s| \le b, \quad \text{where } b < 1 \wedge (3K)^{-2}.$$

Then there is a unique strong solution of the equation

$$X_t(\omega) = \int_0^t f(s, \omega, X.(\omega))\, dZ_s(\omega), \qquad t \ge 0. \tag{14.6}$$

PROOF. Write B for the space of corlol adapted processes $\{Y_t\}$ such that $Y_0 = 0$ a.s. and $Y^* \in L^2(\Omega, P)$, where as usual $Y^*(\omega) = \sup_t |Y_t(\omega)|$.

Define a norm on B by

$$\|Y\| = \|Y^*\|_2 = \int Y^*(\omega)^2\, dP.$$

We prove (14.6) has a solution in B by successive approximation.

For a process $Y \in B$ define the process SY by

$$(SY)_t = \int_0^t f(s, \cdot, Y.(\cdot))\, dZ_s.$$

If $X, Y \in B$

$$(SX)_t - (SY)_t = \int_0^t (f(s, \cdot, X.(\cdot)) - f(s, \cdot, Y.(\cdot)))\, dN_s$$

$$+ \int_0^t (f(s, \cdot, X.(\cdot)) - f(s, \cdot, Y.(\cdot)))\, dA_s.$$

Writing $\Delta = X - Y$ and

$$L_t = \int_0^t (f(s, \cdot, X.(\cdot)) - f(s, \cdot, Y.(\cdot)))\, dN_s,$$

we have

$$[L, L]_\infty = \int_0^\infty (f(s, \cdot, X.(\cdot)) - f(s, \cdot, Y.(\cdot)))^2\, d[N, N]_s$$

$$\le \int_0^\infty K^2 \Delta_s^2\, d[N, N]_s \le K^2 b (\Delta^*)^2.$$

Therefore, by Doob's inequality, Theorem 4.2(5):

$$\|L^*\|_2 \le 2(E[L, L]_\infty)^{1/2} \le 2K\sqrt{b}\|\Delta\|.$$

Writing

$$B_t = \int_0^t (f(s, \cdot, X.(\cdot)) - f(s, \cdot, Y.(\cdot)))\, dA_s,$$

we see that

$$B^* \leq \int_0^\infty K|\Delta_s||dA_s| \leq Kb\Delta^*,$$

so

$$\|B^*\|_2 \leq Kb\|\Delta\|.$$

Consequently

$$\|SX - SY\| \leq K(2\sqrt{b} + b)\|\Delta\|.$$

However, because $b < 1 \wedge (3K)^{-2}$

$$h = K(2\sqrt{b} + b) < 1.$$

Therefore,

$$\|SX - SY\| \leq h\|X - Y\|. \tag{14.7}$$

Again writing 0 for the process which is identically zero we have

$$(SO)_t = \int_0^t f(s, \cdot, 0)\, dN_s + \int_0^t f(s, \cdot, 0)\, dA_s.$$

The first term on the right is a local martingale M_t and

$$[M, M]_\infty = \int_0^\infty f^2(s, \cdot, 0)\, d[N, N]_s \leq C^2 b,$$

because of conditions (J^2) and (J^4). Therefore, by Doob's inequality, Theorem 4.2(5)),

$$\|M\|^2 = \|M^*\|_2^2 \leq 4C^2 b.$$

The second term on the right of (14.7)

$$D_t = \int_0^t f(s, \cdot, 0)\, dA_s$$

is such that

$$D^* \leq \int_0^t |f(s, \cdot, 0)||dA_s| \leq Cb$$

so

$$\|D\|^2 = \|D^*\|_2^2 \leq C^2 b^2.$$

Consequently, because

$$(SO)^* \leq M^* + D^*,$$

we have $\|(SO)\| < \infty$. Substituting $Y = 0$ in (14.7) we can deduce $\|SX\| < \infty$ for $X \in B$.

Consequently, the map S of B into B satisfies (14.7), where $h<1$. Therefore, by the fixed point theorem, the equation $SX=X$ has one, and only one, solution in B. That is, there is a process $X\in B$ such that

$$X_t(\omega)=\int_0^t f(s,\omega,X.(\omega))\,dZ_s(\omega)\quad \text{a.s.}$$

for all $t\geq 0$.

Suppose there is a second solution Y, which does not belong to the space B. Then $Y_0=0$ a.s. and

$$Y_t(\omega)=\int_0^t f(s,\omega,Y.(\omega))\,dZ_s(\omega)\quad \text{a.s.}$$

for all $t\geq 0$. Write $T=\inf\{t:|Y_t|\geq m\}$. Then $|Y_s|\leq m$ on $[\![0,T[\![$ and

$$|Y_T|\leq|Y_{T-}|+|f(T,\cdot,Y.(\cdot))||\Delta Z_T|.$$

Now

$$|Y_{T-}|\leq m$$

and

$$\begin{aligned}|f(T,\cdot,Y.(\cdot))|&\leq|f(T,\cdot,0)|+K|Y_{T-}|\\&\leq C+mK\end{aligned}$$

and

$$|\Delta Z_T|\leq|\Delta A_T|+|\Delta N_T|\leq 2b.$$

Therefore, the stopped process Y^T is bounded, and so belongs to B. However, Y^T is the solution of the equation

$$Y_t^T=\int_0^t f(s,\cdot,Y^T(\cdot))\,dZ_s^T. \tag{14.8}$$

Here Z^T satisfies the same hypotheses as Z, and X^T is the unique solution of (14.8) which belongs to B. Consequently

$$Y_t=X_t\quad \text{for } t\in[\![0,T]\!].$$

Letting m and T tend to infinity the uniqueness follows. The proof of Theorem 14.6 is, therefore, complete.

Lemma 14.14. *Suppose* $f:R^+\times\Omega\times R\to R$ *is such that*

(a) *for each* s,ω

$$|f(s,\omega,x)-f(s,\omega,y)|\leq K|x-y|,$$

(b) *for each* s,x: $f(s,\cdot,x)$ *is* $\mathcal{F}_s$ *measurable,*

(c) *for each* ω,x: $f(\cdot,\omega,x)$ *is collor (i.e. continuous on the left with limits on the right).*

Then for each corlol adapted process X the process $(s, \omega) \to f(s, \omega, X_{s-}(\omega))$ is collor and adapted (and, therefore, predictable and locally bounded.)

PROOF. By hypotheses (a) and (b), for each fixed t the map

$$(\omega, x) \to f(t, \omega, x)$$

is $\mathcal{F}_t \times \mathcal{B}$ measurable. (Here $\mathcal{B}$ is the Borel field in R.) Therefore, the map $\omega \to f(t, \omega, X_{t-}(\omega))$ is $\mathcal{F}_t$ measurable.

By hypotheses (c), the limit

$$f(t+, \omega, x) = \lim_{s \searrow t} f(s, \omega, x)$$

exists. For $s > t$ consider

$$|f(s, \omega, X_{s-}(\omega)) - f(t+, \omega, X_t(\omega))| \leq |f(s, \omega, X_{s-}(\omega)) - f(s, \omega, X_t(\omega))| + |f(s, \omega, X_t(\omega)) - f(t+, \omega, X_t(\omega))|.$$

The first term is less than $K|X_{s-}(\omega) - X_t(\omega)|$ and so tends to 0 because X is right continuous. The second term tends to zero by the definition of $f(t+, \omega, X_t(\omega))$. Therefore, the process $f(s, \omega, X_{s-}(\omega))$ has limits on the right. The left continuity is proved similarly. □

Remark 14.15. The vector form of Theorem 14.6 and Lemma 14.14 combine to show that, if the coefficients σ and g satisfy hypotheses (a), (b) and (c) of Lemma 14.14, then the classical stochastic differential equation

$$X_t = x_0 + \int_0^t \sigma(s, X_{s-})\, dB_s + \int_0^t g(s, X_{s-})\, ds$$

of Example 14.3 has a strong solution. Indeed, the integrals give rise to continuous processes so that X itself is a continuous process, and $X_{s-} = X_s$.

Definition 14.16. Suppose $\{X_t\}$ is an adapted process with values in R^d defined on the filtered probability space $(\Omega, \mathcal{F}, P, \mathcal{F}_t)$, $t \geq 0$. $\{X_t\}$ is said to be a *Markov process* with respect to $\{\mathcal{F}_t\}$ if for all s, and all $t \geq s$,

$$E[f(X_t)|\mathcal{F}_s] = E[f(X_t)|X_s] \quad \text{a.s.}$$

for every bounded real valued Borel function f defined on R^d.

Remarks 14.17. We now discuss the Markov nature of the strong solution of the stochastic differential equation

$$X_t = x_0 + \int_0^t \sigma(u, X_u)\, dB_u + \int_0^t g(u, X_u)\, du,$$

with respect to the filtration generated by the Brownian motion and the initial condition. We shall also consider the above equation starting at an intermediate time s, $0 \le s \le t$,

$$X_t = x_s + \int_s^t \sigma(u, X_u)\, dB_u + \int_s^t g(u, X_u)\, du.$$

Suppose here that the initial condition x_s is an $\mathscr{F}_s$ measurable random variable which is independent of the random variables $\{B_u - B_v, u \ge v \ge s\}$, and denote by $\mathscr{F}_{s,t}$ the σ-field $\mathscr{F}^{x_s} \vee \mathscr{F}^B_{s,t}$, where $\mathscr{F}^{x_s}$ is the σ-field generated by x_s, and $\mathscr{F}^B_{s,t}$ is the completion of the σ-field generated by all the random variables $\{B_u - B_v, s \le v \le u \le t\}$. Finally, write $\mathscr{F}_t$ for $\mathscr{F}_{0,t}$. Our discussion follows that of Kallianpur [55] and Meyer [63].

Lemma 14.18. *Suppose $f: R^d \times \Omega \to R$ is a bounded $\mathscr{B} \times \mathscr{F}^B_{s,t}$ measurable function and Y is a d-dimensional $\mathscr{F}_s$ measurable random variable.*

If $g(x) = E[f(x, \omega)]$ for $x \in R^d$ then

$$E[f(Y, \omega)|\mathscr{F}_s] = g(Y). \tag{14.9}$$

PROOF. Because the random variables $\{B_u - B_v, s \le v \le u \le t\}$ are independent of $\mathscr{F}_s$, for each $x \in R^d$ the random variable $f(x, \cdot)$ is independent of $\mathscr{F}_s$. Write $\mathscr{A}$ for the collection of all $\mathscr{B} \times \mathscr{F}^B_{s,t}$ measurable sets $\{A\}$ whose indicator functions $f = I_A$ satisfy the conclusion of Lemma 14.9. If $A = C \times D$, where $C \in \mathscr{B}$ and $D \in \mathscr{F}^B_{s,t}$ then

$$\begin{aligned} E[I_A(Y, \omega)|\mathscr{F}_s] &= I_C(Y) E[I_D(\omega)|\mathscr{F}_s] \\ &= I_C(Y) E[I_D(\omega)] \\ &= g(Y), \end{aligned}$$

where

$$g(x) = E[I_A(x, \omega)].$$

Therefore, (14.9) holds for $f = I_A$ where A is a finite union of disjoint sets of the form $C_i \times D_i$, $C_i \in \mathscr{B}$, $D_i \in \mathscr{F}^B_{s,t}$, so $\mathscr{A}$ contains all of a family of sets which generates $\mathscr{B} \times \mathscr{F}^B_{s,t}$. Suppose A_n is a monotone sequence with $A_n \in \mathscr{A}$ and with limit A. Then

$$\begin{aligned} E[I_A(Y, \omega)|\mathscr{F}_s] &= \lim_{n\to\infty} E[I_{A_n}(Y, \omega)|\mathscr{F}_s] \\ &= \lim_{n\to\infty} g_n(Y), \end{aligned}$$

where

$$g_n(Y) = \int_\Omega I_{A_n}(x, \omega) P(d\omega) \to \int_\Omega I_A(x, \omega) P(d\omega) = g(x).$$

That is, $\lim g_n(Y) = g(Y)$ so $\mathscr{A}$ is a monotone class, and by Theorem 1.22 $\mathscr{A} = \mathscr{B} \times \mathscr{F}^B_{s,t}$. Consequently, (14.9) holds for any bounded $\mathscr{B} \times \mathscr{F}^B_{s,t}$ measurable function f. $\square$

Corollary 14.19. *Suppose* $f: R^d \times \Omega \to R$ *is a nonnegative* $\mathscr{B} \times \mathscr{F}^B_{s,t}$ *measurable function and* Y *is a* d*-dimensional* $\mathscr{F}_s$ *measurable random variable. Denote by* (PY^{-1}) *the probability measure induced by* Y *on* R^d*. Then*

$$E[f(Y(\omega), \omega)] = \int_{R^d} E[f(x, \omega)](PY^{-1})(dx).$$

PROOF. For $m \in N$ and $x \in R^d$ write

$$f_m = fI_{\{f \le m\}},$$

$$g_m = E[f_m(x, \omega)].$$

Then from Lemma 14.18

$$\begin{aligned} E[f_m(Y(\omega), \omega)] &= E[E[f_m(Y(\omega), \omega) | \mathscr{F}_s]] \\ &= E[g_m(Y(\omega))] = \int_{R^d} g_m(\omega)(PY^{-1})(dx) \\ &= \int_{R^d} E[f_m(x, \omega)](PY^{-1})(dx). \end{aligned}$$

The result follows by applying the monotone convergence theorem to each side. □

We shall now prove a result known as Gronwall's inequality.

Lemma 14.20. *Suppose* $\alpha(t)$ *is a Lebesgue integrable function on* $[a, b]$*, and that* B *and* H *are constants such that*

$$\alpha(t) \le B + H \int_a^t \alpha(s)\, ds \quad \textit{for all } t \in [a, b].$$

Then

$$\alpha(t) \le Be^{H(t-a)}.$$

PROOF. By hypothesis:

$$\begin{aligned} \alpha(t) &\le B + H \int_a^t \left\{ B + \int_a^s \alpha(s_1)\, ds_1 \right\} ds \\ &\le B + H \int_a^t \left\{ B + H \int_a^s \left\{ B + H \int_a^{s_1} \alpha(s_2)\, ds_2 \right\} ds_1 \right\} ds \\ &\le B + HB \int_a^t ds + H^2 B \int_a^t \int_a^s ds_1\, ds + \cdots \\ &\quad + H^n B \int_a^t \int_a^s \cdots \int_a^{s_{n-2}} ds_{n-1} \cdots ds_1\, ds \\ &\qquad + H^{n+1} \int_a^t \int_a^s \cdots \int_a^{s_{n-1}} \alpha(s_n)\, ds_n \cdots ds. \end{aligned}$$

However,

$$\int_a^t \int_a^s \cdots \int_a^{s_{n-1}} \alpha(s_n)\, ds_n \cdots ds = \int_a^t \frac{(t-s_n)^n}{n!} \alpha(s_n)\, ds_n.$$

Therefore,

$$\lim_{n\to\infty} H^{n+1} \int_a^t \int_a^s \cdots \int_a^{s_{n-1}} \alpha(s_n)\, ds_n \cdots ds = 0$$

and

$$\alpha(t) \leq B\left\{1 + H(t-a) + \frac{H^2(t-a)^2}{2!} \cdots\right\}.$$

That is

$$\alpha(t) \leq B\, e^{H(t-a)}. \qquad \square$$

Remarks 14.21. Suppose $g: R^+ \times R^d \to R^d$ is (Borel) measurable, and σ is a $d \times m$ matrix with (Borel) measurable entries $\sigma_{ij}: R^+ \times R^d \to R$ such that, for each $u \geq 0$, and $x, y \in R^d$:

$$\begin{aligned} |g(u,x) - g(u,y)| &\leq K|x-y|, \\ |\sigma(u,x) - \sigma(u,y)| &\leq K|x-y|. \end{aligned} \tag{14.10}$$

By a stopping time argument if necessary we could also assume that

$$|g(u,x)|^2 + |\sigma(u,x)|^2 \leq K(1+|x|^2). \tag{14.11}$$

The coefficient functions g and σ satisfy the conditions of Theorem 14.6 (see Remark 14.15), and for $s \leq t \leq T \leq \infty$ the equation

$$\begin{aligned} dX_t &= g(t, X_t)\, dt + \sigma(t, X_t)\, dB_t, \\ X_s &= x \in R^d \end{aligned} \tag{14.12}$$

has a unique, continuous, strong solution. The following lemmas show that (14.12) has a unique solution $X_t = X_{s,x}(t, \omega)$, which for each $t \geq s$, is $\mathscr{B}(R^d) \times \mathscr{F}^B_{s,t}$ measurable in (x, ω).

Lemma 14.22. *There is a constant C, which depends only on C and T, such that for $x, y \in R^d$ and $s \in [0, T]$:*

$$E[\sup_{s\leq t\leq T} |X_{s,x}(t) - X_{s,y}(t)|^2] \leq C|x-y|^2.$$

Here, $X_{s,x}$ and $X_{s,y}$ are the unique continuous solutions of (14.12) *which have initial values x and y, respectively, at $t = s$.*

Proof. Consider a fixed x, y and s, and for $t \in [s, T]$

$$\phi(t) = E[\sup_{s\leq u\leq T} |X_{s,x}(u) - X_{s,y}(u)|^2].$$

Then

$$\phi(t) \leq 3|x-y|^2 + 3E\left[\sup_{s\leq u\leq t}\left|\int_s^u (g(v, X_{s,x}(v)) - g(v, X_{s,y}(v)))\,dv\right|^2\right]$$

$$+3E\left[\sup_{s\leq u\leq t}\left|\int_s^u (\sigma(v, X_{s,x}(v)) - \sigma(v, X_{s,y}(v)))\,dB_v\right|^2\right]$$

$$\leq 3|x-y|^2 + 3(t-s)E\left[\int_s^t \left|g(v, X_{s,x}(v)) - g(v, X_{s,y}(v))\right|^2 dv\right]$$

$$+12E\left[\int_s^t |\sigma(v, X_{s,x}(v)) - \sigma(v, X_{s,y}(v)|^2\,dv\right],$$

by Doob's inequality, and this is

$$\leq 3|x-y|^2 + (3T+12)K\int_s^t \phi(v)\,dv.$$

Writing, $H = (3T+12)K$ and $C = 3e^{HT}$ we have by Gronwall's inequality, Lemma 14.20, that

$$\phi(T) \leq C|x-y|^2. \qquad \square$$

Lemma 14.23. *There is an R^d valued process*

$$X_s(x, t, \omega): R^d \times [s, T] \times \Omega \to R^d,$$

such that:

(a) *for each $x \in R^d$, $X_s(x, t, \omega)$ is a solution of the stochastic differential equation* (14.12),
(b) *for each $t \in [s, T]$ the restriction of $X_s(x, u, \omega)$ to $R^d \times [s, t] \times \Omega$ is $\mathscr{B}(R^d) \times \mathscr{B}([s, t]) \times \mathscr{F}^B_{s,t}$ measurable.*

PROOF. Consider a point

$$\alpha^k_m = (k_1 2^{-m}, \ldots, k_d 2^{-m}) \in R^d,$$

with dyadic rational coordinates and consider a process $X_{s,\alpha^k_m}(t, \omega)$ which is a solution of (14.12), with initial condition α^k_m at $t = s$, and which is continuous for all ω. Then this solution is progressively measurable, and $X_{s,\alpha^k_m}(u, \omega)$ restricted to $[s, t] \times \Omega$ is $\mathscr{B}([s, t]) \times \mathscr{F}^B_{s,t}$ measurable. For $x = (x_1, \ldots, x_d) \in R^d$ let $\alpha^k_m(x)$ be such that $k_j 2^{-m} \leq x_j \leq (k_j+1)2^{-m}$ $(1 \leq j \leq d)$, and write

$$X^m_s(x, t, \omega) = X_{s,\alpha^k_m(x)}(t, \omega).$$

For each x, ω and m, $X^m_s(x, t, \omega)$ is continuous in t and the function

$$(x, u, \omega) \mapsto X^m_s(x, u, \omega) \quad \text{is } \mathscr{B}(R^d) \times \mathscr{B}([s, t]) \times \mathscr{F}^B_{s,t}$$

measurable. Define

$$X_s(x, t, \omega) = \lim_{m\to\infty} \sup X^m_s(x, t, \omega) \tag{14.13}$$

and let $X_{s.x}(t, \omega)$ be the unique continuous solution of (14.12) with initial condition $x \in R^d$ at $t = s$. Then from Lemma 14.22

$$E[\sup_{s \le t \le T} |X_{s,x}(t) - X_{s,\alpha_m^k(x)}(t)|^2] \le C|x - \alpha_m^k(x)|^2 \le C\,dz^{-m}.$$

By the Borel–Cantelli lemma

$$P[\omega: \sup_{s \le t \le T} |X_{s,x}(t, \omega) - X_{s,\alpha_m^k(x)}(t, \omega)| > 1/m \text{ for infinitely many } m] = 0.$$

Therefore, for each $x \in R^d$,

$$P[\omega: X_s^m(x, t, \omega) \to X_{s,x}(t, \omega) \text{ uniformly on } [s, T]] = 1.$$

Consequently, for almost all $\omega \in \Omega$, and all $(x, t) \in R^d \times [s, T]$

$$X_s(x, t, \omega) = X_{s,x}(t, \omega),$$

and so, for every $x \in R^d$, $X_s(x, t, \omega)$ is a solution of (14.12) with initial condition x at $t = s$, which is continuous for almost all ω. From (14.13) we see the process $X_s(x, t, \omega)$ has the measurability property of statement (b). □

Lemma 14.24. *The stochastic differential equation* (14.12) *has a unique solution* $X_{s,x}(t)$, $t \ge s$, *which is* $\mathscr{B}(R^d) \times \mathscr{F}_{s,t}^B$ *measurable as a function of* (x, ω).

PROOF. This is an immediate consequence of the above lemma. □

Lemma 14.25. *Suppose* $\phi: R^d \to R$ *is a bounded Borel measurable function. Then* $f(x, \omega) = \phi(X_{s,x}(t, \omega))$ *is* $\mathscr{B}(R^d) \times \mathscr{F}_{s,t}^B$ *measurable.*

PROOF. Write $\mathscr{A}$ for the collection of all sets $A \in \mathscr{B}(R^d)$ for which the lemma is true with $\phi = I_A$.

If

$$f(x, \omega) = I_A(X_{s,x}(t, \omega)),$$

then

$$\{(x, \omega): I_A(X) = 1\} = \{(x, \omega): X_{s,x}(t, \omega) \in A\} \in \mathscr{B}(R^d) \times \mathscr{F}_{s,t}^B,$$

so the lemma is true for all $A \in \mathscr{B}(R^d)$. The result follows for general ϕ by approximation with simple functions. □

Definition 14.26. For fixed x, s, t, $s \le t$, and the above notation write

$$P(s, x, t, A) = P\{\omega: X_{s,x}(t, \omega) \in A\}, \text{ for } A \in \mathscr{B}(R^d).$$

Because

$$P(s, x, t, A) = \int_\Omega I_A(X_{s,x}(t, \omega)) P(d\omega),$$

it follows from Lemma 14.25 that for fixed s, t and A, $P(s, x, t, A)$ is $\mathscr{B}(R^d)$ measurable in x. Clearly for fixed x, s, t $P(s, x, t, \cdot)$ is a probability measure on $\mathscr{B}(R^d)$.

$P(s, x, t, A)$ is called the *transition probability function* of the process $X_{s,x}(t, \omega)$.

Theorem 14.27. *The stochastic differential equation*

$$dX_t = g(t, X_t)\,dt + \sigma(t, X_t)\,dB_t, \quad t \in [0, T] \quad \text{and} \quad X_0 = x_0 \in R^d, \tag{14.14}$$

where x_0 is independent of $\mathscr{F}^B_{0,T}$ and the coefficients satisfy the conditions of Remarks 14.21, has a unique solution $\{X_t\}$ which is a continuous Markov process relative to $\{\mathscr{F}_t\}$.

If $P(s, x, t, A)$ is the transition probability function of the process, then for $A \in \mathscr{B}(R^d)$ and $s \le t$:

$$\begin{aligned} P\{X_t \in \mathscr{A} | \mathscr{F}_s\} &= P\{X_t \in A | X_s\} \\ &= P(s, x, t, A). \end{aligned}$$

PROOF. Let $X_{s.x}(t, \omega)$ be the unique solution of (14.14) and write

$$f(x, \omega) = \phi(X_{s,x}(t, \omega))$$

where, as in Lemma 14.25, $\phi: R^d \to R$ is any bounded Borel measurable function. Then for each $x \in R^d$, $f(x, \cdot)$ is $\mathscr{F}^B_{s,t}$ measurable, and so is independent of $\mathscr{F}_s = \mathscr{F}^{x_0} \vee \mathscr{F}^B_s$. Consequently by Lemma 14.18: if Z is an $\mathscr{F}_s$ measurable random variable

$$E[\phi(X_{s,Z}(t, \omega)) | \mathscr{F}_s] = g(Z),$$

where

$$g(x) = \int_\Omega \phi(X_{s,x}(t, \omega)) P(d\omega).$$

Write

$$X_s = X_{0,x_0}(s, \omega),$$

so X_s is $\mathscr{F}_s$ measurable, and take $Z = X_s$ in the above. By the uniqueness of the solution, we have that

$$X_{s,X_s}(t, \omega) = X_{0,x_0}(t, \omega) = X_t$$

so

$$E[\phi(X_t) | \mathscr{F}_s] = g(X_s) \quad \text{a.s.}$$

Similarly, if $s \le u \le t$ we have

$$E[\phi(X_{s,x}(t, \omega)) | \mathscr{F}_u] = g^*(X_{s,x}(u)), \tag{14.15}$$

where now

$$g^*(y) = \int_\Omega \phi(X_{u,y}(t, \omega)) P(d\omega).$$

Now $g(x)$ is $\mathscr{B}(R^d)$ measurable, so $g(X_s)$ is $\mathscr{F}^{X_s} \subset \mathscr{F}_s$ measurable. Therefore, if $t \geq s$:

$$E[\phi(X_t)|\mathscr{F}_s] = g(X_s)$$

$$E[E[\phi(X_s)|\mathscr{F}_s]X_s] = E[\phi(X_s)|X_s] \quad \text{a.s.} \tag{14.16}$$

Consequently, $\{X_t\}$ is a Markov process with respect to $\{\mathscr{F}_t\}$.
From the definition of the transition probability

$$E[\phi(X_{s,x}(t))] = \int_{R^d} \phi(y) P(s, x, t, dy). \tag{14.17}$$

Taking

$$\phi = I_A, \qquad A \in \mathscr{B}(R^d),$$

we have

$$P[X_t \in A|\mathscr{F}_s] = P[X_t \in A|X_s] = P(s, X_s, t, A) \quad \text{a.s.}$$

Now take $\phi(y) = P(t, y, u, A)$ in (14.17), where $u > t$. Then

$$\phi(X_{s,x}(t)) = P(t, X_{s,x}(t), u, A)$$

and so from (14.16)

$$\begin{aligned}
\int_{R^d} P(t, y, u, A) P(s, x, t, dy) &= E[P(t, X_{s,x}(t), u, A)] \\
&= E[P\{X_{s,x}(u) \in A|X_{s,x}(t)\}] \\
&= E[E[I_A(X_{x,s}(u))|X_{s,x}(t)]] \\
&= E[I_A(X_{s,x}(u))] \\
&= P\{X_{s,x}(u) \in A\} = P(s, x, u, A).
\end{aligned}$$

Therefore, for $x \in R^d$, $A \in \mathscr{B}(R^d)$ and $s < t < u$ we have established, as in the case for Markov processes, that the transition probability function $P(s, x, t, A)$ satisfies the Chapman–Kolmogorov equation

$$\int_{R^d} P(t, y, u, A) P(s, x, t, dy) = P(s, x, u, A). \qquad \square$$

Definition 14.28. Suppose $0 \leq s < t \leq T$, $x \in R^d$, and that $X_{s,x}(t, \omega)$ is the unique solution of (14.12). If $B_0(R^d) = B_0$ is space of all real valued, bounded Borel measurable functions on R^d then we can define a family of operators $\{T_s^t\}$ on B_0 by

$$T_s^s = I, \quad \text{the identity operator,}$$

$$T_s^t f(x) = \int_{R^d} f(y) P(s, x, t, dy),$$

where $P(s, x, t, dy)$ is the transition probability of the process $\{X_{s,x}(t, \omega)\}$.

Note that

$$T_s^t(f(X_s)) = E[f(X_t)|\mathcal{F}_s]$$
$$= E[f(X_t)|X_s].$$

Remark 14.29. If $0 \leq s \leq t \leq u \leq T$, it is a consequence of the Chapman–Kolmogorov equation that, for $f \in B_0$,

$$T_s^u f = T_s^t T_t^u f.$$

That is, the family of operators is a semigroup.

Notation 14.30. Write $\mathscr{C}_b$ for the space of bounded real continuous functions on R^d.

Write $\mathscr{C}_b^2 = \mathscr{C}_b^2(R^d)$ for the subspace of functions in $\mathscr{C}_b$ which also have first and second derivatives in $\mathscr{C}_b$.

Suppose $g = (g^1, \ldots, g^d)$ and $\sigma = (\sigma^{ij})$ are as in Remarks 14.21, and write $a = (a^{ij})$ for the $d \times d$ matrix $\sigma\sigma'$, where σ' denotes the transpose of σ.

Write $X_t = X_{0,x_0}$ for the solution of (14.12) and let L_s be the second-order differential operator defined on $\mathscr{C}_b^2$ by:

$$L_s f(x) = \tfrac{1}{2} \sum_{i,j} a^{ij}(s, x) \frac{\partial^2 f}{\partial x^i \, \partial x^j}(x) + \sum_i g^i(s, x) \frac{\partial f(x)}{\partial x^i}.$$

Theorem 14.31. (a) *For each* $f \in \mathscr{C}_b^2$

$L_u f(y)$ is jointly measurable in u and y.

(b) *For every* $x \in R^d$, $0 \leq s < u \leq T$,

$$\int_{R^d} |(L_u f)(y)| P(s, x, u, dy) < \infty.$$

(c) *For every* $x \in R^d$, $0 \leq s < t \leq T$

$$T_s^t f(x) = f(x) + \int_s^t T_s^u L_u f(x) \, du.$$

(d) *If g and σ are continuous then, for every $x \in R^d$ and $0 \leq s < T$:*

$$\lim_{h \searrow 0} h^{-1}(T_s^{s+h} f(x) - f(x)) = L_s f(x).$$

PROOF. The process $X_{s,x}(t)$ is the solution of

$$X_{s,x}(t) = x + \int_s^t g(u, X_{s,x}(u)) \, du + \int_0^t \sigma(u, X_{s,x}(u)) \, dB_u.$$

Applying the differentiation rule, Theorem 12.21, to $f \in \mathscr{C}_b^2$ and taking expectations:

$$E[f(X_{s,x}(t))] = f(x) + \int_0^t E[L_u f(X_{s,x}(u))]\, du.$$

The final integral exists if $f \in \mathscr{C}_b^2$ and we assume the growth conditions (14.11) on g and σ, because $E[\sup_{s\le t\le T}|X_{s,x}(t)|^2] < \infty$. This proves part (c) of the theorem, and the remarks also establish part (b). The measurability properties of part (a) are immediate from the definitions.

Finally, when g and σ are continuous

$$T_s^u L_u f(x) = E[L_u f(X_{s,x}(u))]$$

is a continuous function of $u \in [s, T]$, and this implies part (d). □

Theorem 14.32. *Suppose $f \in \mathscr{C}_b^2$. Then*

$$M_t^f = f(X_t) - f(X_0) - \int_0^t L_u f(X_u)\, du$$

is an $\mathscr{F}_t$ martingale.

PROOF. From the remarks in the proof of Theorem 14.31 above

$$\int_0^T E|L_u f(X_u)|\, du < \infty,$$

so

$$E[|M_t^f|] < \infty, \qquad t \in [0, T].$$

Because $\{X_t\}$ is Markov, M_t^f is certainly $\mathscr{F}_t$ measurable. Now for $0 \le x \le t \le T$

$$M_t^f - M_s^f = f(X_t) - f(X_s) - \int_s^t L_u f(X_u)\, du.$$

Therefore,

$$E[M_t^f - M_s^f | \mathscr{F}_s] = E[f(X_t)|\mathscr{F}_s] - f(X_s) - E\left[\int_s^t L_u f(X_u)\, du | \mathscr{F}_s\right]$$

$$= T_s^t f(X_s) - f(X_s) - \int_s^t (T_s^u L_u f(X_s)\, du = 0 \quad \text{a.s.},$$

by Theorem 14.31(c). □

Definition 14.33. Because the operator $\{L_u\}$ defined on $\mathscr{C}_b^2$ satisfies the identity of part (c) of Theorem 14.31, $\{L_u\}$ is called the *extended generator* of the Markov process $\{X_t\}$. The right-hand side of the identity of part (d) of Theorem 14.31 defines the *weak infinitesimal generator* of the Markov process $\{X_t\}$.

We have seen that, if b and σ are continuous, these operators coincide on $\mathscr{C}_b^2$.

Definition 14.34. The Markov process $\{X_t\}$ has *stationary transition probabilities* if $P(s, x, t, A) = P(0, x, t-s, A)$. $\{X_t\}$ is then said to be a *homogeneous Markov process.*

It then follows that,

$$T_s^t = T_0^{t-s},$$

so writing

$$P(0, x, t-s, A) = Q(t-s, x, A),$$

$$T_0^{t-s} = T_{t-s},$$

the semigroup $\{T_t\}$, $t \geq 0$, is now defined by

$$T_t f(x) = \int_{R^d} f(y) Q(t, x, dy), \qquad f \in B_0.$$

The extended generator is the linear operator A on $\mathscr{C}_b^2$ such that

$$T_t f(x) = f(x) + \int_0^t T_u A f(x)\, du.$$

Remark 14.35. Note that if the coefficients g and σ of (14.14) are functions only of X_t, and not of t, then the unique solution process $\{X_t\}$ is a homogeneous Markov process.

CHAPTER 15

Random Measures

This chapter is concerned with random measures. However, to illustrate and motivate some of the ideas of the general situation (and, indeed some concepts presented in earlier chapters), the first half of the chapter discusses, in some detail, a very basic stochastic process which has just one random jump. Random measures are associated with such a process in an elementary way, and related martingales can be considered. This section is based on the work of Chou and Meyer [10], Davis [12], Elliott [30], [32] and Jacod [49]. The second part of the chapter will then discuss general random measures following Jacod [50].

The Single Jump Process 15.1

In this section we shall consider a process $\{X_t\}$, $t \geq 0$, which takes its values in a Lusin space $(E, \mathscr{E})$ and which remains at its initial point $z_0 \in E$ until a random time T, when it jumps to a new random position z. The underlying probability space can be taken to be

$$\Omega = [0, \infty] \times E,$$

with the σ-field $\mathscr{B} \times \mathscr{E}$. (As usual, $\mathscr{B}$ denotes the Borel field on $[0, \infty]$.) A sample path of the process is

$$X_t(\omega) = \begin{cases} z_0, & \text{if } t < T(\omega), \\ z(\omega), & \text{if } t \geq T(\omega). \end{cases}$$

Suppose a probability measure μ is given on Ω, and that

$$\mu([0, \infty] \times \{z_0\}) = 0 = \mu(\{0\} \times E),$$

so that the probabilities of a zero jump and a jump at time zero are zero.

Write $\mathcal{F}_t$ for the completed σ-field generated by $\{X_s\}$, $s \leq t$, so that $\mathcal{F}_t$ is generated by $\mathcal{B}([0, t]) \times \mathcal{E}$. Note that $]t, \infty] \times E$ is an atom in $\mathcal{F}_t$. For $A \in \mathcal{E}$ write

$$F_t^A = \mu([t, \infty] \times A),$$

so that F_t^A is the probability that $T > t$ and $z \in A$. Furthermore, write

$$F_t = F_t^E$$

and $c = \inf\{t: F_t = 0\}$.

F_t is right continuous and monotonic decreasing, so there are only countably many points of discontinuity $\{u\} = D$, where $\Delta F_u = F_u - F_{u-} \neq 0$. Any constant time is a predictable stopping time, so each time u, where $\Delta F_u \neq 0$, is predictable. The set of ω where $T(\omega) = u$ and $\Delta F_u \neq 0$, therefore, corresponds to the accessible part of T. Clearly F_t is continuous if and only if T is totally inaccessible.

Lemma 15.2. *Suppose τ is an $(\mathcal{F}_t)$ stopping time. Then there is a $t_0 \in [0, \infty[$ such that $\tau \wedge T = t_0 \wedge T$ a.s.*

PROOF. As observed above, modulo any null sets $]t, \infty[\times E$ is an atom in $\mathcal{F}_t$. Suppose τ takes two values t_1, t_2 on $\{\tau \leq T\}$ with positive probability. Then for $t \in]t_1, t_2[$

$$\{\tau \leq t\} \cap]t, \infty] \times E \neq]t, \infty] \times E,$$

so $\{\tau \leq t\} \notin \mathcal{F}_t$. Therefore, for some $t_0 \in [0, \infty[$, $\{\tau \leq T\} \subset \{t_0 \leq T\}$. A similar argument gives the reverse inclusion and the result follows. □

The Stieltjes measure on $([0, \infty], \mathcal{B})$ given by F_t^A is absolutely continuous with respect to that given by F_t, so there is a Radon–Nikodym derivative $\lambda(A, s)$ such that

$$F_t^A - F_0^A = \int_{]0,t[} \lambda(A, s)\, dF_s.$$

Because $(E, \mathcal{E})$ is a Lusin space $\lambda(\cdot, s)$ is a regular version of a conditional probability measure on $(E, \mathcal{E})$; for $A \in \mathcal{E}$, $\lambda(A, s)$ is the probability that $z \in A$ given that the jump occurs at time s.

Definition 15.3. The pair (λ, Λ) is the *Lévy system* for the jump process where

$$\Lambda(t) = -\int_{]0,t]} \frac{dF_s}{F_{s-}}.$$

If $\tilde{\Lambda}(t) = \Delta(t \wedge T)$ then, roughly, $d\tilde{\Lambda}(t)$ is the probability that the jump occurs at time t, given that it has not happened so far.

For $A \in \mathscr{E}$ write

$$p(t, A) = I_{t \geq T} I_{z \in A},$$

$$\tilde{p}(t, A) = -\int_{]0, t \wedge T]} \lambda(A, s) \frac{dF_s}{F_{s-}}$$

$$= \int_{]0,t]} \lambda(A, s)\, d\tilde{\Lambda}(s).$$

Clearly $t \wedge T$ is left continuous and $-\int_{]0,t]} \lambda(A, s)\, d\tilde{\Lambda}(s)$ is a Borel function, so $\tilde{p}(t, A)$ is predictable. In fact $\tilde{p}(t, A)$ is the dual predictable projection of $p(t, A)$, as the following result shows.

Theorem 15.4.

$$q(t, A) = p(t, A) - \tilde{p}(t, A)$$

is an $\mathscr{F}_t$ martingale.

Proof. For $t > s$

$$E[p(t, A) - p(s, A) | \mathscr{F}_s] = I_{\{T > s\}} (F_s^A - F_t^A) F_s^{-1},$$

$$E\left[\int_{]s,t]} \lambda(A, u)\, d\tilde{\Lambda}(u) \middle| \mathscr{F}_s\right]$$

$$= -I_{\{T>s\}} \left[\frac{F_t}{F_s} \int_{]s,t]} \lambda(A, u) F_{u-}^{-1}\, dF_u - \frac{1}{F_s} \int_{]s,t]} \int_{]s,r]} \lambda(A, u) F_{u-}^{-1}\, dF_u\, dF_r\right].$$

However,

$$\int_{]s,t]} \int_{]s,r]} \lambda(A, u) F_{u-}^{-1}\, dF_u\, dF_r = \int_{]s,t]} \lambda(A, u) F_{u-}^{-1} \int_{[u,t]} dF_r\, dF_u$$

$$= F_t \int_{]s,t]} \lambda(A, u) F_{u-}^{-1}\, dF_u + F_t^A - F_s^A.$$

Therefore,

$$E[q(t, A) - q(s, A) | \mathscr{F}_s] = 0 \quad \text{a.s.} \qquad \square$$

Remarks 15.5. The jump of $q(t, A)$ at a discontinuity u of F_t is

$$\Delta q(u, A) = I_{T=u} I_{z \in A} + \lambda(A, u) \frac{\Delta F_u}{F_{u-}} I_{T \geq u}.$$

However,

$$E[I_{T=u} I_{z \in A} | \mathscr{F}_{u-}] = E[I_{z \in A} | T = u] . P(T = u | \mathscr{F}_{u-})$$

$$= -\lambda(A, u) F_u^{-1}\, \Delta F_u I_{T \geq u},$$

so

$$E[\Delta q(u, A)|\mathscr{F}_{u-}]=0.$$

Therefore, from Theorem 9.29,

$$q^{\Delta u}=\Delta q(u, A)I_{t\geq u}$$

is a square integrable martingale orthogonal to every square integrable martingale which is continuous at u. Furthermore, from Theorem 9.29, the predictable quadratic variation of $q^{\Delta u}$ is:

$$\begin{aligned}\langle q^{\Delta u}, q^{\Delta u}\rangle_t &= E[\Delta q(u, A)^2|\mathscr{F}_{u-}]I_{t\geq u}\\ &= -\lambda(A, u)F_{u-}^{-1}\,\Delta F_u(1+\lambda(A, u)F_{u-}^{-1}\,\Delta F_u)I_{T\geq u}I_{t\geq u}.\end{aligned}$$

Theorem 15.6. *The predictable quadratic variation* $\langle q, q\rangle(t, A)$ *of* $q(t, A)$ *is*

$$\tilde{p}(t, A)-r(t, A),$$

where

$$r(t, A)=\sum_{0<u\leq t\wedge T}\lambda(A, u)^2\frac{\Delta F_u^2}{F_{u-}^2}.$$

PROOF. Decompose F_t into the sum of its continuous part F_t^c and its sum of jumps $F_t^d=\sum_{0<u\leq t}\Delta F_u$. Then $p(t, A)$ can be similarly decomposed as a sum of

$$\tilde{p}^c(t, A)=-\int_{]0,t\wedge T]}\lambda(A, s)F_{s-}^{-1}\,dF_s^c,$$

and

$$\tilde{p}^d(t, A)=-\sum_{0<u\leq t\wedge T}\lambda(A, u)F_{u-}^{-1}\,\Delta F_u.$$

If, as above $D=\{u\colon \Delta F_u\neq 0\}$, write

$$p^d(t, A)=\sum_{0<u\leq t\wedge T}I_{T=u}I_{z\in A}I_{u\in D},$$

and

$$p^c(t, A)=p(t, A)-p^d(t, A).$$

Then

$$q(t, A)=q^c(t, A)+q^d(t, A),$$

where

$$q^c(t, A)=p^c(t, A)-\tilde{p}^c(t, A)$$

and

$$\begin{aligned}q^d(t, A)&=p^d(t, A)-\tilde{p}^d(t, A)\\ &=\sum_{u\in D}q^{\Delta u}.\end{aligned}$$

Now if $u \neq u'$ the martingale $q^{\Delta u}$ is orthogonal to $q^{\Delta u'}$, and $q^d(t, A)$ is orthogonal to $q^c(t, A)$. The product of two orthogonal martingales is, by definition, a martingale, so the predictable quadratic variation of a sum of orthogonal martingales is the sum of their quadratic variations. Therefore,

$$\langle q, q\rangle(t, A) = \langle q^c, q^c\rangle(t, A) + \langle q^d, q^d\rangle(t, A)$$

and

$$\langle q^d, q^d\rangle(t, A) = \sum_{q \in D} \langle q^{\Delta u}, q^{\Delta u}\rangle(t, A) = \tilde{p}^d(t, A) - r(t, A).$$

Because F_t^c is continuous direct integration, as in Theorem 15.4, (or see Lemma 3.3 of [30]), and application of the differentiation rule shows that

$$\langle q^c, q^c\rangle(t, A) = \tilde{p}^c(t, A).$$

Therefore,

$$\begin{aligned}\langle q, q\rangle(t, A) &= \tilde{p}^c(t, A) + \tilde{p}^d(t, A) - r(t, A) \\ &= \tilde{p}(t, A) - r(t, A).\end{aligned}$$

□

Remark 15.7. The above form of predictable quadratic variation, with second-order terms arising from the accessible part $D = \{u : \Delta F_u \neq 0\}$, is very typical of what happens in more complicated situations.

Notation 15.8. Consider the set $L^1(\Omega, \mathscr{B} \times \mathscr{E}, \mu)$. Because $p(t, A)$ and $\tilde{p}(t, A)$ are countably additive in A we can define for $g \in L^1(\mu)$ the Stieltjes integrals:

$$\int_\Omega g(s, x) p(ds, dx) = g(T, z),$$

$$\int_\Omega g(s, x) \tilde{p}(ds, dx) = \int_{]0,T]} \int_E g(s, x) \lambda(s, dx)\, d\Lambda(s).$$

Put

$$\int_\Omega g\, dq = \int_\Omega g\, dp - \int_\Omega g\, d\tilde{p}.$$

Lemma 15.9. *Writing $\|g\|$ for the L^1 norm $\int |g|\, d\mu$ of $g \in L^1(\mu)$ we have*

$$\|g\| = E\left[\int_\Omega |g|\, dp\right] = E\left[\int_\Omega |g|\, d\tilde{p}\right].$$

Proof. We have

$$\int_\Omega |g|\, dp = |g(T, z)|$$

so the first identity is immediate. Now

$$\int_{\Omega} |g|\, d\tilde{p} = \int_{]0,T]\times E} F_{s-}^{-1} |g(s,x)|\, d\mu(s,x),$$

so

$$E\int_{\Omega} |g|\, d\tilde{p} = -\int_{]0,\infty]}\int_{]0,t]\times E} F_{s-}^{-1} |g(s,x)|\, d\mu(s,x)\, dF_t$$

$$= \int_{]0,\infty]\times E}\left(-\int_{[s,\infty]} dF_t\right) F_{s-}^{-1} |g(s,x)|\, d\mu$$

$$= \int_{\Omega} |g|\, d\mu = \|g\|. \qquad \square$$

Notation 15.10. Write $L^1(p)$ (resp. $L^1(\tilde{p})$) for the set of measurable functions $g:\Omega\to R$ such that $E[\int_\Omega |g|\, dp]<\infty$ (resp. $E[\int_\Omega |g|\, d\tilde{p}]<\infty$).

Write $L^1_{\text{loc}}(\mu)$ for the set of measurable functions $g:\Omega\to R$ such that $I_{s\le t}\, g(s,x)\in L^1(\mu)$ for all $t<c$.

Lemma 15.11.

$$L^1_{\text{loc}}(p) = L^1_{\text{loc}}(\tilde{p}) = L^1_{\text{loc}}(\mu).$$

PROOF. The identity of $L^1_{\text{loc}}(p)$ and $L^1_{\text{loc}}(\tilde{p})$ follows from Lemma 15.9 above. Suppose $g\in L^1_{\text{loc}}(p)$. Then there is an increasing sequence of stopping times $\{T_k\}$ such that $\lim T_k=\infty$ a.s. and $I_{s<T_k} g\in L^1(p)$ for each k. Let $\{t_k\}$ be the corresponding sequence, given by Lemma 15.2 such that

$$T_k \wedge T = t_k \wedge T.$$

Because $\lim T_k=\infty$ a.s. we see that $\lim t_k = c$ a.s. Now

$$\int_{\Omega} I_{s<T_k} |g|\, dp = g(T,z) I_{T_k>T}.$$

However,

$$\{T_k > T\} = [0, t_k[\times E$$

so

$$E\left[\int_{\Omega} I_{s<T_k} |g|\, dp\right] = \int_{[0,t_k[\times E} |g|\, d\mu.$$

Because $\lim t_k = c$ a.s. we see that $g\in L^1_{\text{loc}}(\mu)$.

Conversely, suppose $g\in L^1_{\text{loc}}(\mu)$. Construct the following sequence of stopping times $\{T_k\}$:

(a) if $c=\infty$ put $T_k=k$,
(b) if $c<\infty$ and $F_{c-}>0$ put $T_k=\infty$ for all k,
(c) if $c<\infty$ and $F_{c-}=0$ consider a sequence t_k such that $\lim t_k = c$ and put

$$T_k = k I_{\{T\le t_k\}} + t_k I_{\{T>t_k\}}.$$

Clearly $\lim T_k = \infty$ a.s. and $I_{s<T_k} g \in L^1(p)$, so

$$L^1_{\text{loc}}(p) = L^1_{\text{loc}}(\mu). \qquad \square$$

Lemma 15.12. *Suppose $\{M_t\}$ is a uniformly integrable $\{\mathscr{F}_t\}$ martingale such that $M_0 = 0$ a.s. Then there is an $\mathscr{F} = \bigvee_t \mathscr{F}_t$ measurable function $h: \Omega \to R$ such that $h \in L^1(\mu)$ and*

$$M_t = h(T, z) I_{\{t \geq T\}} - I_{\{t<T\}} F_t^{-1} \int_{]0,t] \times E} h(s, x)\, d\mu(s, x) \quad a.s.$$

PROOF. If M is a uniformly integrable martingale then from Corollary 4.10

$$M_t = E[h | \mathscr{F}_t]$$

for some $\mathscr{F} = \bigvee_t \mathscr{F}_t$ measurable random variable h. From the definition of $\mathscr{F}$, h is of the form $h(T, z)$. However,

$$E[h(T, z) | \mathscr{F}_t] = h(T, z) I_{\{t \geq T\}} + I_{\{t<T\}} F_t^{-1} \int_{]t,\infty] \times E} h(s, x)\, d\mu(s, x).$$

Because $M_0 = E[h] = \int_\Omega h\, d\mu = 0$ the result follows. $\square$

Lemma 15.13. *Suppose $\{M_t\}$, $t \geq 0$, is a local martingale of $\{\mathscr{F}_t\}$.*

(a) *If $c = \infty$, or $c < \infty$ and $F_{c-} = 0$, then $\{M_t\}$ is a martingale on $[0, c[$.*
(b) If *$c < \infty$ and $F_{c-} > 0$ then $\{M_t\}$ is a uniformly integrable martingale.*

PROOF. (a) Let $\{T_k\}$ be an an increasing sequence of stopping times such that $\lim T_k = \infty$ a.s. and $\{M_{t \wedge T_k}\}$ is a uniformly integrable martingale. If there is a k such that $T_k \geq T$ a.s. then $M_t = M_{t \wedge T_k}$ a.s. so $\{M_t\}$ is a uniformly integrable martingale. Otherwise, suppose for each k that $P\{T_k < T\} > 0$. Then there is a sequence $\{t_k\}$ such that $T_k \wedge T = t_k \wedge T$ for each k, and because $P\{T > t_k\} > 0$ we have $t_k \leq c$. In fact $t_k < c$, otherwise we should have $T_k \geq T$. Because $\lim T_k = \infty$ we see that $\lim_k P\{T > t_k\} = 0$, so $\lim t_k = c$. Now $\{M_t\}$ is stopped at time T so $M_{t \wedge T_k} = M_{t \wedge t_k}$. Consequently $\{M_t\}$, $t \leq t_k$, is a uniformly integrable martingale, and $\{M_t\}$ is certainly a martingale on $[0, c[$.

(b) Suppose now that $c < \infty$ and $F_{c-} > 0$. Then $P\{T = c\} > 0$. Because $\lim T_k = \infty$ a.s. there is a k such that $P\{\omega: T(\omega) = c,\ T_k(\omega) > c\} > 0$. Consequently, for such a k: $T_k \geq T$ a.s. and the process $\{M_{t \wedge T_k}\} = \{M_t\}$ is a uniformly integrable martingale. $\square$

Theorem 15.14. *$\{M_t\}$ is a local $\mathscr{F}_t$ martingale with $M_0 = 0$ a.s. if and only if $M_t = M_t^g$ for some $g \in L^1_{\text{loc}}(\mu)$, where $M_t^g = \int_\Omega I_{s \leq t} g(s, x) q(ds, dx)$.*

PROOF. Suppose $g \in L^1_{\text{loc}}(\mu)$. Then if $\{T_k\}$ is the sequence of stopping times introduced in Lemma 15.11 calculations similar to Theorem 15.4 show that $M^g_{t \wedge T_k}$ is a uniformly integrable $\{\mathscr{F}_t\}$ martingale. Conversely, suppose

$\{M_t\}$ is a local $\{\mathcal{F}_t\}$ martingale with $M_0 = 0$ a.s. We consider two situations

(i) $c < \infty$, $F_{c-} > 0$. From Lemma 15.13 $\{M_t\}$ is then a uniformly integrable martingale, and so is of the form

$$M_t = h(T, z)I_{\{t \geq T\}} - I_{\{t < T\}}F_t^{-1} \int_{]0,t] \times E} h(s, x)\, d\mu(s, x), \qquad (15.1)$$

where $h(T, z) = M_\infty$. Define

$$g(t, z) = h(t, z) + I_{\{t < c\}}F_t^{-1} \int_{]0,t] \times E} h(s, x)\, d\mu(s, x) \quad \text{if } t < \infty \qquad (15.2)$$

and $g(\infty, z) = 0$. Then

$$M_t^g = \int_\Omega I_{s \leq t} g\, dq$$

$$= I_{\{t \geq T\}}\left[g(T, z) - \int_{]0,t] \times E} g(s, x)F_{s-}^{-1}\, d\mu\right] - I_{\{t < T\}} \int_{[0,t] \times E} F_{s-}^{-1} g\, d\mu. \qquad (15.3)$$

From (15.1) and (15.3) we see that $M_t = M_t^g$ if

$$h(t, x) = g(t, x) - \int_{]0,t] \times E} F_{s-}^{-1} g(s, x)\, d\mu. \qquad (15.4)$$

However, if g is given by (15.2) and $t < c$

$$\begin{aligned}
\int_{]0,t] \times E} F_{s-}^{-1} g(s, x)\, d\mu &= \int_{]0,t] \times E} F_{s-}^{-1} h(s, x)\, d\mu(s, x) \\
&\quad - \int_{]0,t]} F_s^{-1} F_{s-}^{-1} \int_{]0,s] \times E} h(u, x)\, d\mu(u, x)\, dF_s \\
&= \int_{]0,t] \times E} F_{s-}^{-1} h\, d\mu \\
&\quad + \int_{]0,t] \times E} \left(-\int_{[u,t]} F_s^{-1} F_{s-}^{-1}\, dF_s\right) h(u, x)\, d\mu \\
&= \int_{]0,t] \times E} F_{s-}^{-1} h\, d\mu + \int_{]0,t] \times E} \left(\frac{1}{F_t} - \frac{1}{F_{u-}}\right) h(u, x)\, d\mu \\
&= \frac{1}{F_t} \int_{]0,t] \times E} h\, d\mu.
\end{aligned}$$

Therefore, (15.4) is satisfied if $t < c$. A similar calculation shows that the coefficients of $I_{\{t < T\}}$ agree in (15.1) and (15.3), so $M_t = M_t^g$ if $t < c$. However, both M_t and M_t^g are stopped at T and $P\{T > c\} = 0$ so it remains only to show that $M_c = M_c^g$ when $T(\omega) = c$. This is verified by a similar calculation to that above.

We now check that $g \in L^1(\mu)$. Because $\{M_t\}$ is uniformly integrable

$$\int_\Omega |h|\, d\mu < \infty.$$

Therefore,

$$\begin{aligned}\int_\Omega |g|\, d\mu &\leq \int |h|\, d\mu - \int_{]0,c[} F_t^{-1} \int_{]0,t]\times E} |h|\, d\mu\, dF_t \\ &\leq \int |h|\, d\mu - F_{c-}^{-1} \int_{]0,c[} \int_{]0,t]\times E} |h|\, d\mu\, dF_t \\ &= \int |h|\, d\mu + F_{c-}^{-1} \int_{]0,c[\times E} (F_t - F_{c-})|h|\, d\mu \\ &\leq (1+F_{c-}^{-1}) \int |h|\, d\mu < \infty.\end{aligned}$$

Consequently, $g \in L^1(\mu)$.

(ii) Now suppose $c = \infty$, or $c < \infty$ and $F_{c-} = 0$. Then from Lemma 15.13 $\{M_t\}$ is a martingale on $[0, c[$, and so uniformly integrable on $[0, t]$ for any $t < c$. Therefore M_t is of the form (15.1) for some h satisfying

$$\int_{]0,t]\times E} |h(s,x)|\, d\mu(s,x) < \infty$$

for all $t < c$. Calculations as in (i) above show that, for g given by (15.2) and $t < c$, $M_t = M_t^g$. Also,

$$\begin{aligned}\int_{]0,t]\times E} |g|\, d\mu &\leq \int_{]0,t]\times E} |h|\, d\mu - \int_{]0,t]} F_s^{-1} \int_{]0,s]\times E} |h|\, d\mu\, dF_s \\ &\leq \int_{]0,t]\times E} |h|\, d\mu \,.\left(1 - \int_{]0,t]} F_s^{-1}\, dF_s\right) \\ &< \infty \quad \text{if } t < c.\end{aligned}$$

Therefore $g \in L^1_{\text{loc}}(\mu)$ and the proof is complete. □

Remarks 15.15. With the notation of Theorem 15.6 $q(t, A) = q^c(t, A) + \sum_{u\in D} q^{\Delta u}$ so for $g \in L^1(\mu)$

$$M_t^g = \int_\Omega I_{s\leq t} g\, dq^c + \sum_{u\in D} \int_\Omega I_{s\leq t} g\, dq^{\Delta u}.$$

For example, if $u \in D$:

$$\begin{aligned}\int_\Omega I_{s\leq t} g\, dq^{\Delta u} &= I_{t\geq u} I_{T=u}\, g(u, z) \\ &\quad + I_{\{T\wedge t \geq u\}} \int_E g(u,x)\lambda(dx, u) F_{u-}^{-1} \Delta F_u\end{aligned}$$

and

$$E[g(u,x)I_{\{T=u\}}|\mathscr{F}_{u-}] = -\int_E g(u,x)\lambda(dx,u)F_{u-}^{-1}\Delta F_u \,.\, I_{\{T\geq u\}}.$$

The q^c and $q^{\Delta u}$ martingales are orthogonal, and a calculation similar to Theorem 15.6 establishes the following result (see [33]):

Theorem 15.16.

$$\langle M^g, M^g\rangle_t = \int_\Omega I_{S\leq t} g^2 d\tilde{p} - \sum_{\substack{0<u\leq t\wedge T \\ u\in D}} \left(\int_E g(u,x)\lambda(dx,u)\right)^2 \frac{\Delta F_u^2}{F_{u-}^2}.$$

Remarks 15.17. $p(ds, dx)$ and $\tilde{p}(ds, dx)$ are examples of random measures. $\tilde{p}$ is, in fact, the dual predictable projection of p. The results above are taken from work of Chou and Meyer [10], Davis [12] and Elliott [32, 33]. The results are extended to a sequence of jumps with a single time accumulation point in [12], and to general right constant processes with values in E in [30]. Jacod, discussing multivariate point processes, obtained related theorems in [49]. Further results for the single jump process are given in Chapter 17, where the optimal control of jump processes is discussed.

Having motivated some of the ideas by discussing the basic example of a single jump process we now define a general random measure. Jacod was the first to discuss such measures in a detailed and rigorous manner, and our presentation is based on his work. For further results see [50].

General Random Measures

Convention 15.18. As before, we suppose we are working on a probability space $(\Omega, \mathscr{F}, P)$ which has a complete, right continuous filtration $\{\mathscr{F}_t\}$, $t\geq 0$. E will be a Lusin space, that is, E is homeomorphic to a Borel subset of a compact metric space. E is given its Borel σ-field $\mathscr{E}$ and a distance d which is compatible with its topology. More generally, we could take E to be a Blackwell space ([12]); however, such generality will not be required as the applications we have in mind are when $E\subset R^n$ or $E\subset R^N$.

Notation 15.19. Write

$$\tilde{\Omega} = \Omega\times R^+\times E,$$

$$\tilde{\mathscr{F}} = \mathscr{F}\times\mathscr{B}(R^+)\times\mathscr{E},$$

$$\tilde{\Sigma}_x = \Sigma_x\times\mathscr{E}, \qquad x = o, a, p.$$

$\tilde{\Sigma}_x$ will also be used to denote the space of $\tilde{\Sigma}_x$ measurable process.

Definition 15.20. A *random measure* μ is a family $\{\mu(\omega,\cdot), \omega\in\Omega\}$ of σ-finite measures on $(R^+\times E, \mathscr{B}(R^+)\times\mathscr{E})$.

Example 15.21. A random measure can be associated with any process A of finite variation by taking E to be a one point set $\{\Delta\}$ and defining $\mu(\omega, dt \times \{\Delta\}) = dA_t(\omega)$.

Remarks 15.22. Because each measure $\mu(\omega, \cdot)$ is σ-finite it has a Jordan–Hahn decomposition

$$\mu(\omega, \cdot) = \mu^+(\omega, \cdot) - \mu^-(\omega, \cdot)$$

and an absolute value $|\mu|(\omega, \cdot) = \mu^+(\omega, \cdot) + \mu^-(\omega, \cdot)$.

Notation 15.23. Suppose $W: \tilde{\Omega} \times R$ is such that each section $W(\omega, \cdot): R^+ \times E \to R$ is a Borel function. Then write

$$(W * \mu)_t(\omega) = \int_{[0,t]\times E} W(\omega, s, x)\mu(\omega, ds, dx),$$

if this integral exists (possibly equal to $+\infty$ or $-\infty$).

If $(W * \mu)_t$ exists for all $t \in R^+$ one can talk of the *process* $W * \mu$.

Also, the random measure $W \cdot \mu$ is defined by

$$(W \cdot \mu)(\omega, dt, dx) = W(\omega, t, x)\mu(\omega, dt, dx).$$

Clearly for each ω the Radon–Nikodym derivative of $W \cdot \mu$ with respect to μ is $W(\omega, \cdot)$, and, if the appropriate integrals exist,

$$W' * (W \cdot \mu) = (W'W) * \mu. \tag{15.5}$$

Definition 15.24. The random measure μ is *optional* (resp. *predictable*) if for each positive process $W \in \tilde{\Sigma}_x$, $x = o$ (resp. p) the processes $W * \mu^+$ and $W * \mu^-$ are optional (resp. predictable).

Remarks 15.25. If μ is optional (resp. predictable), then clearly the random measures μ^+, μ^- and $|\mu|$ are optional (resp. predictable). Furthermore, for each $W \in \tilde{\Sigma}_x$, $x = o$ (resp. p) for which the process $W * \mu$ exists

$$W * \mu = W^+ * \mu^+ + W^- * \mu^- - W^+ * \mu^- - W^- * \mu^+,$$

and so $W * \mu$ is then optional (resp. predictable).

Lemma 15.26. *Suppose μ is an optional (resp. predictable) random measure and $W \in \tilde{\Sigma}_x$, $x = o$ (resp. p). Then, if the random measure $W \cdot \mu$ is defined, it is optional (resp. predictable).*

PROOF. Clearly

$$(W \cdot \mu)^+ = W^+ \cdot \mu^+ + W^- \cdot \mu^-,$$

$$(W \cdot \mu)^- = W^- \cdot \mu^+ + W^+ \cdot \mu^-,$$

and, as observed above,

$$W' * (W \cdot \mu) = (W'W) * \mu. \qquad \square$$

Remarks 15.27. We now consider spaces of random measures which generalize the spaces $\mathscr{A}$, $\mathscr{A}_{\text{loc}}$ and $\mathscr{V}$. Again we should consider the equivalence classes of random measures which are P-almost surely equal, but we use the same notation, μ, for an equivalence class and a member of the equivalence class, that is, a *version* if μ. A random measure is then *optional* if its equivalence class contains an optional member.

Definition 15.28. A random measure μ is said to be *integrable* if, with 1 denoting the process which is identically 1,

$$E[1 * |\mu|_\infty] < \infty.$$

Notation 15.29. $\tilde{\mathscr{A}}$ will denote the set of (equivalence classes of) optional and integrable random measures.

$\tilde{\mathscr{V}}$ (resp. $\tilde{\mathscr{A}}_\sigma$) will denote the set of (equivalence classes of) random measures μ for which there exists a $\tilde{\Sigma}_x$ measurable, $x = o$ (resp. p) partition $\{B_n\}$ of $\tilde{\Omega}$, such that $I_{B_n} \cdot \mu \in \tilde{\mathscr{A}}$ for each n.

Clearly each measure in $\tilde{\mathscr{V}}$ is optional and $\tilde{\mathscr{A}} \subset \tilde{\mathscr{A}}_\sigma \subset \tilde{\mathscr{V}}$. $\tilde{\mathscr{A}}^+$ (resp. $\tilde{\mathscr{A}}_\sigma^+$, $\tilde{\mathscr{V}}^+$) will denote the positive measures in $\tilde{\mathscr{A}}$ (resp. $\tilde{\mathscr{A}}_\sigma$, $\tilde{\mathscr{V}}$).

Remarks 15.30. Clearly, $\mu \in \tilde{\mathscr{A}}$ if and only if $1 * \mu \in \mathscr{A}$.

$1 * \mu \in \mathscr{V}$ (resp. $\mathscr{A}_{\text{loc}}$) if and only if $\mu \in \tilde{\mathscr{V}}$ (resp. $\tilde{\mathscr{A}}_\sigma$) and if also there is an increasing sequence $\{T_n\}$ of stopping times, with $\lim T_n = \infty$ a.s., such that $I_{D_n} \cdot \mu \in \mathscr{A}$, where $D_n = [\![0, T_n[\![$ (resp. $[\![0, T_n]\!]$). When $1 * \mu \in \mathscr{V}$ one can take for $\{T_n\}$ the sequence of stopping times which localizes $1 * \mu$.

Definition 15.31. Suppose $\mu \in \tilde{\mathscr{V}}$. Let $\{B_n\}$ be a $\tilde{\Sigma}_o$ measurable partition of $\tilde{\Omega}$ such that $I_{B_n} \cdot \mu \in \tilde{\mathscr{A}}$ for each n, and let W be a bounded $\tilde{\mathscr{F}}$ measurable process. Write

$$M_\mu(WI_{B_n}) = E[WI_{B_n} * \mu_\infty].$$

This defines a σ-finite measure M_μ on $(\tilde{\Omega}, \tilde{\Sigma}_o)$ which is independent of the partition $\{B_n\}$. M_μ is the *Doléans measure* associated with μ.

If μ is associated with $A \in \mathscr{A}_{\text{loc}}$ as in Example 15.21, then

$$M_\mu = \mu_A * \delta_\Delta,$$

where μ_A is the measure associated with A as in Theorem 7.13.

Theorem 15.32. *Suppose m is a measure on $(\tilde{\Omega}, \tilde{\mathscr{F}})$. There is a random measure $\mu \in \tilde{\mathscr{V}}$ (resp. a predictable random measure $\mu \in \tilde{\mathscr{A}}_\sigma$), such that $m = M_\mu$ if and only if:*

(a) *m is $\tilde{\Sigma}_x$ σ-finite, $x = 0$ (resp. p),*
(b) *$m(N \times E) = 0$ for every P-evanescent set N,*
(c) *for every $A \in \tilde{\Sigma}_x$, $x = o$ (resp. p) such that $|m|(A) < \infty$, and every bounded*

$\mathscr{F}$-measurable process X

$$m(I_A X) = m(I_A \Pi_o(X))$$
$$(\textit{resp.} = m(I_A \Pi_p(X)).$$

μ is then unique.

Furthermore, $\mu \in \tilde{\mathscr{A}}$ (resp. $\tilde{\mathscr{A}}_\sigma$, ≥ 0) if and only if m is finite (resp. $\tilde{\Sigma}_p$ σ-finite (resp. positive)).

PROOF. (i) Suppose $\mu \in \tilde{\mathscr{V}}$ (resp. $\mu \in \tilde{\mathscr{A}}_\sigma$ is a predictable random measure), and $m = M_\mu$. Properties (a) and (b) are immediate. If $A \in \tilde{\Sigma}_x$, $x = o$ (resp. p) is such that $|m|(A) < \infty$ then the processes $\alpha^+ = I_A * \mu^+$ and $\alpha^- = I_A * \mu^-$ are in $\mathscr{A}$ (resp. are predictable processes in $\mathscr{A}$). Then for any bounded $\mathscr{F}$-measurable process X

$$\begin{aligned} m(I_A X) &= E[XI_A * \mu^+_\infty] - E[XI_A * \mu^-_\infty] \\ &= E\left[\int_0^\infty X\,d\alpha^+\right] - E\left[\int_0^\infty X\,d\alpha^-\right] \\ &= (X, \alpha^+) - (X, \alpha^-). \end{aligned}$$

Property (c) then follows from Corollary 7.23. The necessity of conditions (a), (b) and (c) is, therefore, established.

(ii) We now show that $\mu \in \tilde{\mathscr{V}}$ is determined by its Doléans measure M_μ. Suppose $\{B_n\}$ is a $\tilde{\Sigma}_o$ measurable partition of $\tilde{\Omega}$ such that $I_{B_n} \cdot \mu \in \tilde{\mathscr{A}}$. It is sufficient to show that each measure $I_{B_n} \cdot \mu$ is determined by its Doléans measure $I_{B_n} \cdot M_\mu$; that is, it is sufficient to prove the result when $\mu \in \tilde{\mathscr{A}}$.

In that case, because $\mathscr{E}$ is separable, the measure μ is determined by the values $\mu([0, t] \times D)$, for t a positive rational number, and $D \in \mathscr{E}$ a member of a countable family which generates $\mathscr{E}$. However, for any set $G \in \mathscr{F}$

$$M_\mu(G \times [0, t] \times D) = E[I_G(\mu([0, t] \times D))].$$

Therefore, P-almost surely, the measure M_μ determines each random variable $\mu([0, t] \times D)$, and so, P-almost surely, the random measure μ itself.

(iii) We now show that conditions (a), (b), and (c) are sufficient.

Initially we suppose that m is a positive, finite measure and that for every set $H \in \tilde{\Sigma}_x$ and every bounded measurable process X,

$$m(I_H \Pi_x(X)) = m(I_H X), \qquad x = o \text{ (resp. } p).$$

Write $\hat{m}$ for the measure on $([O, \infty[\times \Omega, \mathscr{B} \times \mathscr{F})$ defined by $\hat{m}(B \times G) = m \cdot (B \times G \times E)$. Then $\hat{m}$ satisfies the conditions of Theorem 7.13 and so there is an optional (resp. predictable), process $A \in \mathscr{A}^+$ such that $\hat{m} = \mu_A$. Because E is a Lusin space, by taking Radon–Nikodym derivatives, one can factorize the measure m with respect to $\hat{m}$, that is, there is a family of random transition probability measures $\{n\}$ from $([0, \infty[\times \Omega, \Sigma_x)$ on $(E, \mathscr{E})$, such that on $(\tilde{\Omega}, \tilde{\Sigma}_x)$

$$m(dt, d\omega, dx) = \hat{m}(dt, d\omega) n(t, \omega, dx).$$

Therefore,

$$\mu(\omega, dt, dx) = dA_t(\omega) n(\omega, t, dx)$$

is a random measure in $\tilde{\mathscr{A}}^+$. Suppose X is a positive, bounded measurable process and $D \in \mathscr{E}$. Then by the above definition of μ:

$$M_\mu(XI_D) = \mu_A(X \cdot n(D)) = \hat{m}(X \cdot n(D)).$$

The process $n(D)$ is optional (resp. predictable), so from Theorem 7.21 and Corollary 6.44 this is equal to

$$\hat{m}(\Pi_x(X) \cdot n(D)) = m(\Pi_x(X) \cdot I_D).$$

By a monotone class argument, the measures m and M_μ are seen to be equal. In the case $x = p$, μ is seen to be a predictable random measure.

(iv) Suppose m is now a signed finite measure satisfying conditions (a), (b), and (c) of the theorem, and that $m = m^+ - m^-$ is its Jordan–Hahn decomposition.

Define finite positive measures n^+, n^- on $(\tilde{\Omega}, \tilde{\mathscr{F}})$ by putting for every $D \in \mathscr{E}$ and positive bounded measurable process X;

$$n^+(I_D \cdot X) = m^+(I_D \Pi_x(X)), \qquad x = o \text{ (resp. } p),$$

$$n^-(I_D \cdot X) = m^-(I_D \Pi_x(X)), \qquad x = o \text{ (resp. } p).$$

If $n = n^+ - n^-$ then

$$n(I_D \cdot X) = m(I_D \Pi_x(X))$$

from (c), so $n = m$.

If $N \in \mathscr{F}$ is a P-evanescent set $\Pi_x(I_N)$ is indistinguishable from the zero process, so from their definition and property (c) we then have:

$$n^+(N \times E) = 0 = n^-(N \times E).$$

Consequently, the measures n^+ and n^- satisfy the conditions of (iii) above, and so there are measures $\upsilon, \eta \in \tilde{\mathscr{A}}^+$ such that $M_\upsilon = n^+$ and $M_\eta = n^-$. The measure $\mu = \upsilon - \eta$ then belongs to $\tilde{\mathscr{A}}$ and satisfies $M_\mu = m$. In the case $x = p$, μ is also a predictable random measure.

(v) Suppose m is now a general measure satisfying conditions (a), (b), and (c) of the theorem. Let $\{B_n\}$ be a Σ_x, $x = o$ (resp. p), measurable partition of $\tilde{\Omega}$ such that m restricted to each B_n is finite. If $m_n = I_{B_n} \cdot m$, m_n is a finite measure satisfying the conditions of (iv). Consequently, for each n there is a measure $\mu_n \in \tilde{\mathscr{A}}$ such that $M_{\mu_n} = m_n$. Because m_n is concentrated on B_n, $m_n(B_n^c) = 0$ and $I_{B_n} \cdot \mu_n = \mu_n$ so the measure $\mu = \Sigma_n \mu_n$ is in $\tilde{\mathscr{V}}$ (resp. a predictable random measure in $\tilde{\mathscr{A}}_\sigma$ if $x = p$ above). By construction, $M_\mu = m$.

(vi) Clearly $\mu \in \tilde{\mathscr{A}}$ (resp. $\tilde{\mathscr{A}}_\sigma$) if and only if M_μ is finite (resp. M_μ is $\tilde{\Sigma}_p$ σ-finite. From the construction of μ in (iii) and (v), and the uniqueness established in (ii) we see $\mu \geq 0$ if and only if $M_\mu \geq 0$. □

Corollary 15.33. *Suppose μ and υ are two random measures in $\tilde{\mathscr{V}}$ (resp. $\tilde{\mathscr{A}}_\sigma$). If the measures M_μ, M_υ are equal on $\tilde{\Sigma}_x$, $x = o$ (resp. p), then $\mu = \upsilon$.*

We also observe from the theorem the following result:

Corollary 15.34. *The spaces $\tilde{\mathscr{A}}, \tilde{\mathscr{A}}_\sigma, \tilde{\mathscr{V}}$* (resp. the predictable random measures in each of these spaces of equivalence classes of random measures), form vector spaces of equivalence classes of random measures.

Theorem 15.35. *Suppose $\mu \in \tilde{\mathscr{A}}_\sigma$. There is then a unique predictable random measure $\Pi_p^*(\mu) \in \tilde{\mathscr{A}}_\sigma$ which satisfies the two following equivalent conditions:*

(a) *the measures M_μ and $M_{\Pi_p^*(\mu)}$ coincide on $\tilde{\Sigma}_p$,*
(b) *for every function $W \in \tilde{\Sigma}_p$ such that $W * \mu \in \mathscr{A}_{\text{loc}}$, $\Pi_p^*(W * \mu) = W * \Pi_p^*(\mu)$.*

Proof. Suppose $\{B_n\}$ is a $\tilde{\Sigma}_p$ measurable partition of $\tilde{\Omega}$ such that for each $n \in Z^+, I_{B_n} \cdot \mu \in \tilde{\mathscr{A}}$.

Define a $\tilde{\Sigma}_p$ σ-finite measure m on $(\tilde{\Omega}, \tilde{\Sigma}_p)$ by putting, for each $D \in \mathscr{E}$ and each bounded measurable process X,

$$m(XI_DI_{B_n}) = M_\mu(\Pi_P^*(X)I_DI_{B_n}).$$

Then m satisfies the conditions of Theorem 15.32, and

$$m(I_AX) = m(I_A\Pi_p^*(X))$$

for every $A \in \tilde{\Sigma}_p$. Therefore, there is a predictable random measure $\Pi_p^*(\mu) \in \tilde{\mathscr{A}}_\sigma$ such that $m = M_{\Pi_p^*(\mu)}$. Clearly M_μ and $M_{\Pi_p^*(\mu)}$ are equal on $\tilde{\Sigma}_p$. We now prove that (a)⇒(b):

Suppose $\Pi_p^*(\mu)$ is a predictable random measure in $\tilde{\mathscr{A}}_\sigma$ and $W \in \tilde{\Sigma}_p$ is such that $W * \mu \in \mathscr{A}_{\text{loc}}$. Then

$$M_{W*\mu}(X) = M_\mu(WX)$$

and

$$M_{W*\Pi_p^*(\mu)}(X) = M_{\Pi_p^*(\mu)}(WX)$$

for every bounded measurable process X. From their definition the measures $M_{W*\mu}$ and $M_{\Pi_p^*(W*\mu)}$ coincide on Σ_p, so if X is also predictable, $XW \in \tilde{\Sigma}_p$ and

$$M_\mu(WX) = M_{\Pi_p^*(\mu)}(WX),$$

$$M_{W*\mu}(X) = M_{\Pi_p^*(W*\mu)}(X).$$

Therefore, $\Pi_p^*(W * \mu) = W * \Pi_p^*(\mu)$. Finally we prove that (b)⇒(a).

For every $A \in \tilde{\Sigma}_p$ the process $I_{A \cap B_n} * \mu$ is in $\mathscr{A} \subset \mathscr{A}_{\text{loc}}$. Therefore, from (b)

$$\Pi_p^*(I_{A \cap B_n} * \mu) = (I_{A \cap B_n}) * \Pi_p^*(\mu).$$

Consequently,

$$M_\mu(A \cap B_n) = E[(I_{A \cap B_n} * \mu)_\infty],$$

$$E[(I_{A \cap B_n} * \Pi_p^*(\mu))_\infty] = M_{\Pi_p^*(\mu)}(A \cap B_n)$$

and so $\Pi_\pi^*(\mu)$ satisfies (a). □

Definition 15.36. $\Pi_p^*(\mu)$ is called the *dual predictable projection* of μ.

Corollary 15.37. *Suppose $\mu \in \tilde{A}_\sigma$ and $W \in \tilde{\Sigma}_p$ are such that $W \cdot \mu \in \tilde{A}_\sigma$. Then $\Pi_p^*(W \cdot \mu) = W \cdot \Pi_p^*(\mu)$. In particular, if $\mu \in \tilde{\mathscr{A}}_\sigma$ is predictable then $\Pi_p^*(\mu) = \mu$.*

Remarks 15.38. From Theorem 9.29 and the characterization Corollary 7.31 of the dual predictable projection we see that if T is a predictable stopping time and $A \in \mathscr{A}_{\text{loc}}$ then

$$\Delta\Pi_p^*(A)_T = E[\Delta A_T | \mathscr{F}_{T-}] \quad \text{on } \{T < \infty\}.$$

Therefore, if $W \in \tilde{\Sigma}_p$ is such that

$$\int_E W(T, x)\mu(\{T\} \times dx) I_{\{T<\infty\}}$$

exists, then

$$\int_E W(T, x)\Pi_p^*(\mu)(\{T\} \times dx)$$
$$= E\left[\int_E W(T, x)\mu(\{T\} \times dx) \middle| \mathscr{F}_{T-}\right] \quad \text{on } \{T < \infty\}, \qquad (15.6)$$

if the conditional expectation exists.

Definition 15.39. $\mu \in \tilde{\mathscr{A}}_\sigma$ is said to be a *martingale random measure* if the restriction of M_μ to $\tilde{\Sigma}_p$ is zero.

We see that if $\mu \in \tilde{\mathscr{A}}_\sigma$ then $\Pi_p(\mu)$ is the unique predictable measure in $\tilde{\mathscr{A}}_\sigma$ such that $\mu - \Pi_p(\mu)$ is a martingale random measure.

Definition 15.40. A random measure μ is said to be *integer valued* if

(a) $\mu(\omega, \{t\} \times E) \le 1$ for all (ω, t),
(b) for every $A \in \mathscr{B}(R^+) \times \mathscr{E}$, $\mu(A)$ has values in $\{0, 1, 2, \ldots\} \cup \{\infty\}$. Note that μ is then a positive random measure.

Remarks 15.41. If μ is an integer valued random measure write

$$D=\{(\omega, t): \mu(\omega, \{t\}\times E)=1\}.$$

From Definition 15.40(b), if $(\omega, t)\in D$ there is a unique point $\varepsilon_t(\omega)\in E$ such that

$$\mu(\omega, \{t\}\times dx)=\delta_{\varepsilon_t(\omega)}(dx).$$

Here $\delta_{\varepsilon_t(\omega)}(dx)$ denotes the unit mass at $\varepsilon_t(\omega)$.

If $(\omega, t)\notin D$, define $\varepsilon_t(\omega)=e$, where e is an additional point, not in E. Then we have a process $\varepsilon=\{\varepsilon_t\}, t\geq 0$, with values in $E\cup\{e\}$, and the measure μ can be written

$$\mu(\omega, dt, dx)=\sum_{(\omega,s)\in D}\delta_{(s,\varepsilon_s(\omega))}(dt, dx)$$

$$=\sum_{s\geq 0} I_{\{\varepsilon_s\in E\}}\delta_{(s,\varepsilon_s(\omega))}(dt, dx). \tag{15.7}$$

Because μ is σ-finite each section of D,

$$D_\omega=\{t\in R^+: (\omega, t)\in D\}$$

is at most countable. That is, in the terminology of [22], D is "minced".

Conversely, if ε is a process with values in $E\cup\{e\}$ such that the set $\{\varepsilon\in E\}$ is minced, then the above expression (15.7) defines a random measure μ. These random measures are sometimes called *point processes*, the points being the $(s, \varepsilon_t(\omega))\in R^+\times\mathscr{E}$ such that $\varepsilon_t(\omega)\in E$.

Lemma 15.42. *Suppose μ is an integer valued random measure given by* (15.7). *Then $\mu\in\tilde{\mathscr{V}}$ if and only if $\{\varepsilon\}$ is (P-indistinguishable from) an optional process, and then D is (P-indistinguishable from) an optional set.*

Proof. Suppose $\{B_n\}$ is a $\tilde{\Sigma}_o$ measurable partition of $\tilde{\Omega}$ such that $M_\mu(B_n)<\infty$. Then, for every $C\in\mathscr{E}$

$$\{\varepsilon\in C\}=\bigcup_n\{\Delta(I_C I_{B_n}*\mu)=1\}\in\Sigma_o.$$

Consequently D is an optional set and ε is an optional process.

Conversely, suppose $\{\varepsilon\}$ is an optional process. Let $\{T_n\}$ be a sequence of stopping times with disjoint graphs such that $D=\bigcup_n [\![T_n]\!]$. If W is a positive $\tilde{\Sigma}_o$ measurable process then

$$W*\mu=\sum_n W(T_n, \varepsilon_{T_n})I_{[\![T_n,\infty]\!]}$$

is optional, so μ is optional. Furthermore, writing $B_n=[\![T_n\times E]\!]$ and $B=\tilde{\Omega}\backslash(\bigcup_n B_n)$, the measures $I_B\cdot\mu$ and $I_{B_n}\cdot\mu$ are of finite total mass, and so $\mu\in\tilde{\mathscr{V}}$. □

Notation 15.43. $\tilde{\mathscr{V}}^1$ denotes the members of $\tilde{\mathscr{V}}$ which have a representation which is an integer valued random measure and $\tilde{\mathscr{A}}_\sigma^1 = \tilde{\mathscr{V}}^1 \cap \tilde{\mathscr{A}}_\sigma$. When discussing elements in $\tilde{\mathscr{V}}^1$ we always consider a representative which has integer values.

Example 15.44. Suppose X is a corlol process with values in E. Write

$$\mu(dt, dx) = \sum_{s>0} I_{\{X_{s-} \neq X_s\}} \delta_{(s, X_s)}(dt, dx).$$

Then μ is an integer valued random measure, with

$$D = \{(\omega, t): X_{t-}(\omega) \neq X_t(\omega)\} \cap [\![0]\!]^c$$

and $\varepsilon = X$ on D.

If X is optional then, from Lemma 15.42 above, so is μ. In fact define the following stopping times:

$$T(0, m) = 0,$$

$$T(n+1, m) = \inf\left\{t > T(n, m): d(X_{t-}, X_t) \in \left[\frac{1}{m}, \frac{1}{m-1}\right]\right\}.$$

Then the measure M_μ is finite on each of the $\tilde{\Sigma}_p$ measurable sets

$$\left\{(\omega, t, x): t \leq T(n, m)(\omega), d(X_{t-}(\omega), x) \in \left[\frac{1}{m}, \frac{1}{m-1}\right[\right\},$$

whose union is $\tilde{\Omega}$, so in fact $\mu \in \tilde{\mathscr{A}}_\sigma^1$.

Definition 15.45. The dual predictable projection $\Pi_p^*(\mu)$ of the measure μ of this example is called the *Lévy system* of the process X.

Example 15.46. Again suppose X is a corlol process with values in E, but now suppose E is a subset of an additive group, for example, E might even be a vector space. Consider the following random measure

$$\mu^X(dt, dx) = \sum_{s>0} I_{\{\Delta X_s \neq 0\}} \delta_{(s, \Delta X_s)}(dt, dx).$$

Here

$$D = \{(\omega, t): X_{t-}(\omega) \neq X_t(\omega)\} \cap [\![0]\!]^c$$

and $\varepsilon = \Delta X$ on D. As in the above example we can show that if X is adapted then $\mu^X \in \tilde{\mathscr{A}}_\sigma^1$. $\Pi_p^*(\mu^X)$ is also often called the Lévy system of X, though care should be taken to indicate which measure is under discussion. In fact

$$\mu(A) = \int \mu^X(ds, dx) I_A(s, X_{s-} + x)$$

and

$$(\Pi_p^*(\mu))(A) = \int \Pi_p^*(\mu^X)(ds, dx) I_A(s, X_{s-} + x).$$

Theorem 15.47. *If* $\mu \in \tilde{\mathscr{A}}^1$ *there is a version of* $\Pi_p^*(\mu)$ *such that for all* ω

$$\Pi_p^*(\mu)(\omega, \cdot) \geq 0 \quad \text{and} \quad \Pi_p^*(\mu)(\omega, \{t\} \times E) \leq 1.$$

PROOF. Suppose υ is a positive version of $\Pi_p^*(\mu)$ and write

$$a_t(\omega) = \upsilon(\omega, \{t\} \times E).$$

If $\{B_n\}$ is a $\tilde{\Sigma}_p$ measurable partition of $\tilde{\Omega}$ such that $I_{B_n} \cdot \mu \in \tilde{\mathscr{A}}^+$ then

$$a_t = \sum_n \Delta(I_{B_n} * \upsilon),$$

so the process $\{a_t\}$ is predictable. Therefore, the random measure $\upsilon' = I_{\{a \leq t\}} \cdot \upsilon$ is a predictable positive random measure, which satisfies the conditions of the theorem and which is a version of $\Pi_p^*(\mu)$ if the set $\{a > 1\}$ is evanescent. However, from (15.6), for every predictable stopping time T, $a_T \leq 1$ almost surely. Therefore, applying the Section Theorem 6.24 to the predictable set $\{a > 1\}$ we see this set is evanescent.

Notation 15.48. If $\mu \in \tilde{\mathscr{A}}_\sigma^1$ we shall always consider a version of $\Pi_p^*(\mu)$ such that

$$\Pi_p^*(\mu)(\omega, \{t\} \times E) \leq 1 \quad \text{for all } \omega.$$

Write:

$$a_t(\omega) = \Pi_p^*(\mu)(\omega, \{t\} \times E),$$

$$J = \{a > 0\}.$$

If W is a positive $\tilde{\Sigma}_o$ measurable process

$$\hat{W}_t(\omega) = \begin{cases} \int_E W(\omega, t, x) \Pi_p(\mu)(\omega, \{t\} \times dx), & \text{if this integral exists,} \\ +\infty, & \text{or otherwise.} \end{cases}$$

Clearly $\hat{W} = 0$ on J^c. Formula (15.6) can be written

$$\hat{W}_T = E[W(T, \varepsilon_T) I_D(T) | \mathscr{F}_{T-}] \quad \text{on } \{T < \infty\}. \tag{15.8}$$

Note that $a = \hat{1}$.

Remarks 15.49. We now define the local characteristics of a semimartingale.

Suppose $X = (X^j)$, $1 \leq j \leq m$, is a process with values in R^m. When we speak of X being a martingale, a special semimartingale (with a canonical decomposition), etc., we are referring to each component of X. Suppose X is in fact a semimartingale and write

$$\tilde{X}_t = \sum_{s \leq t} (\Delta X_s) I_{\{|\Delta X_s| > 1\}} I_{]\!]0,\infty[\![}.$$

Then the process $X - \tilde{X} - X_0$ is a semimartingale with bounded jumps and so, by Corollary 12.40, it has a canonical decomposition

$$X - \tilde{X} - X_0 = M + B,$$

where M is a process whose components are in $\mathscr{L}$ and B is a predictable process whose components are in $\mathscr{V}_0$.

Definition 15.50. The *local characteristics* of the semimartingale X are the members of the triplet (B, C, υ) where:

(a) B is the process defined above,
(b) $C=(C^{jk})$, $1\le j,\ k\le m$, is the $m\times m$ matrix process with components

$$C_t^{jk}=\langle (X^j)^c, (X^k)^c\rangle_t,$$

(that is, the predictable quadratic variation process of the continuous martingale parts of X^j and X^k),
(c) $\upsilon=\Pi_p^*(\mu^X)$, the dual predictable projection of the measure μ^X associated with X as in Example 15.46.

Remarks 15.51. $\upsilon\in\tilde{\mathscr{A}}^1$, but while υ and C are intrinsic, the process B depends on the size of the jumps considered in the process $\tilde{X}$. For example, if $0<\alpha<\infty$ and we consider a process

$$\tilde{X}_t^\alpha=\sum_{s\le t}(\Delta X_s)I_{\{|\Delta X_s|>\alpha\}}I^{]\!]0,\infty[\![},$$

then $X-\tilde{X}^\alpha-X_0$ is again special but the process B in its decomposition would be different.

Theorem 15.52. *A version of (B, C, υ) can be chosen such that for each ω:*

(a) *$B(\omega)$ is corlol and of finite variation on each compact set,*
(b) *$C(\omega)$ is continuous and, if $s\le t$, the matrix $C_t(\omega)-C_s(\omega)$ is symmetric and nonnegative,*
(c) *$\upsilon(\omega, R^+\times\{0\})=0$ and for every $t\in R^+$, $\upsilon(\omega,\{t\}\times R^m)\le 1$,*

$$\int(1\wedge|x|^2)\upsilon(\omega,[0,t]\times dx)<\infty \quad \textit{and} \quad \sum_{s\le t}\left|\int I_{\{|x|\le 1\}}\upsilon(\omega,\{s\}\times dx\right|<\infty,$$

(d) *for every $t\in R^+$*

$$\Delta B_t(\omega)=\int I_{\{|x|\le 1\}}\upsilon(\omega,\{t\}\times dx).$$

PROOF. It is enough to show that the above properties are satisfied, except on a set of measure zero, by any version (B, C, υ) of the local characteristics. Clearly this is so for (a).

If $\alpha\in R^m$ then

$$\sum_{j,k=1}^m \alpha^j\alpha^k C^{jk}=\left\langle\left(\sum_{j=1}^m \alpha^j X^j\right)^c,\left(\sum_{j=1}^m \alpha^j X^j\right)^c\right\rangle,$$

so the process $\sum_{j,k=1}^m \alpha^j\alpha^k C^{jk}$ is increasing. Now by definition the matrix C_t-C_s is symmetric, and by the above observation it is nonnegative if $s\le t$. Furthermore, C is a continuous process, and so (b) is satisfied.

By definition, $\mu^X(\omega, R^+ \times \{0\}) = 0$. The second statement of (c) is the result of Theorem 15.47.

Write

$$H = \{|\Delta X| > 1\}.$$

Now because X is a semimartingale it is corlol and

$$A_t(\omega) = \sum_{s \le t} I_{\{(\omega,s) \in H\}} \in \mathscr{V},$$

while

$$A'_t(\omega) = \sum_{s \le t} (\Delta X_s)^2(\omega) I_{\{(\omega,s) \in H^c\}} \le [X, X]_t(\omega)$$

and so is in $\mathscr{V}$. In fact, because A and A' both have jumps bounded by 1 they are in $\mathscr{A}_{\text{loc}}$. However,

$$A_t + A'_t = \int (1 \wedge |x|^2) \mu^X([0, t] \times dx)$$

by the definition of μ^X. The dual predictable projection of this process is

$$\int (1 \wedge |x|^2) v([0, t] \times dx),$$

which is, therefore, almost surely finite for every $t < \infty$.

Now, because B is predictable

$$\begin{aligned} \Pi_p(\Delta B) &= \Pi_p(\Delta X - \Delta \tilde{X} - \Delta M) \\ &= \Pi_p(\Delta X \cdot I_{H^c}) \end{aligned}$$

while from (15.8)

$$\Pi_p(\Delta X \cdot I_{H^c})_t = \int_{R^m} I_{\{|x| \le 1\}} v(\{t\} \times dx),$$

which implies the last part of (c), and (d). □

CHAPTER 16

The Optimal Control of a Continuous Process

Introduction 16.1

In this chapter and the next we show how some of the results of previous chapters can be used to discuss the optimal control of stochastic processes. Only the completely observed case is considered in detail, and two basic situations are treated: the optimal control of a process described by a stochastic differential equation and, in the following chapter, the optimal control of a jump process. The main features of this approach are:

(a) the idea that the effect of a control is to change the measure on the space of sample paths of the state process,
(b) the principle of optimality,
(c) the Doob–Meyer decomposition,
(d) the representation of martingales as stochastic integrals, and the stochastic minimum principle.

Earlier work is reviewed in [37].

Controlled Stochastic Differential Equations 16.2

Write Ω for the space of continuous functions defined on $[0, 1]$ with values in R^n. Let $\{B_t\}$ be the family of coordinate functions and $\mathscr{F}_t^0 = \sigma\{B_s : s \le t\}$. Suppose P is Wiener measure on $(\Omega, \mathscr{F}_1^0)$ and write $\mathscr{F}_t$ for the completion of $\mathscr{F}_t^0$ by the null sets of $\mathscr{F}_1^0$. For $x_t \in \mathscr{C}([0, 1], R^n) = \Omega$ write, as before,

$$|x|_t^* = \sup_{0 \le s \le t} |x_s|.$$

Suppose $\sigma: [0, 1] \times \Omega \to R^{n^2}$ is a matrix valued function such that

(a) $\sigma_{ij}(\ ,\ )$ is $\mathscr{F}_t$ predictable,
(b) $|\sigma_{ij}(t, x) - \sigma_{ij}(t, y)| \le M|x - y|_t^*$,

(c) for each $(t, x) \in [0, 1] \times \Omega$ $\sigma(t, x)$ is nonsingular and $|\sigma^{-1}(t, x)| \leq M$.
Here M is a fixed constant, independent of t, i, j.

Then, by Theorem 14.6, for any given initial condition $x_0 \in R^n$, there is a unique strong solution to the stochastic differential equation

$$dx_t = \sigma(t, x)\, dB_t.$$

The control values will be chosen from a compact metric space U (which may be a subset of some Euclidean space).

Let $f: [0, 1] \times \Omega \times U \to R^n$ be a function such that:

(a) for each $(t, x) \in [0, 1] \times \Omega$, $f(t, x, \cdot)$ is continuous in $u \in U$,
(b) for each u, f is $\mathscr{F}_t$ predictable in (t, x),
(c) $|f(t, x, u)| \leq M(1 + |x|_t^*)$.

Controls 16.3. The set of admissible controls $\mathscr{U}$ is the set of $\mathscr{F}_t$ predictable U-valued processes.

For $u \in \mathscr{U}$ write

$$\Lambda_{r,t}(f^u) = \Lambda_{r,t}(u) = \exp\left(\int_r^t (\sigma^{-1}(s, x) f(s, x, u_s))'\, dB_s - \tfrac{1}{2}\int_r^t |\sigma_s^{-1} f_s^u|^2\, ds\right).$$

The boundedness of σ^{-1} and the growth condition on f imply (see Lemma 13.37), that $E[\Lambda_{0,1}(u)] = 1$, and we can define a measure P_u on $(\Omega, \mathscr{F}_1)$ by putting

$$\frac{dP_u}{dP} = \Lambda_{0,1}(u).$$

Theorem 16.4. *Under P_u the process x_t satisfies*

$$dx_t = f(t, x, u_t)\, dt + \sigma(t, x)\, dW_t^u, \tag{16.1}$$

with initial condition x_0, where $\{W_t^u\}$ is the vector P_u Brownian motion given by

$$dW_t^u = dB_t - \sigma^{-1}(t, x) f(t, x, u_t)\, dt. \tag{16.2}$$

PROOF. It was shown in Corollary 13.25 that W^u is a Brownian motion under P_u and the result follows by multiplying both sides of (16.2) by $\sigma(t, x)$ and integrating, because stochastic integrals under P or P_u give the same process (see Theorem 13.15). □

Remarks 16.5. We shall be discussing the optimal control of a stochastic system $\{x_t\}$ whose dynamics are described by an equation of the form (16.1). For each admissible control $u \in \mathscr{U}$ the controlled dynamics are given by the above weak solution. That is, for each $u \in \mathscr{U}$ we have a probability measure P_u under which the process $\{x_t\}$ satisfies (16.1). The advantage of using weak solutions is that we do not need the dependence of f on x to be smooth (for example, that f be Lipschitz is the usual requirement for strong solutions, see Theorem 14.6). Consequently, "bang-bang", and other noncontinuous controls can be discussed.

Example 16.6. Note also that to define $\Lambda(u)$, σ must be nonsingular. However, Davis [14] showed how the following important class of degenerate systems can be treated.

Suppose $x_t = (x_t^1, x_t^2) \in R^{n+m}$ is defined by

$$\begin{aligned} dx_t^1 &= f^1(t, x_t^1, x_t^2)\,dt, \\ dx_t^2 &= f^2(t, x_t^1, x_t^2, u_t)\,dt + \bar{\sigma}(t, x_t^1, x_t^2)\,dW_t, \end{aligned} \tag{16.3}$$

with initial condition $x_0 \in R^{n+m}$, and where $\bar{\sigma}$ is nonsingular, satisfying the above conditions, f^2 satisfies the same conditions as f, and f^1 is Lipschitz in x_t^1 uniformly in (t, x_t^2) Then for each trajectory x^2 there is a unique solution $x_t^1 = \phi_t(x^2)$ of the first of the above equations, and the second equation can be written

$$dx_t^2 = f^2(t, \phi_t(x^2), x^2, u)\,dt + \bar{\sigma}(t, \sigma_t(x^2), x_t^2)\,dW_t.$$

This equation is now of the form (16.1) and so has a weak solution for each $u \in \mathcal{U}$. If a scalar nth-order stochastic differential equation is written as a first-order system, then a degenerate family of equations like (16.3) is obtained.

Cost 16.7. Suppose the running cost is determined by a real, bounded, measurable function $c(t, x, u)$, which satisfies the same conditions as f, and that the terminal cost is given by a real, bounded, measurable function $g(x)$.

Then if control $u \in \mathcal{U}$ is used the total expected cost is

$$J(u) = E_u\left[\int_0^1 c(t, x_t, u_t)\,dt + g(x_1)\right],$$

where E_u denotes expectation with respect to P_u. We wish to investigate how the control u should be selected so that the total expected cost is minimized. For any $t \in [0, 1]$ and $u \in \mathcal{U}$ the conditional remaining cost from time t is

$$\varphi_t^u = E_u\left[\int_t^1 c(s, x_s, u_s)\,ds + g(x_1)\middle|\mathcal{F}_t\right].$$

Suppose control $u \in \mathcal{U}$ is built up from two controls $v, w \in \mathcal{U}$ by concatenation, that is:

$$u(\omega, s) = \begin{cases} w(\omega, s), & 0 \le s \le t, \\ v(\omega, s), & t < s \le 1. \end{cases}$$

Then

$$\Lambda_{0,1}(u) = \Lambda_{0,t}(w)\Lambda_{t,1}(v)$$

and by [60], 27.4:

$$\varphi_t^u = \frac{E[\Lambda_{0,t}(w)\Lambda_{t,1}(v)\{\int_t^1 c(s, x_s, u_s)\,ds + g(x_1)\}|\mathcal{F}_t]}{E[\Lambda_{0,t}(w)\Lambda_{t,1}(v)|\mathcal{F}_t]},$$

where E denotes expectation with respect to the original measure P. Now $\Lambda_{0,t}(w)$ is $\mathcal{F}_t$ measurable and, as an exponential, is almost surely never zero. $\Lambda_{0,t}(w)$, therefore, cancels; also

$$E[\Lambda_{t,1}(v)|\mathcal{F}_t]=1 \quad \text{a.s.}$$

Consequently,

$$\varphi_t^u = E\left[\Lambda_{t,1}(v)\left\{\int_t^1 c(s, x_s, u_s)\, ds + g(x_1)\right\}\middle|\mathcal{F}_t\right],$$

showing that φ_t^u depends only on the values of the control u on the interval $]t, 1]$. As a conditional expectation, φ_t^u is defined only almost everywhere. However, all the measures P_u are equivalent, so the null sets up to which φ_t^u is defined are also independent of the control u. Because c and g are bounded $\varphi_t^u \in L^1(\Omega, \mathcal{F}_t, P)$ for each $u \in \mathcal{U}$.

From Dunford and Schwartz [29], p. 302, L_1 is a complete lattice, so the lattice infimum

$$V_t = \bigwedge_{u\in\mathcal{U}} \varphi_t^u$$

exists and is $\mathcal{F}_t$ measurable. See also Lemma 16.A.2 of the appendix to this chapter.

Definition 16.8. V_t is the *value function.*

Remarks 16.9. To continue our discussion we need some results on lattices and conditional expectations due to Striebel [76]. These are proved in the appendix at the end of this chapter.

Suppose $(\Omega, \mathcal{F}, P)$ is a probability space and $\{\phi_\gamma: \gamma \in \Gamma\}$ is a subset of $L^1(\Omega, \mathcal{F}, P)$ such that $\phi_\gamma(\omega) \geq 0$ a.s. for each $\gamma \in \Gamma$.

Definition 16.10. The set $\{\phi_\gamma\}$ has the *ε-lattice property*, for $\varepsilon > 0$, if given $\gamma_1, \gamma_2 \in \Gamma$ there is a $\gamma_3 \in \Gamma$ such that

$$\phi_{\gamma_3}(\omega) \leq \phi_{\gamma_1}(\omega) \wedge \phi_{\gamma_2}(\omega) + \varepsilon \quad \text{a.s.}$$

The following result is then proved in the appendix (Lemma 16.A.5).

Lemma 16.11. *Suppose that $\mathcal{G}$ is a sub-σ-field of $\mathcal{F}$ and that the set $\{\phi_\gamma: \gamma \in \Gamma\} \subset L^1(\Omega, \mathcal{F}, P)$ of nonnegative functions has the ε-lattice property for every $\varepsilon > 0$. Then*

$$E\left[\bigwedge_\gamma \phi_\gamma \middle| \mathcal{G}\right] = \bigwedge_\gamma E[\phi_\gamma|\mathcal{G}] \quad \text{a.s.}$$

That is, the lattice infimum and conditional expectation operations commute.

Remarks 16.12. The above result extends immediately to the situation where $\|\phi_\gamma\|_\infty < K < \infty$ for every $\gamma \in \Gamma$. (Because then $\psi_\gamma = \phi_\gamma + K \geq 0$ for every $\gamma \in \Gamma$. The result can be applied to the set $\{\psi_\gamma\}$ and K subtracted from each side.)

The lemma will now be applied to the random variables $\{\varphi_t^u : u \in \mathcal{U}\}$.

Notation 16.13. Write $\mathcal{U}_r^t$ for the set of restrictions of admissible controls to the time interval $[r, t] \subset [0, 1]$.

Lemma 16.14. *Suppose, as above, that for $u \in \mathcal{U}_t^1$*

$$\varphi_t^u = E_u\left[\int_t^1 c(s, x_s, u_s)\, ds + g(x_1) \Big| \mathcal{F}_t\right].$$

Then the set of random variables $\{\varphi_t^u, u \in \mathcal{U}_t^1\} \subset L^1(\Omega, \mathcal{F}, P)$ has the ε-lattice property for $\varepsilon = 0$ (and so for every $\varepsilon > 0$).

PROOF. For $u_1, u_2 \in \mathcal{U}_t^1$ write

$$A = \{\omega : \varphi_t^{u_1} \leq \varphi_t^{u_2}\}.$$

Note that $A \in \mathcal{F}_t$. Because the admissible controls are $\mathcal{F}_t$ predictable, and so $\mathcal{F}_t$ adapted, they depend on complete information about the process. We can, therefore, define for $s \in]t, 1]$:

$$u_3(\omega, s) = \begin{cases} u_1(\omega, s), & \omega \in A, \\ u_2(\omega, s), & \omega \in A^c. \end{cases}$$

Then

$$\varphi_t^{u_3} \leq \varphi_t^{u_1} \wedge \varphi_t^{u_2} \quad \text{a.s.}$$

□

Corollary 16.15. *For any $u \in \mathcal{U}$ and $0 \leq r \leq t \leq 1$ the value function V_t satisfies the following principle of optimality:*

$$V_r \leq E_u\left[\int_r^t c(s, x_s, u_s)\, ds + V_t \Big| \mathcal{F}_r\right] \quad a.s.$$

PROOF.

$$\begin{aligned} V_r &= \bigwedge_{v \in \mathcal{U}_r^1} \varphi_r^u \\ &= \bigwedge_{v \in \mathcal{U}_r^1} E_v\left[\int_r^1 c_s^v\, ds + g(x_1) \Big| \mathcal{F}_r\right], \end{aligned}$$

where $c_s^v = c(s, x_s, v_s)$. Consider any fixed $u \in \mathcal{U}_r^t$ and the set $\mathcal{V}_r^1 \subset \mathcal{U}_r^1$ of all admissible controls which are equal to u on $]r, t]$. Then certainly

$$\begin{aligned} V_r &\leq \bigwedge_{v \in \mathcal{V}_r^1} E_v\left[\int_r^1 c_s^v\, ds + g(x_1) \Big| \mathcal{F}_r\right] \\ &= \bigwedge_{v \in \mathcal{U}_t^1} E_u\left[\int_r^t c_s^u\, ds + E_v\left[\int_t^1 c_s^v\, ds + g(x_1) \Big| \mathcal{F}_t\right] \Big| \mathcal{F}_r\right] \\ &= E_u\left[\int_r^t c_s^u\, ds \Big| \mathcal{F}_r\right] + \bigwedge_{v \in \mathcal{U}_t^1} E_u[\varphi_t^v | \mathcal{F}_r]. \end{aligned}$$

From Lemmas 16.11 and 16.14 we can interchange the order of the final infimum and expectation so that

$$\bigwedge_{v \in \mathcal{U}_t^1} E_u[\varphi_t^v | \mathcal{F}_r] = E_u\left[\bigwedge_{v \in \mathcal{U}_t^1} \varphi_t^u \middle| \mathcal{F}_r\right]$$
$$= E_u[V_t | \mathcal{F}_r],$$

and the result follows. □

Definition 16.16. For any $u \in \mathcal{U}$ the process

$$M_t^u = \int_0^t c_s^u \, ds + V_t$$

is the *cost process.*

Theorem 16.17. $\{M_t^u\}$ *is a* P_u *submartingale. Furthermore,* $\{M_t^u\}$ *is a* P_u *martingale if and only if control u gives the minimum expected cost, i.e. if and only if u is optimal.*

PROOF. For any $u \in \mathcal{U}$ we have from Corollary 16.15 if $0 \le r \le t \le 1$;

$$V_r \le E_u\left[\int_0^t c_s^u \, ds \middle| \mathcal{F}_r\right] + E_u[V_t | \mathcal{F}_r] \quad a.s.$$

Now $\int_0^r c_s^u \, ds$ is $\mathcal{F}_r$ measurable so adding this to each side above we have

$$M_r^u \le E_u[M_t^u | \mathcal{F}_r] \quad \text{a.s.}$$

That is, $\{M_t^u\}$ is a P_u submartingale.

However, for any $u \in \mathcal{U}$ we have that

$$M_1^u = \int_0^1 c_s^u \, ds + g(x_1),$$

that is M_1^u is the "sample cost" obtained by using control u. Also

$$M_0^u = V_0 = \inf_{v \in \mathcal{U}_0^1} \varphi_0^v,$$

the minimum expected cost. Consequently, if $\{M_t^u\}$ is a P_u martingale

$$E_u[M_1^u] = E_u[M_0^u] = V_0$$

and u is optimal. Conversely, if u is optimal then for any t:

$$V_0 = E_u\left[\int_0^t c_s^u \, ds + \varphi_t^u\right].$$

However, from Corollary 16.15,

$$V_0 \le E_u\left[\int_0^t c_s^u \, ds + V_t\right]$$

so

$$E_u[V_t - \varphi_t^u] \geq 0 \quad \text{a.s.}$$

By definition $V_t \leq \varphi_t^u$ a.s. and so $V_t = \varphi_t^u$ a.s. Consequently

$$M_t^u = E_u[M_1^u | \mathscr{F}_t]$$

and $\{M_t^u\}$ is a P_u martingale if u is optimal. □

Remark 16.18. The following result, that there is a corlol version of the cost process, is often glossed over. However, we give a careful proof based on that of Memin [61].

Theorem 16.19. *There is a corlol version of the P_u-submartingale $\{M_t^u\}$.*

Proof. Because the filtration $\mathscr{F}_t$ is right continuous (being generated by a continuous process), the result will follow from Theorem 4.6(3) if we can prove that $E_u[M_t^u]$ is right continuous.

For an admissible control $u \in \mathscr{U}$ write $\Pi_t u \in \mathscr{U}_0^t$ for the restriction of u to $[0, t]$. From Lemmas 16.11 and 16.14

$$\begin{aligned}
E_u[M_t^u] &= \int_\Omega \left\{ \int_0^t c_s^u \, ds + V_t \right\} dP_u \\
&= \int_\Omega E\left[\Lambda_{0,t}(u) \Lambda_{t,1}(u) \left\{ \int_0^t c_s^u \, ds + V_t \right\} \Big| \mathscr{F}_t \right] dP \\
&= \int_\Omega \Lambda_{0,t}(u) \left\{ \int_0^t c_s^u \, ds + V_t \right\} dP \\
&= \int_\Omega \left\{ \int_0^t c_s^u \, ds + V_t \right\} dP_{\Pi_t u} \\
&= \int_\Omega \int_0^t c_s^u \, ds \, dP_{\Pi_t u} + \inf_{\substack{v \in \mathscr{U} \\ \Pi_t v = \Pi_t u}} \int_\Omega E_v\left[\int_t^1 c_s^v \, ds + g(x_1) \Big| \mathscr{F}_t \right] dP_{\Pi_t u}.
\end{aligned}$$

For any $\varepsilon > 0$ there is, therefore, a control $v_1 \in \mathscr{U}$ such that $\Pi_t v_1 = \Pi_t u$ and

$$\int_\Omega \int_0^t c_s^u \, ds \, dP_{\Pi_t u} + \int_\Omega E_{v_1}\left[\int_t^1 c_s^{v_1} \, ds + g(x_1) \Big| \mathscr{F}_t \right] dP_{\Pi_t u} \leq E_u[M_t^u] + \varepsilon.$$

For any $h > 0$ consider the control v_2 defined by

$$v_2(s) = \begin{cases} u(s), & \text{for } 0 \leq s \leq t + h, \\ v_1(s), & \text{for } t + h < s \leq 1. \end{cases}$$

Then v_1 differs from v_2 only on $]t, t+h]$.
We establish that for small enough h

$$\Delta_{t,t+h} = \left| \int_\Omega \left\{ \int_0^1 c_s^{v_1} \, ds + g(x_1) \right\} dP_{v_1} - \int_\Omega \left\{ \int_0^1 c_s^{v_2} \, ds + g(x_1) \right\} dP_{v_2} \right| < \varepsilon.$$

Now

$$\Delta_{t,t+h} = \left| \int_\Omega \int_0^t c_s^u \, ds \, dP_{\Pi_t u} + \int_\Omega E_{v_1}\left[\int_t^1 c_s^{v_1} \, ds + g(x_1) \middle| \mathscr{F}_t \right] dP_{\Pi_t u} \right.$$

$$\left. - \int_\Omega \int_0^{t+h} c_s^u \, ds \, dP_{\Pi_{t+h} u} - \int_\Omega E_{v_2}\left[\int_{t+h}^1 c_s^{v_2} \, ds + g(x_1) \middle| \mathscr{F}_{t+h} \right] dP_{\Pi_{t+h} u} \right|.$$

However,

$$\int_\Omega E_{v_1}\left[\int_t^1 c_s^{v_1} \, ds + g(x_1) \middle| \mathscr{F}_t \right] dP_{\Pi_t u}$$

$$= \int_\Omega \Lambda_{0,t}(u) E\left[\Lambda_{t,1}(v_1) \left\{ \int_t^1 c_s^{v_1} \, ds + g(x_1) \right\} \middle| \mathscr{F}_t \right] dP$$

and

$$\int_\Omega E_{v_2}\left[\int_{t+h}^1 c_s^{v_2} \, ds + g(x_1) \middle| \mathscr{F}_{t+h} \right] dP_{\Pi_{t+h} u}$$

$$= \int_\Omega \Lambda_{0,t+h}(u) E\left[\Lambda_{t+h,1}(v_1) \left\{ \int_{t+h}^1 c_s^{v_1} \, ds + g(x_1) \right\} \middle| \mathscr{F}_{t+h} \right] dP.$$

Also,

$$\int_\Omega \int_0^{t+h} c_s^u \, ds \, dP_{\Pi_{t+h} u} = \int_\Omega \int_0^t c_s^u \, ds \, dP_{\pi_t u} + \int_\Omega \int_t^{t+h} c_s^u \, ds \, dP_{\Pi_{t+h} u}.$$

Therefore,

$$\Delta_{t,t+h} \leq \left| \int_\Omega \int_t^{t+h} c_s^u \, ds \, dP_{\Pi_{t+h} u} \right| + \left| \int_\Omega \Lambda_{0,t}(u) \Lambda_{t,1}(v_1) \left\{ \int_t^1 c_s^{v_1} \, ds + g(x_1) \right\} dP \right.$$

$$\left. - \int_\Omega \Lambda_{0,t+h}(u) \Lambda_{t+h,1}(v_1) \left\{ \int_{t+h}^1 c_s^{v_1} \, ds + g(x_1) \right\} dP \right|. \tag{16.4}$$

The cost functions c and g are bounded; suppose $|c|$ and $|g|$ are less than K. Then

$$\left| \int_\Omega \int_t^{t+h} c_s^u \, ds \, dP_{\Pi_{t+h} u} \right| \leq Kh. \tag{16.5}$$

Consider the second term in (16.4). This is less than

$$\left| \int_\Omega \Lambda_{0,t}(u) \Lambda_{t,1}(v_1) \int_t^{t+h} c_s^{v_1} \, dP \right| + \Gamma_{t,t+h},$$

where

$$\Gamma_{t,t+h} = \left| \int_\Omega \Lambda_{0,t}(u) (\Lambda_{t,t+h}(\nu_1) - \Lambda_{t,t+h}(u)) \Lambda_{t+h,1}(v_1) \left\{ \int_{t+h}^1 c_s^{v_1} \, ds + g(x_1) \right\} dP \right|.$$

As above,

$$\left|\int_\Omega \Lambda_{0,t}(u)\Lambda_{t,1}(v_1)\int_t^{t+h} c_s^{v_1}\,ds\,dP\right| \le Kh. \tag{16.6}$$

Now for $0 \le s \le 1-t$

$$\Lambda_{0,t}(u)\Lambda_{t,t+s}(v_1) \quad \text{and} \quad \Lambda_{0,t}(u)\Lambda_{t,t+s}(u)$$

are right continuous, uniformly integrable martingales, and

$$\lim_{s\to 0} \Lambda_{t,t+s}(v_1) = \lim_{s\to 0} \Lambda_{t,t+s}(u) = 1 \quad \text{a.s.}$$

Therefore,

$$\begin{aligned}\lim_{h\to 0}\Gamma_{t,t+h} &= \lim_{h\to 0}\left|\int_\Omega \Lambda_{0,t}(u)(\Lambda_{t,t+h}(v_1)-\Lambda_{t,t+h}(u))\right.\\ &\qquad \left.\times \Lambda_{t+h,1}(v_1)\left\{\int_{t+h}^1 c_s^{v_1}\,ds + g(x_1)\right\}dP\right|\\ &\le \lim_{h\to 0} 2K\int_\Omega |\Lambda_{0,t}(u)\Lambda_{t,t+h}(v_1)-\Lambda_{0,t}(u)\Lambda_{t,t+h}(u)|\,dP = 0,\end{aligned}$$

so there is an $h_1 > 0$ such that, if $0 \le h < h_1$, then

$$\Gamma_{t,t+h} \le \varepsilon/3. \tag{16.7}$$

Consequently, from (16.5), (16.6) and (16.7) we see there is an h_2 such that if $0 \le h < h_2$

$$\Delta_{t,t+h} \le \varepsilon.$$

Because $\{M_t^u\}$ is a P_u submartingale

$$\int_\Omega M_{t+h}^u\,dP_u \ge \int_\Omega M_t^u\,dP_u.$$

However, by the choice of v_1 and v_2,

$$\int_\Omega M_{t+h}^u\,dP_u \le \int_\Omega\left\{\int_0^1 c_s^{v_2}\,ds + g(x_1)\right\}dP_{v_2}$$

and

$$\int_\Omega M_t^u\,dP_u \ge \int_\Omega\left\{\int_0^1 c_s^{v_1}\,ds + g(x_1)\right\}dP_{v_1} - \varepsilon.$$

Therefore,

$$\begin{aligned}\int_\Omega M_{t+h}^u - M_t^u\,dP &= \left|\int_\Omega M_{t+h}^u - M_t^u\,dP\right|\\ &\le \Delta_{t,t+h} + \varepsilon,\end{aligned}$$

so for $0 \leq h < h_2$

$$|E_u[M^u_{t+h}] - E[M^u_t]| \leq 2\varepsilon.$$

Consequently, $E_u[M^u_t]$ is right continuous.

From Theorem 4.6(3) there is, therefore, a right continuous modification of $\{M^u_t\}$ and from Theorem 4.3 this modification is corlol. □

Corollary 16.20. *There is a corlol version of the process V_t.*

Remarks 16.21. Because both the running cost c and terminal cost g are bounded by K the submartingale $\{M^u_t\}$ is of class D and so has a Doob–Meyer decomposition. That is, there is a unique predictable increasing process $\{A^u_t\}$, with $A^u_0 = 0$, and a uniformly integrable $P_u$martingale $\{N^u_t\}$, with $N^u_0 = 0$, such that

$$M^u_t = W_0 + A^u_t + N^u_t. \tag{16.8}$$

The next step is to represent $\{N^u_t\}$ as a stochastic integral with respect to the Brownian motion $\{W^u_t\}$, (16.2). If the σ-field were generated by $\{W^u_t\}$, and if $\{N_t\}$ were square integrable, we could immediately apply Theorem 12.33 and Remark 12.34. However, our solution $\{x_t\}$, (16.1), is only a weak solution, so $\{W^u_t\}$ does not necessarily generate $\{\mathscr{F}_t\}$. Nevertheless, it was shown by Fujisaki *et al.* [41] that all square integrable $\{\mathscr{F}_t\}$ martingales are stochastic integrals of $\{W^u_t\}$. We now give the proof of this result following Davis and Varaiya [21] in the following general situation.

First note because $dB_t = \sigma(t, x)^{-1}\, dx_t$, and σ is Lipschitz that $\sigma\{B_s : s \leq t\} = \sigma\{x_s : s \leq t\}$, so we can suppose the filtration $\{\mathscr{F}_t\}$ is generated by the process $\{x_t\}$. For the proof of the next theorem we shall assume that on a probability space $(\Omega, \mathscr{F}, P)$ we have a process $\{x_t\}$, $0 \leq t \leq 1$, such that, if $\{\mathscr{F}_t\}$ is the filtration generated by $\{x_t\}$, $\{W_t\}$ is an $\{\mathscr{F}_t\}$ Brownian motion and that under P, for $x_0 \in R^n$,

$$x_t = x_0 + \int_0^t f(s, x)\, ds + \int_0^t \sigma(s, x)\, dW_s \quad \text{a.s.}$$

Here f and σ satisfy measurability and growth conditions similar to those of f and σ above.

We can then prove the following result.

Theorem 16.22. *Suppose $\{N_t\}$, $N_0 = 0$, is a square integrable P martingale with respect to the filtration $\{\mathscr{F}_t\}$. Then there is an $\mathscr{F}_t$ predictable process $\{\gamma_t\}$ such that*

$$\int_0^1 E_u|\gamma_t|^2\, dt < \infty$$

and

$$N_t = \int_0^t \gamma_s \, dW_s \quad a.s.$$

PROOF. For $n \in Z^+$ define

$$T_n = \min\left\{1, \inf\left(t: \int_0^t |\sigma_s^{-1} f_s|^2 \, ds \geq n\right\}\right).$$

Clearly T_n is an $\mathscr{F}_t$ stopping time and $\lim T_n = 1$ a.s. from the conditions on σ^{-1} and f. Write

$$\Lambda_t^* = \exp\left(-\int_0^t \sigma_s^{-1} f_s \, dW_s - \tfrac{1}{2}\int_0^t |\sigma_s^{-1} f_s|^2 \, ds\right),$$

and define a measure P_n^* by

$$\frac{dP_n^*}{dP} = \Lambda_{t \wedge T_n}^*.$$

From Girsanov's theorem (Corollary 13.25), P_n^* is a probability measure for each n and the process

$$z_t^n = W_t + \int_0^{t \wedge T_n} \sigma_s^{-1} f_s \, ds$$

is a Brownian motion under P_n^*. Write

$$\mathscr{Z}_t^n = \sigma\{z_s^n : 0 \leq s \leq t\}.$$

From Theorem 12.33 and Remark 12.34 if $\{\tilde{N}_t\}$ is a square integrable, centred P_n^* martingale with respect to the filtration $\{\mathscr{Z}_t^n\}$ there is a $\{\mathscr{Z}_t^n\}$ predictable process $\{\phi_t^n\}$ such that

$$\tilde{N}_t = \int_0^t \phi_s^n \, dz_s^n \quad \text{a.s.}$$

However, for $t < T_n$

$$z_t^n = \int_0^t \sigma_s^{-1}(\sigma_s \, dW_s + f_s \, ds) = \int_0^t \sigma_s^{-1} \, dx_s,$$

so

$$\mathscr{Z}_t^n = \mathscr{F}_{t \wedge T_n}.$$

Consequently we have established that if $\{\tilde{N}_t\}$ is a square integrable, centred P_n^* martingale with respect to the filtration $\{\mathscr{F}_{t \wedge T_n}\}$ then

$$\tilde{N}_{t \wedge T_n} = \int_0^{t \wedge T_n} \phi_s^n \sigma_s^{-1} \, dx_s \quad \text{a.s.} \tag{16.9}$$

By the optional stopping Theorem 4.12 and Lemma 13.10, if $\{N_t\}$ is a square integrable, centred P martingale with respect to the filtration $\{\mathscr{F}_t\}$ then $\{\tilde{N}_t\}$ is a square integrable, centred P_n^* martingale with respect to the filtration $\{\mathscr{F}_{t\wedge T_n}\}$ where

$$\tilde{N}_t = (\Lambda^*_{t\wedge T_n})^{-1} N_{t\wedge T_n}.$$

In this situation we certainly have that $\tilde{N}_t = \tilde{N}_{t\wedge T_n}$, so from (16.9)

$$\begin{aligned}\tilde{N}_t &= \int_0^{t\wedge T_n} \phi_s^n \sigma_s^{-1}\, dx_s \\ &= \int_0^{t\wedge T_n} \phi_s^n\, dW_s + \int_0^{t\wedge T_n} \phi_s^n \sigma_s^{-1} f_s\, ds.\end{aligned}$$

Also, $\Lambda_t^* = \mathscr{E}(-\int_0^t \sigma_s^{-1} f_s\, dW_s)$ so from Theorem 13.5

$$\Lambda^*_{t\wedge T_n} = 1 - \int_0^{t\wedge T_n} \Lambda_s^* \sigma_s^{-1} f_s\, dW_s.$$

Applying the Ito product formula

$$\begin{aligned}N_{t\wedge T_n} &= \tilde{N}_t \,.\, \Lambda^*_{t\wedge T_n} \\ &= \int_0^{t\wedge T_n} \tilde{N}_s\, d\Lambda_s^* + \int_0^{t\wedge T_n} \Lambda_s^*\, d\tilde{N}_s + \langle \tilde{N}, \Lambda^*\rangle_{t\wedge T_n} \\ &= -\int_0^{t\wedge T_n} \tilde{N}_s \Lambda_s^* \sigma_s^{-1} f_s\, dW_s + \int_0^{t\wedge T_n} \Lambda_s^* \phi_s^n \sigma_s^{-1}\, dx_s \\ &\quad -\int_0^{t\wedge T_n} \Lambda_s^* \phi_s^n \sigma_s^{-1} f_s\, ds \\ &= \int_0^{t\wedge T_n} \gamma_s^n\, dW_s,\end{aligned}$$

where $\gamma_s^n = \Lambda_s^*(\phi_s^n - \tilde{N}_s \sigma_s^{-1} f_s)$. Furthermore,

$$E[N^2_{t\wedge T_n}] = \int_0^t E[(I_{[\![0,T_n]\!]}\gamma_s^n)^2]\, ds \le E[N_1^2] < \infty.$$

This representation is unique, so that if $m \ge n$

$$\gamma_s^n = \gamma_s^m \quad \text{for } s \le T_n.$$

Define γ_s to be the process such that on $[\![0, T_n]\!]$ $\gamma_s = \gamma_s^n$. Then

$$N_{t\wedge T_n} = \int_0^{t\wedge T_n} \gamma_s\, dW_s,$$

so

$$N_t = \int_0^t \gamma_s\, dW_s$$

for $t < T_n$. However,

$$\lim T_n = 1 \quad \text{a.s.} \quad \text{so} \quad \int_0^1 E_u[|\gamma_s|^2]\, ds < \infty$$

and the result follows. □

Corollary 16.23. *The uniformly integrable martingale $N_t^u, N_0 = 0$, which occurs in the Doob–Meyer decomposition* (16.8) *has a representation*

$$N_t^u = \int_0^t \gamma_s \, dW_s^u \quad a.s.$$

Here $\{\gamma_s\}$ is a predictable process for which there is an increasing sequence of stopping times $\{T_n\}$ with $\lim T_n = 1$ *a.s. and $E[\int_0^{T_n} |\gamma_s|\, ds] < \infty$.*

PROOF. First note that the character of a stopping time is independent of the probability measure, so because $\{\mathscr{F}_t\}$ is generated by a Brownian motion the proof of Theorem 12.33 shows that all $\{\mathscr{F}_t\}$ stopping times are predictable. Consequently the filtration is quasi-left continuous: $\mathscr{F}_{T-} = \mathscr{F}_T$ for any predictable stopping time T. Therefore, as in Theorem 12.33

$$\begin{aligned} N_{T-}^u &= E[N_T^u | \mathscr{F}_{T-}] = E[N_T^u | \mathscr{F}_T] \\ &= N_T^u \end{aligned}$$

for any stopping time and so $\{N_t^u\}$ has continuous paths. Consider the sequence of stopping times $\{T_n\}$ defined by

$$T_n = \inf\{t: |N_t^u| \geq n\}.$$

Then $\{T_n\}$ is an increasing sequence and $\lim T_n = 1$ a.s. If $N_t^n = N_{t \wedge T_n}^u$ then $\{N_t^n\}$ is a square integrable martingale so

$$N_t^n = N_{t \wedge T_n}^u = \int_0^t \gamma_s^n \, dW_s^u$$

for a predictable integrand $\{\gamma_s^n\}$ such that $\int_0^1 E[|\gamma_s^n|^2]\, ds < \infty$. Because N_t^n is constant after T_n, γ_s^n is zero after T_n and this integral is in fact $E[\int_0^{T_n} |\gamma_s^n|^2 \, ds]$. Again this representation is unique, so if $m \geq n$, $\gamma_s^n = \gamma_s^m$ for $s \leq T_n$ and an integrand with the stated properties is obtained by putting

$$\gamma_s = \gamma_s^n \quad \text{on } [\![0, T_n]\!].$$ □

Remarks 16.24. Note that

$$N_1^n = N_{T_n}^u \quad \text{and that as } n \to \infty,$$

$$N_{T_n}^u \to N_1^u \quad \text{a.s.}$$

Consequently we can represent

$$N_1^u \quad \text{as} \int_0^1 \gamma_s \, dW_s^u.$$

Furthermore, $\int_0^1 \gamma_s^2 \, ds < \infty$ a.s.

We can now establish a necessary condition for a control $u^* \in \mathcal{U}$ to be optimal. In fact, similarly to the Pontrjagin minimum principle [40] of deterministic control theory we show that if a control u^* is optimal it minimizes almost surely (Lebesgue $\times P$) a certain Hamiltonian. See [31].

Theorem 16.25. *Suppose $u^* \in \mathcal{U}$ is optimal. Then, if $\{\gamma_t\}$ is the process of Corollary 16.23 above, for almost every $(t, x) \in [0, 1] \times \Omega$*

$$\gamma_t \sigma_t^{-1} f_t^{u^*} + c_t^{u^*} = \inf_{u \in U} \gamma_t \sigma_t^{-1} f_t^u + c_t^u.$$

PROOF. We have above for any $u \in \mathcal{U}$ the representation

$$\begin{aligned} M_t^u &= \int_0^t c_s^u \, ds + W_t \\ &= W_0 + A_t^u + N_t^u \\ &= W_0 + A_t^u + \int_0^t \gamma_s \, dW_s^u. \end{aligned}$$

If $u \in \mathcal{U}$ is any other admissible control and $H_s(u_s) = \gamma_s \sigma_s^{-1} f_s^u + c_s^u$:

$$\begin{aligned} M_t^v &= \int_0^t c_s^v \, ds + W_t \\ &= W_0 + \left[A_t^u + \int_0^t (H_s(v_s) - H_s(u_s)) \, ds \right] + \int_0^t \gamma_s \, dW_s^v \end{aligned} \tag{16.10}$$

by the above representation for W_t. However, the submartingale $\{M_t^v\}$ also has a Doob–Meyer decomposition

$$M_t^v = W_0 + A_t^v + N_t^v.$$

Certainly $\{M_t^v\}$ is a special semimartingale, in the sense of Definition 12.37, and so from Theorem 12.38 the local martingale and predictable bounded variation processes in these two decompositions must be the same. That is

$$\begin{aligned} N_t^v &= \int_0^t \gamma_s \, dW_s^v, \\ A_t^v &= A_t^u + \int_0^t (H_s(v_s) - H_s(u_s)) \, ds. \end{aligned} \tag{16.11}$$

Here A_t^v is a predictable increasing process.

Suppose now that $u^* \in \mathcal{U}$ is optimal. Then from Theorem 16.17 $\{M_t^{u^*}\}$ is a P_{u^*} martingale so certainly $A_t^{u^*} = 0$. That is, the increasing process A_t^v has a representation

$$A_t^v = \int_0^t (H_s(v_s) - H_s(u_s^*)) \, ds$$

and consequently the integrand must, almost surely (Lebesgue $\times P$), be positive:

$$H_s(u_s^*) \leq H_s(v_s) \quad \text{a.s.}$$

for any other admissible $v \in \mathscr{U}$. □

Remark 16.26. We note that a further consequence of the unique decomposition and representation (16.11) of N_t^v is that the integrand process $\{\gamma_s\}$ is independent of the control. A sufficient condition for optimality is now established.

Theorem 16.27. *For a control $u^* \in \mathscr{U}$ consider the P_{u^*} martingale*

$$J_t = E_{u^*}\left[\left[\int_0^1 c_s^{u^*}\, ds + g(x_1)\middle|\mathscr{F}_t\right]\right] = E_{u^*}[M_1^{u^*} | \mathscr{F}_t] = \int_0^t c_s^{u^*}\, ds + \varphi_t^{u^*}.$$

Then u^* is optimal if for any other $v \in \mathscr{U}$ the process

$$I_t^v = \int_0^t c_s^v\, ds + \varphi_t^{u^*}$$

is a P_v submartingale.

PROOF. This is immediate because then

$$J_0 = J(u^*) = \varphi_0^{u^*} = E_v[I_0^v]$$
$$\leqq E_v[I_1^v] = J(v).$$

That is, u^* minimizes the total expected cost. □

Corollary 16.28. *We can localize this condition as follows:*

The (square integrable) P_{u^*} martingale J_t has a representation as a stochastic integral

$$J_t = J(u^*) + \int_0^t \tilde{\gamma}_s\, dW_s^{u^*}$$

by Theorem 16.22. Write

$$\tilde{H}_s(u_s) = \tilde{\gamma}_s \sigma_s^{-1} f_s^u + c_s^u \quad \text{for } u \in \mathscr{U}.$$

Then if, almost surely Lebesgue $\times P$, for any $v \in \mathscr{U}$

$$\tilde{H}_s(u_s^*) \leq \tilde{H}_s(v_s) \quad a.s.$$

u^* is optimal.

PROOF. Clearly

$$J_1 = \int_0^1 c_s^{u^*}\, ds + g(x_1)$$

$$= J(u^*) + \int_0^1 \tilde{\gamma}_s\, dW^{u_s^*}$$

$$= J(u^*) + \int_0^1 \tilde{g}_s\, dW_s^v + \int_0^1 \tilde{\gamma}_s \sigma_s^{-1}(f_s^v - f_s^{u^*})\, ds.$$

We can, therefore, write for any $v \in \mathcal{U}$:

$$I_1^v = \int_0^1 c_s^v \, ds + g(x_1)$$

$$= J(u^*) + \int_0^1 \tilde{\gamma}_s \, dW_s^v + \int_0^1 (\tilde{H}_s(v_s) - \tilde{H}_s(u_s^*)) \, ds.$$

However, we only know the integrals in the final expression are, respectively, local martingales and locally of integrable variation. That is, there is an increasing sequence of stopping times $\{T_n\}$, $\lim T_n = 1$ a.s. such that $\{I_{t \wedge T_n}^v\}$ is a P_v submartingale and

$$E_v[I_{1 \wedge T_n}^v] \geq J(u^*).$$

However, the process I_t^v is uniformly bounded by $2K$ and $I_{1 \wedge T_n}^v$ converges almost surely to I_1^v. Therefore,

$$\lim_{n \to \infty} E_v[I_{1 \wedge T_n}^v] = J(v).$$

That is, $J(v) \geq J(u^*)$ and u^* is optimal. □

Remark 16.29. In this case

$$J_t = M_t^{u^*}$$

and $\gamma_s = \tilde{\gamma}_s$ a.s.

As noted above, the integrand $\{\gamma_s\}$ in the representation of $\{N_t^u\}$ is independent of the control $u \in \mathcal{U}$. This suggests that an optimal control can be constructed by minimizing the Hamiltonian $H_s(u_s)$. To establish this result we first quote two results of Beneš. The first, Lemma 1 of [3], is an implicit function result.

Lemma 16.30. *Suppose $(M, \mathcal{M})$ is a measure space, A and U are separable metric spaces with their Borel fields, and that U is compact. Let $k: M \times U \to A$ be continuous in its second argument for each value of the first, and $\mathcal{M}$ measurable in the first for each value of the second. Let $y: M \to A$ be $\mathcal{M}$ measurable with*

$$y(x) \in k(x, U) \quad \textit{for all } x \in \mathcal{M}.$$

Then there is an $\mathcal{M}$ measurable function

$$u: M \to U,$$

such that $y(x) = k(x, u(x))$.

Notation 16.31. Write Φ for the set of functions from $[0, 1] \times \Omega \to R^n$ such that, if $\phi \in \Phi$:

(a) ϕ is $\mathcal{F}_t$ predictable,
(b) $|\phi(t, x)| \leq K(1 + |x|_t^*)$.

For $\phi \in \Phi$ we can construct the random variables

$$\Lambda_{r,t}^{\phi} = \exp\left\{\int_r^t (\sigma^{-1}(s,x)\phi(s,x))\,dB_s - \tfrac{1}{2}\int_r^t |\sigma_s^{-1}\phi(s,x)|^2\,ds\right\}.$$

Again the growth condition on ϕ implies that $E[\Lambda_{0,1}^{\phi}] = 1$, so that a new probability measure P_ϕ can be defined on $(\Omega, \mathscr{F}_1)$ by putting

$$\frac{dP_\phi}{dP} = \Lambda_{0,1}^{\phi}.$$

Write $\mathscr{D}$ for the set of random variables of the form $\Lambda_{0,1}^{\phi}$, $\phi \in \Phi$. Lemma 1 of Beneš [4] shows that $\mathscr{D}$ is a uniformly integrable subset of $L^1(\Omega, \mathscr{F}, P)$, and so $\mathscr{D}$ is relatively compact in the weak topology. Theorem 2 of Duncan and Varaiya [28] shows that $\mathscr{D}$ is convex and strongly closed. Consequently we have the following result:

Lemma 16.32. *$\mathscr{D}$ is a weakly compact subset of $L^1(\Omega, \mathscr{F}, P)$.*

Remarks 16.33. For $p \in R^n$ and $(t, x, u) \in [0, 1] \times \Omega \times U$ consider the Hamiltonian

$$\mathscr{H}'(t, x, p, u) = p \cdot f(t, x, u) + c(t, x, u).$$

Because f and c are continuous in $u \in \mathscr{U}$ for each (t, x) there is a control value in $u'(t, x, p) \in U$ which minimizes $\mathscr{H}'(t, x, p, u)$. That is

$$\mathscr{H}'(t, x, p, u'(t, x, p)) = \min_{u \in \mathscr{U}} \mathscr{H}'(t, x, p, u).$$

Consider now the predictable integrand process $\{\gamma_t\}$ of Corollary 16.23. Our candidate for an optimal control is the process

$$u_t^* = u'(t, x, \gamma_t(x)).$$

We first show that u^* is an admissible control by establishing that u' can be chosen in a suitably measurable way.

Lemma 16.34.

$$u_t^* = u'(t, x, \gamma_t(x))$$

can be selected so that it is an admissible control.

PROOF. For fixed (t, x, p), $\mathscr{H}'(t, x, p, u)$ is continuous in $u \in U$. For fixed u and $p \in R^n$, $\mathscr{H}'(t, x, p, u)$ is $\mathscr{F}_t$ predictable, so with $\mathscr{B}$ denoting the Borel field on R^n, for fixed u, $\mathscr{H}'(t, x, p, u)$ is measurable with respect to $\mathscr{M} = \Sigma_p \times \mathscr{B}$. Suppose S is a countable dense subset of U. Then, writing $H(t, x, p) = \min_{u \in U} \mathscr{H}'(t, x, p, u)$,

$$\mathscr{H}(t, x, p) = \inf_{u \in S} \mathscr{H}'(t, x, p, u).$$

Therefore, for any $a \in R$

$$\{(t, x, p): \mathcal{H}(t, x, p) < a\} = \bigcup_{u \in S} \{(t, x, p): \mathcal{H}'(t, x, p, u) < a\}.$$

Consequently, $\mathcal{H}(t, x, p)$ is $\mathcal{M}$ measurable. Certainly for each (t, x, p)

$$\mathcal{H}(t, x, p) \in \mathcal{H}'(t, x, p, U),$$

so from Lemma 16.30 there is on $\mathcal{M}$ measurable function

$$u' : [0, 1] \times \Omega \times R^n \to U,$$

such that

$$\mathcal{H}(t, x, p) = pf(t, x, u'(t, x, p)) + c(t, x, u'(t, x, p)).$$

The map

$$\varphi : ([0, 1] \times \Omega, \Sigma_p) \to ([0, 1] \times \Omega \times R^n, \mathcal{M})$$

defined by

$$\varphi(t, x) = (t, x, \gamma_t(x))$$

is measurable, so

$$u_t^* = u'(t, x, \gamma_t(x))$$

is $\mathcal{F}_t$ predictable. That is, $u^* \in \mathcal{U}$ and

$$\mathcal{H}(t, x, \gamma_t(x)) = \mathcal{H}'(t, x, p, u'(t, x, \gamma_t(x)). \qquad \square$$

Theorem 16.35. *The admissible control* $u_t^* = u'(t, x, \gamma_t(x))$ *is optimal.*

PROOF. For any other admissible control $v \in \mathcal{U}$

$$H_s(v_s) \geq H_s(u_s^*) \quad \text{a.s.}$$

From Theorem 16.25 and equation (16.10)

$$\begin{aligned} M_t^v &= W_0 + A_t^{u^*} + \int_0^t (H_s(v_s) - H_s(u_s^*))\, ds + \int_0^t \gamma_s\, dW_s^v \\ &\geq W_0 + A_t^{u^*} + \int_0^t \gamma_s\, dW_s^v. \end{aligned}$$

Taking the expectation at $t = 1$:

$$J(v) \geq W_0 + E_v[A_1^{u^*}]. \tag{16.12}$$

We shall show that $A_1^{u^*} = 0$ a.s. Suppose $\{u_n\}$ is a sequence of admissible controls such that

$$\lim_n J(u_n) = W_0 = \inf_{u \in U} J(u). \tag{16.13}$$

Then for any positive integer N

$$E_{u_n}[A_1^{u^*} \wedge N] = E[\Lambda_{0,1}(u_n)(A_1^{u^*} \wedge N)].$$

From (16.12) and (16.13) this must converge to zero, because $A_t^{u^*}$ is a positive increasing process, so

$$E_{u_n}[A_1^{u^*} \wedge N] \geq 0 \quad \text{for all } n.$$

However, by Lemma 16.32 there is a subsequence of $\{\Lambda_{0,1}(u_n)\}$ which converges weakly to some $\Lambda \in \mathscr{D} \subset L^1(\Omega, \mathscr{F}, P)$. Therefore,

$$\lim_n E_{u_n}[A_1^{u^*} \wedge N] = E[\Lambda(A_1^{u^*} \wedge N].$$

Certainly $\Lambda > 0$ a.s. so this implies that $A_1^{u^*} \wedge N = 0$ a.s. Consequently $A_1^{u^*} = 0$ a.e. and $A_t^{u^*} = 0$ a.e. for $t \in [0, 1]$.

The process $\{M_t^{u^*}\}$ can be written

$$\begin{aligned} M_t^{u^*} &= W_0 + A_t^{u^*} + N_t^{u^*} \\ &= W_0 + N_t^{u^*}, \end{aligned}$$

so taking the expectation at $t = 1$ we see that

$$J(u^*) = W_0$$

and u^* is optimal. □

Remarks 16.36. The above result due to Davis [11] is much stronger than any obtained so far in deterministic control theory. This is because the noise "smooths out" the process. The existence of an optimal control was originally established by Beneš [4], and Duncan and Varaiya [28], under the hypothesis that the set $f(t, x, U)$ is convex. This implies the convexity, and weak closure, of the set $\{\Lambda_{0,1}(u)\}$ of attainable densities.

In the deterministic case the running cost can be combined with a purely terminal cost by introducing an additional state variable x^0. However, in the stochastic case above an extra, independent Brownian motion $\{W_t^0\}$ must also be introduced, so that

$$dx_t^0 = c(t, x, u)\, dt + dW_t^0.$$

However, this must not be done before the proof of results such as the above Theorem 16.35 because there is then no way of showing that the optimal control u^* does not depend on $\{W_t^0\}$.

To actually construct the optimal control u^* we must determine the integrand process $\{\gamma_t\}$. This is very difficult in general, but some progress has been made by Haussmann [44–46]. Suppose u^* is the optimal control and that the random variable

$$M_1^{u^*} = \int_0^1 c(s, x, u^*(s, x))\, ds + g(x_1)$$

is Frechet differentiable in $x \in \Omega$. Then for each $x \in \Omega$ there is, by the Riesz representation theorem, an R^n valued Radon measure μ_x such that for $y \in \Omega$

$$M_1^{u^*}(x+y) = M_1^{u^*}(x) + \int_{[0,1]} y(s)\mu_x(ds) + O(\|y\|).$$

Because u^* is optimal we also have

$$M_t^u = J(u^*) + \int_0^t \gamma_s \, dW_s^{u^*}.$$

Under additional smoothness assumptions Haussmann [44, 45], proves that

$$\gamma_t = E_{u^*}\left[\left[\int_{]t,1]} \mu'_x(ds)\Phi(s,t)\,\middle|\,\mathscr{F}_t\right]\right],$$

where $\Phi(s, t)$ is the (random) fundamental solution of a linearized form of the dynamical equations with control u^*.

Partial Information 16.37

Suppose the state vector $\{x_t\}$ is made up of two sets of components $x_t = (y_t, z_t)$, where $y_t \in R^m$, $z_t \in R^{n-m}$, and that only the first can be observed by the controller. Write

$$\mathscr{Y}_t = \sigma\{y_s : s \le t\}.$$

The set of admissible controls $\mathscr{K}$ now comprises those which are predictable (and so certainly adapted), with respect to the filtration $\mathscr{Y}_t$.

We shall state, without proof, a necessary condition for optimality obtained in [34]. Suppose $u^* \in \mathscr{K}$ is optimal and again write

$$\varphi_t^{u^*} = E_{u^*}\left[\left[\int_t^1 c_s^{u^*}\, ds + g(x_1)\,\middle|\,\mathscr{F}_t\right]\right],$$

$$M_t^u = \int_0^t c_s^u \, ds + \varphi_t^{u^*} \quad \text{for } u \in \mathscr{K}.$$

Then $\{M_t^{u^*}\}$ is an $(\mathscr{F}_t, P_{u^*})$ martingale so

$$E_{u^*}[M_t^{u^*} | \mathscr{Y}_t] \text{ is a } (\mathscr{Y}_t, P_{u^*})$$

martingale. Using the principle of optimality

$$E_{u^*}[M_t^u | \mathscr{Y}_t] \le E_{u^*}[E_u[M_{t+h}^u | \mathscr{F}_t] \mathscr{Y}_t] \quad \text{a.e.} \tag{16.14}$$

for any $u \in \mathscr{K}$ and any $h > 0$. $\{M_t^{u^*}\}$ has a representation as a stochastic integral

$$M_t^{u^*} = J(u^*) + \int_0^t \gamma_s \, dW_s^{u^*}, \tag{16.15}$$

where $\{\gamma_s\}$ is a $\mathcal{F}_t$ predictable process. As in Theorem 16.25 we can then write

$$M_t^u = J(u^*) + \int_0^t \gamma_s \, dW_s^u + \int_0^t \Delta H_s^*(u) \, ds, \tag{16.16}$$

where

$$\Delta H_s^*(u) = H_s(u_s) - H_s(u_s^*)$$
$$= (\gamma_s f(s, x, u_s) + c(s, x, u_s)) - (\gamma_s f(s, x, u_s^*) + c(s, x, u_s^*)).$$

From (16.14) and (16.16) we have that

$$(1/h) E_{u^*}\left[E_u\left(\int_t^{t+h} \Delta H_s^*(u) \, ds \middle| \mathcal{F}_t \right) \middle| \mathcal{Y}_t \right] \geq 0 \quad \text{a.e.}$$

A delicate argument in [34] shows that we can take the limit in this expression as $h \to 0$ to obtain the following result:

Theorem 16.38. *Suppose $u^* \in \mathcal{K}$ is an optimal (partially observable) control. Then there is a set $T \subset [0, 1]$ of zero Lebesgue measure such that for $t \notin T$:*

$$E_{u^*}[\gamma_t f(t, x, u_t^*) + c(t, x, u_t^*) | \mathcal{Y}_t] \leq E_{u^*}[\gamma_t f(t, x, u_t) + c(t, x, u_t) | \mathcal{Y}_t] \quad \textit{a.e.}$$

for any other $u \in \mathcal{K}$. Here $\{\gamma_t\}$ is the process of (16.15).

Remarks 16.39. Compared with [21] the advantage of this result is that the condition is stated in terms of a single "adjoint process" $\{\gamma_t\}$, rather than a different such process for each control $u \in \mathcal{K}$.

A sufficient condition for $u^* \in \mathcal{K}$ to be optimal, discussed in [36], is that (using the above notation),

$$E_u[H_t(u_t^*) | \mathcal{Y}_t] \leq E_u[H_t(u_t) | \mathcal{Y}_t] \quad \text{a.e.}$$

for every $u \in \mathcal{K}$.

A disadvantage of the above formulation is that in a partially observed control problem the control variables should really be functions of the conditional distribution of the state $\{x_t\}$ given the observations so far, $\mathcal{Y}_t$. This suggests the so-called "separation principle" approach, in which the best estimate, $\hat{x}_t = E[x_t | \mathcal{Y}_t]$, of the state given the observations is first derived, using filtering theory, and then a new "completely observable" control problem, with state $\hat{x}_t$, is discussed. Whilst this approach is successful for partially observable problems with linear dynamics and quadratic cost [(40, 78]), serious technical difficulties arise in the general case. Indeed, the existence of an optimal control in the general partially observed situation is still an open problem, Using a wider class of randomized controls some results have been obtained by Fleming [38], Fleming and Pardoux [39] and Haussmann [47]. Reference [17] is a survey by Davis of the separation principle approach to the partially observed control problem.

Example 16.40 (The Predicted Miss Problem). As an example of the above martingale formulation for the optimal control problem we now describe the treatment of Davis and Clark [18] of the "predicted miss" problem. Here the dynamics are described by a linear system, and the control values are restricted to the product interval $[-1, 1]^r$. The objective is to steer the system to a given hyperplane at the fixed terminal time 1. There is a natural candidate for the optimal control: in Beneš' [5] phrase it is full "bang" to reduce predicted miss. Because this candidate optimal control is not smooth the classical approach, of studying the Bellman dynamic programming equation, cannot be used. However, the above martingale techniques can be applied.

Dynamics 16.41. Suppose $A(t)$, $B(t)$, $C(t)$, are, respectively, $n\times n$, $r\times n$ and $n\times n$ valued functions for $t\in[0, 1]$, with $C(t)C'(t)$ positive definite. As above, $\Omega=\mathscr{C}([0, 1], R^n)$. Let the control set U be $[-1, 1]^r\subset R^r$. Then, with admissible controls $\mathscr{U}$ as defined in Section 16.3, for any $u\in\mathscr{U}$ the system equation is

$$\begin{aligned} dx_t &= A(t)x_t\,dt+B(t)u_t\,dt+C(t)\,dW_t, \\ x_0 &= \xi\in R^n. \end{aligned} \tag{16.17}$$

Here $\{W_t\}$ is an n-dimensional Brownian motion.

For any $u\in\mathscr{U}$

$$W_t^u=\int_0^t C^{-1}(s)\,dx_s-\int_0^t C^{-1}(s)(A(s)x_s+B(s)u_s)\,ds$$

is an n-dimensional Brownian motion under the probability measure P_u.

Cost 16.42. Suppose $k\in R^n$ and $\lambda: R\to R^+$ is such that

(a) $\lambda(\rho)=\lambda(-\rho)$ for all $\rho\in R$,
(b) λ is continuously differentiable and

$$\frac{d\lambda(\rho)}{d\rho}\ge 0 \quad \text{for all } \rho\ge 0,$$

(c) $\lambda(\rho)=0(\exp\mu|\rho|)$ for some $\mu>0$.

Corresponding to control $u\in\mathscr{U}$ the total expected cost is then

$$J(u)=E_u[\lambda(\langle k, x_1\rangle)].$$

Here and below, $\langle a, b\rangle$ denotes the Euclidean inner product in R^n or R^r. That is, the objective of the controller is to minimize the distance of x_t from the hyperplane $\{X: \langle k, x\rangle=0\}$ at the final time $t=1$. Condition (c) implies $J(u)$ is finite for all $u\in\mathscr{U}$. Write Φ for the transition function corresponding to A. Then if $x_t=x\in R^n$ and the zero control is used over time interval $[t, 1]$ the expected miss distance is

$$E[\langle k, x_1\rangle|x_t=x]=\langle k, \Phi(1, t)x\rangle.$$

Define

$$s(t) = \Phi'(1, t)k \in R^n.$$

Then

$$\frac{ds(t)}{dt} = -A'(t)s(t) \quad \text{and} \quad s(1) = k. \tag{16.18}$$

Define

$$m_t = E[\langle k, x_1\rangle | x_t] = \langle s(t), x_t\rangle.$$

Then, writing $b(t) = B'(t)s(t) \in R^r$,

$$dm_t = \langle b(t), u_t\rangle\, dt + \langle C'(t)s(t), dW_t^u\rangle \tag{16.19}$$

and

$$m_0 = \langle s(0), \xi\rangle,$$

from (16.17) and (16.18). The cost can be written

$$J(u) = E_u[\lambda(m_1)].$$

Because $-1 \le u_t^i \le 1$ for $1 \le i \le r$ this expression suggests that the optimal control might be

$$u_t^{*i} = -\text{sign}(m_t b^i(t)).$$

This control is certainly admissible. If used, the drift term in (16.19) becomes

$$\langle b(t), u_t^*\rangle = -\text{sign}(m) \sum_{i=1}^{r} |b^i(t)|.$$

Furthermore, write

$$v_t = \int_0^t \langle C'(r)s(r), dW_r^{u^*}\rangle$$

and

$$\tau(t) = \inf\left\{\sigma : \int_0^{\sigma} \langle C'(r)s(r), C'(r)s(r)\rangle\, dr \ge t\right\}.$$

Then

$$\tilde{v}_t = v_{\tau(t)}$$

is a standard Brownian motion. If $\tilde{m}_t = m_{\tau(t)}$ then

$$\begin{aligned} d\tilde{m}_t &= \phi(\tau(t), \tilde{m}_t)\dot{\tau}(t)\, dt + d\tilde{v}_t \\ &= -\sum_i |b^i(\tau(t))|(|C'(\tau(t))s(\tau(t))|)^{-2} \,\text{sign}\, \tilde{m}_t\, dt + d\tilde{v}_t \\ &= -\tilde{a}(t)\, \text{sign}\, \tilde{m}_t\, dt + dv_t, \quad \text{say.} \end{aligned}$$

Writing $T = \tau(1)$, $\tilde{a}(t)$ is bounded on $[0, T]$ because CC' is positive definite. Because $\{v_t\}$ is a Brownian motion under a deterministic time change, and $\tilde{m}$ then satisfies a similar equation, we shall, without loss of generality, suppose that $\{v_t\}$ is a standard Brownian motion.

Equation (16.19) can now be written

$$dm_t = \langle b(t), u_t^* \rangle \, dt + dv_t.$$

Consider the square integrable martingale

$$J_t = E_{u^*}[\lambda(m_1) | \mathscr{F}_t].$$

Now $\{m_t\}$ is a Markov process under P_{u^*}, so if $\mathscr{F}_t^m = \sigma\{m_s : s \leq t\}$ then

$$J_t = E_{u^*}[\lambda(m_1) | \mathscr{F}_t^m].$$

Consequently $\{J_t\}$ has a representation as a stochastic integral with respect to $\{v_t\}$:

$$J_t = J(u^*) + \int_0^t \tilde{\gamma}_s \, dv_s \tag{16.20}$$

for some $\mathscr{F}_t^m$ predictable process $\{\tilde{\gamma}_s\}$.

Theorem 16.43. *Suppose*

$$\operatorname{sign} \tilde{\gamma}_t = \operatorname{sign} m_t \quad a.e. \quad dP \times dt.$$

Then u^* is optimal.

Proof. Note the measures P and P_{u^*} have the same null sets. Consider any other control $u \in \mathscr{U}$. Then

$$\begin{aligned} dv_t &= dm_t - \langle b(t), u_t^* \rangle \, dt + \langle b(t), u_t \rangle \, dt - \langle b(t), u_t \rangle \, dt \\ &= \langle C'(t) s(t), dW_t^u \rangle + (\langle b(t), u_t \rangle - \langle b(t), u_t^* \rangle) \, dt. \end{aligned}$$

Therefore,

$$\begin{aligned} J_t &= J(u^*) + \int_0^t \tilde{\gamma}_\sigma \langle C'(\sigma) s(\sigma), dW^u \rangle + \int_0^t \tilde{\gamma}_\sigma (\langle b(\sigma), u \rangle - \langle b(\sigma), u_\sigma^* \rangle) \, d\sigma \\ &= J(u^*) + M_t^u + A_t^u, \quad \text{say,} \end{aligned} \tag{16.21}$$

where $\{M_t^u\}$ is a P_u-local martingale and $\{A_t^u\}$ is a predictable process of integrable variation. Under the hypothesis of the theorem $\{A_t^u\}$ is, in fact, an increasing process. Consequently, by using an increasing sequence of stopping times we can show, as in Corollary 16.28 above, that

$$J(u^*) \leq E_u[J_1] = E_u[(m_1)] = J(u),$$

and so u^* is optimal. □

Remarks 16.44. The significance of this result is that it provides a criterion for optimality which involves only properties of the process $\{m_t\}$ under measure P_{u^*}. Other measures P_u and the process $\{x_t\}$ are not involved.

Consider the Girsanov exponential

$$L_t = \exp\left\{-\int_0^t \langle b(s), u_s^*\rangle\, dv_s - \tfrac{1}{2}\int_0^t \langle b(s), u_s^*\rangle^2\, ds\right\}$$

and define a new measure $\tilde{P}$ on $\mathscr{F}_1^m$ by putting

$$\frac{d\tilde{P}}{dP_{u^*}} = L_1.$$

Because $\langle b(s), u_s^*\rangle$ is bounded $\tilde{P}$ is a probability measure, and under $\tilde{P}\{m_t\}$ is a Brownian motion. Furthermore

$$dL_t = -L_t\langle b(t), u_t^*\rangle\, dv_t, \qquad L_0 = 1. \tag{16.22}$$

Using the Ito differentiation rule for the product of (16.21) and (16.22)

$$J_tL_t = J(u^*)L_0 + \int_0^t L_s(\tilde{\gamma}_s - J_s\langle b(s), u_s^*\rangle)\, dv_s + \int_0^t \tilde{\gamma}_s L_s\Big(\sum_i |b^i(s)|\Big)\, \text{sign}(m_s)\, ds. \tag{16.23}$$

As before, $\{J_t\}$ is a submartingale under $\tilde{P}$ if and only if $\{J_tL_t\}$ is a submartingale under P_{u^*}. We can, therefore, state the following necessary condition:

Lemma 16.45. *Suppose $\{J_t\}$ is a $\tilde{P}$ submartingale. Then*

$$\text{sign}(m_t) = \text{sign}(\tilde{\gamma}_t) \quad a.e. \qquad dP_{u^*}\times dt.$$

PROOF. Equation (16.33) gives the unique decomposition of $\{J_tL_t\}$ into the sum of a local martingale and a predictable process of bounded variation. If $\{J_tL_t\}$ is a submartingale the second process must be increasing; because $L_s(\sum_i |b^i(s)|) > 0$ this implies that $\tilde{\gamma}_s$ sign$(m_s) \geq 0$ a.e. □

Appendix 16.A

Essential Infima and Conditional Expectations

We now establish the result of Striebel [76], that if the ε-lattice property is satisfied for every $\varepsilon > 0$ the essential infimum and conditional expectation operations commute.

Let $(\Omega, \mathscr{F}, P)$ be a probability space and $\{\phi_\gamma : \gamma \in \Gamma\}$ a subset of $L^1(\Omega, \mathscr{F}, P)$ such that $\phi_\gamma(\omega) \geq 0$ a.s. for each $\gamma \in \Gamma$. Write

$$\phi = \bigwedge_\gamma \phi_\gamma$$

for the lattice infimum and λ_γ for the measure defined by

$$\frac{d\lambda_\gamma}{dP} = \phi_\gamma.$$

Lemma 16.A.1. *For $A \in \mathscr{F}$ define*

$$\lambda(A) = \inf \sum_{i=1}^{n} \lambda_{\gamma_i}(A_i),$$

where the infimum is taken over n, $\gamma_1, \ldots, \gamma_n \in \Gamma$, and finite disjoint partitions $A_1, \ldots, A_n$ of A. Then λ is a finite nonnegative measure on $\mathscr{F}$.

Proof. Because $\phi_\gamma(\omega) \geq$ a.s. for all $\gamma \in \Gamma$ each measure λ_γ is nonnegative and so is λ. Similarly, because ϕ_γ is in L^1, λ is finite. Consider disjoint $A_\alpha \in \mathscr{F}$ such that

$$\bigcup_{\alpha=1}^{m} A_\alpha = A \in \mathscr{F}.$$

For any $\varepsilon > 0$ and each α there are $\gamma_{\alpha,1}, \ldots, \gamma_{\alpha,n_\alpha}$ and $A_{\alpha,1}, \ldots, A_{\alpha,n_\alpha}$ such that

$$A_\alpha = \bigcup_{i=1}^{n_\alpha} A_{\alpha,i}$$

and, writing $\lambda_{\alpha,i}$ for $\lambda_{\gamma_{\alpha,i}}$,

$$\sum_i \lambda_{\alpha,i}(A_{\alpha,i}) \leq \lambda(A_\alpha) + \frac{\varepsilon}{m}.$$

Clearly

$$\bigcup_{\alpha=1}^{m} \bigcup_{i=1}^{n_\alpha} A_{\alpha,i} = A,$$

so

$$\begin{aligned} \lambda(A) &\leq \sum_{\alpha=i}^{m} \sum_{i=1}^{n_\alpha} \lambda_{\alpha,i}(A_{\alpha,i}) \\ &\leq \sum_{\alpha=1} \lambda(A_\alpha) + \varepsilon. \end{aligned} \tag{16.A.1}$$

Now take $\gamma_1, \ldots, \gamma_n$ and disjoint $A'_1, \ldots, A'_n$ such that

$$\bigcup_{i=1}^{m} A'_i = A$$

and

$$\sum_{i=1}^{n} \lambda_{\gamma_i}(A_i') \leq \lambda(A) + \varepsilon. \tag{16.A.2}$$

However, for each α

$$\bigcup_{i=1}^{n} (A_i' \cap A_\alpha) = A \cap A_\alpha = A_\alpha$$

and for all i

$$\bigcup_{\alpha=1}^{m} (A_i' \cap A_\alpha) = A_i' \cap A = A_i'. \tag{16.A.3}$$

Therefore,

$$\lambda(A_\alpha) \leq \sum_{i=1}^{n} \lambda_{\gamma_i}(A_i' \cap A_\alpha),$$

so from (16.A.1) and (16.A.2)

$$\sum_{\alpha=1}^{m} \lambda(A_\alpha) \leq \sum_{i=1}^{n} \sum_{\alpha=1}^{m} \lambda_{\gamma_i}(A_i' \cap A_\alpha)$$

$$= \sum_{i=1}^{n} \lambda_{\gamma_i}(A_i') \leq \lambda(A) + \varepsilon.$$

As ε is arbitrary we see that λ is finitely additive.

Consider a decreasing sequence of sets $\{A_n\} \in \mathscr{F}$ such that $\bigcap A_n = \varnothing$. Then because each λ_γ is a measure $\lim_n \lambda_\gamma(A_n) = 0$. Consequently,

$$0 \leq \lambda(A_n) \leq \lambda_\gamma(A_n)$$

so

$$\lim_n \lambda(A_n) = 0.$$

Therefore, λ is continuous at $\varnothing$ and so is a measure. □

Lemma 16.A.2. *Suppose, as above, that* $\{\phi_\gamma, \gamma \in \Gamma\}$ *is a subset of* $L^1(\Omega, \mathscr{F}, P)$ *and that* $\phi_\gamma(\omega) \geq 0$ *a.s. for each* $\gamma \in \Gamma$. *If* λ *is the measure defined in Lemma* 16.A.1, *then*

$$\frac{d\lambda}{dP} = \phi = \bigwedge_\gamma \phi_\gamma.$$

PROOF. First note that if $A \in \mathscr{F}$ is such that $P(A) = 0$ then for any $\gamma \in \Gamma$

$$0 = \lambda_a(A) = \int_A \phi_\alpha \, dP \geq \lambda(A) \geq 0.$$

Therefore, λ is absolutely continuous with respect to P and its Radon–Nikodym derivative $d\lambda/dP$ is $\mathscr{F}$ measurable. Furthermore, for every $A \in \mathscr{F}$ and every $\gamma \in \Gamma$:

$$\lambda(A) \geq \lambda_\gamma(A).$$

Therefore,

$$\frac{d\lambda}{dP} \leq \frac{d\lambda_\gamma}{dP} = \phi_\gamma \quad \text{a.s.}$$

for every $\gamma \in \Gamma$.

Suppose g is any other $\mathscr{F}$ measurable function such that for any $\gamma \in \Gamma$ $g(\omega) \leq \phi_\gamma(\omega)$ a.s. Then for all $A \in \mathscr{F}$ and $\gamma \in \Gamma$:

$$\int_A g \, dP \leq \lambda_\alpha(A).$$

For any $A \in \mathscr{F}$ and $\varepsilon > 0$ consider disjoint $A_1, \ldots, A_n$ and $\gamma_1, \ldots, \gamma_n$ such that

$$A = \bigcup_{i=1}^n A_i$$

and

$$\sum_{i=1}^n \lambda_{\gamma_i}(A_i) \leq \lambda(A) + \varepsilon.$$

Then

$$\int_A g \, dP = \sum_{i=1}^n \int_{A_i} g \, dP \leq \sum_{i=1}^n \lambda_{\gamma_i}(A_i) \leq \lambda(A) + \varepsilon$$

so

$$\int_A g \, dP \leq \lambda(A)$$

for all $A \in \mathscr{F}$. Consequently

$$g \leq \frac{d\lambda}{dP} \quad \text{a.s.}$$

and the result follows. □

Definition 16.A.3. The set $\{\phi_\gamma\} \subset L^1(\Omega, \mathscr{F}, P)$ has the *ε-lattice property* for $\varepsilon > 0$ if, given $\gamma_1, \gamma_2 \in \Gamma$ there is a $\gamma_3 \in \Gamma$ such that

$$\phi_{\gamma_3}(\omega) \leq \phi_{\gamma_1}(\omega) \wedge \phi_{\gamma_2}(\omega) + \varepsilon \quad \text{a.s.}$$

Lemma 16.A.4. *Suppose the set $\{\phi_\gamma, \gamma \in \Gamma\} \subset L^1(\Omega, \mathscr{F}, P)$ of nonnegative functions has the ε-lattice property for every $\varepsilon > 0$. If λ is the measure defined in Lemma* 16.A.1 *then, for every $A \in \mathscr{F}$,*

$$\lambda(A) = \inf_\gamma \lambda_\gamma(A).$$

PROOF. From the definition of $\lambda(A)$ we certainly have that

$$\lambda(A) \le \inf_\gamma \lambda_\gamma(A).$$

For $A \in \mathscr{F}$ and $\varepsilon > 0$ consider disjoint $A_1, \ldots, A_n$ and $\gamma_1, \ldots, \gamma_n$ such that

$$\bigcup_{i=1}^n A_i = A$$

and

$$\sum_{i=1}^n \lambda_{\gamma_i}(A_i) \le \lambda(A) + \frac{\varepsilon}{2}.$$

From the ε-lattice property there is a $\gamma_0 \in \Gamma$ such that

$$\phi_{\gamma_0} \le \min_{i=1,\ldots,n} \phi_{\gamma_i} + \frac{\varepsilon}{2}.$$

Therefore,

$$\lambda_{\gamma_0}(A_i) \le \lambda_{\gamma_i}(A_i) + \frac{\varepsilon}{2} P(A_i)$$

for each i and

$$\inf_\gamma \lambda_\gamma(A) \le \lambda_{\gamma_0}(A) \le \sum_{i=1}^n \lambda_{\gamma_i}(A_i) + \frac{\varepsilon}{2} P(A)$$

$$\le \lambda(A) + \varepsilon, \quad \text{so the result follows.}$$ □

Lemma 16.A.5. *Suppose that $\mathscr{G}$ is a sub-σ-field of $\mathscr{F}$ and that the set $\{\phi_\gamma, \gamma \in \Gamma\} \subset L^1(\Omega, \mathscr{F}, P)$ of nonnegative functions has the ε-lattice property for every $\varepsilon > 0$. Then*

$$E\left[\bigwedge_\gamma \phi_\gamma \middle| \mathscr{G}\right] = \bigwedge_\gamma E[\phi_\gamma | \mathscr{G}] \quad a.s.$$

PROOF. It is easy to check that the set $\{E[\phi_\gamma|\mathscr{G}], \gamma \in \Gamma\}$ has the ε-lattice property. For $B \in \mathscr{G}$ and $\gamma \in \Gamma$ define

$$\lambda'_\gamma(B) = \int_B E[\phi_\gamma|\mathscr{G}]\, dP,$$

$$\lambda'(B) = \inf_\gamma \lambda'_\gamma(B).$$

Using the ε-lattice property and Lemmas 16.A.2 and 16.A.4,

$$\frac{d\lambda'}{dP} = \bigwedge_\gamma E[\phi_\gamma|\mathscr{G}]$$

so $d\lambda'/dP$ is $\mathscr{G}$ measurable. However, if λ_γ and λ are the measures discussed above, for $B \in \mathscr{G}$

$$\lambda'_\gamma(B) = \lambda_\gamma(B)$$

so

$$\lambda'(B) = \lambda(B).$$

Therefore

$$\frac{d\lambda'}{dP} = E\left[\frac{d\lambda}{dP}\middle|\mathscr{G}\right].$$

From Lemma 16.A.2, $d\lambda/dP = \bigwedge_\gamma \phi_\gamma$ so the result is proved. □

CHAPTER 17

The Optimal Control of a Jump Process

The second situation we discuss is the optimal control of a jump process. Initially we consider a single jump process, as in Chapter 15, whose jump time and position are described by controlled families of Radon–Nikodym derivatives. A large class of jump processes can be constructed by the concatenation of single jumps, and we finally indicate how the results extend to multi-jump problems.

It appears necessary in these situations to require all measures corresponding to admissible controls to be mutually absolutely continuous. Otherwise, various conditional expectations are defined only up to control-dependent sets. (The nature of the Girsanov exponential density removes this difficulty in the continuous case discussed above.) We first discuss the form of absolutely continuous measures describing the process, and prove that the absolute continuity of the measures implies the absolute continuity of the Lévy systems, and, more significantly, that the converse holds.

The results of this section were first presented in the paper [19] of M. H. A. Davis and the author. They extend earlier work of Boel and Varaiya [7]. Pliska [70] has obtained related results in terms of infinitesimal generators for the optimal control of Markov jump processes. The form of the infinitesimal generators is closely related to that of the Hamiltonian function $H(t, u)$ of Corollary 17.22 below.

The Single Jump Process 17.1

We first consider a single jump process, as described in Chapter 15. That is, we consider a process $\{z_t\}$, $t \geq 0$, with values in a Lusin space $(E, \mathscr{E})$ which remains at its initial point $z_0 \in E$ until a random time T, when it jumps to a new random position z. The underlying probability space is again taken to be

$$\Omega=[0,\infty]\times E,$$

with the σ-field $\mathscr{B}\times\mathscr{E}$. A sample path of the process is

$$z_t(\omega)=\begin{cases} z_0, & \text{if } t<T(\omega),\\ z(\omega), & \text{if } t\geq T(\omega).\end{cases}$$

A probability measure μ is given on $(\Omega, \mathscr{B}\times\mathscr{E})$ and we suppose

$$\mu([0,\infty]\times\{z_0\})=0=\mu(\{0\}\times E).$$

Notation 17.2. The definitions and notation from Chapter 15 for the single jump process will now be used without further explanation.

In the sequel we shall discuss the same concepts associated with a second measure $\bar{\mu}$ on Ω. Such functions, etc., will be denoted by $\bar{F}_t$, $\bar{\lambda}$ and so on.

Suppose $\bar{\mu}$ is absolutely continuous with respect to μ. Then there is a Radon–Nikodym derivative $L=d\bar{\mu}/d\mu$. Write $L_t=E[L|\mathscr{F}_t]$. From Theorem 15.12

$$\begin{aligned} L_t &= L(T,z)I_{t\geq T}+I_{t<T}F_t^{-1}\int_{]t,\infty]\times X} L(s,x)\,d\mu(s,x)\\ &= L(T,z)I_{t\geq T}+I_{t<T}\bar{F}_t/F_t \end{aligned}$$

and the martingale L_t-1 has a representation as a stochastic integral

$$L_t-1=\int_\Omega I_{s\leq t}g(s,x)\,dq, \tag{17.1}$$

where

$$g(s,x)=L(s,x)-(\bar{F}_s/F_s)I_{s<c}.$$

With $\bar{c}=\inf\{t: F_t=0\}$ we see $\bar{c}\leq c$ and $g(s,x)=0$ for $s>\bar{c}$.

Theorem 17.3. *Write*

$$\phi(s,x)=(F_{s-}/\bar{F}_{s-})(L(s,x)-(\bar{F}_s/F_s)) \quad \textit{for } s<\bar{c},$$

$$\phi(s,x)=0 \quad \textit{for } s>\bar{c}, \textit{ and for } s=c \textit{ if } \bar{F}_{\bar{c}-}=0 \textit{ or } \bar{c}=\infty,$$

$$\phi(s,x)=(F_{\bar{c}-}/\bar{F}_{\bar{c}-})L(c,x) \quad \textit{if } \bar{c}<\infty \textit{ and } \bar{F}_{\bar{c}-}\neq 0.$$

Then

$$\begin{aligned} L_t=&\left[\exp\left(-\int_\Omega I_{s\leq t}\phi(s,x)\,d\tilde{p}^c\right)\right]\\ &\times\left(1+\phi(T,z)I_{t\geq T}+I_{t\geq T}\int_E \phi(T,x)\lambda(dx,T)\frac{\Delta F_T}{F_{T-}}\right)\\ &\times\prod_{\substack{u\leq t\wedge T\\ u\neq T}}\left(1+\int_E\phi(u,x)\lambda(dx,u)\frac{\Delta F_u}{F_{u-}}\right). \end{aligned}$$

PROOF. If $s \leq T$ then $L_{s-} = \bar{F}_{s-}/F_{s-}$. Therefore, for $s \leq T$ and $s < \bar{c}$ (and, if $\bar{F}_{\bar{c}-} \neq 0$, for $s \leq T \wedge \bar{c}$), $L_{s-} > 0$ so

$$\phi(s, x) = g(s, x) L_{s-}^{-1} = (F_{s-}/\bar{F}_{s-})(L(s, x) - (\bar{F}_s/F_s)).$$

Writing

$$M_t = \int_\Omega I_{s \leq t} \phi(s, x)\, dq$$

we notice that (17.1) can be written

$$L_t = 1 + \int_0^t L_{s-}\, dM_s.$$

Consequently, by the exponential formula of Theorem 13.5

$$L_t = e^{M_t} \prod_{u \leq t} (1 + \Delta M_u)\, e^{-\Delta M_u}.$$

At the discontinuities of F_u:

$$\Delta M_u = \int_E \phi(u, x) \lambda(dx, u) \frac{\Delta F_u}{F_{u-}},$$

and at the jump time T

$$\Delta M_T = \phi(T, x) + \int_E \phi(T, x) \lambda(dx, T) \frac{\Delta F_T}{F_{T-}}.$$

Substituting, we see that L has the stated form. □

Theorem 17.4. *Suppose* $(\bar{\lambda}, \bar{\Lambda})$ *is the Lévy system of* $\bar{\mu}$. *Then a.s.* $d\bar{F}$:

$$\bar{\lambda}(A, s) = \int_A \left(1 + \phi + \frac{\Delta F_s}{F_{s-}} \int_E \phi\, d\lambda\right) d\lambda \Big/ \int_E \left(1 + \phi + \frac{\Delta F_s}{F_{s-}} \int_E \phi\, d\lambda\right) d\lambda$$

and

$$\bar{\Lambda}_t = \int_{]0,t]} \int_E \left(1 + \phi + \frac{\Delta F_s}{F_{s-}} \int_E \phi\right) \lambda(dx, s)\, d\Lambda_s.$$

PROOF. For $t > 0$ and $A \in \mathscr{E}$:

$$\begin{aligned} \bar{F}_t^A &= \bar{\mu}(]t, \infty] \times A) \\ &= \int_{]t,\infty] \times A} L\, d\mu \\ &= -\int_{]t,\infty]} \int_A L(s, x) \lambda(dx, s)\, dF_s. \end{aligned}$$

However,

$$\bar{F}_t^A = -\int_{]t,\infty]} \bar{\lambda}(A, s)\, d\bar{F}_s$$
$$= -\int_{]t,\infty]} \bar{\lambda}(A, s) \frac{d\bar{F}_s}{dF_s}\, dF_s.$$

So a.s. dF_s:

$$\bar{\lambda}(A, s)\frac{d\bar{F}_s}{dF_s} = \int_A L(s, x)\lambda(dx, s) = \int_A \left(\frac{\bar{F}_{s-}}{F_{s-}}\phi(s, x) + \frac{\bar{F}_s}{F_s}\right)\lambda(dx, s).$$

Therefore, for $s < \bar{c}$, and, if $\bar{F}_{\bar{c}-} \neq 0$, for $s \leq \bar{c}$:

$$\bar{\lambda}(A, s)\frac{F_s}{\bar{F}_{s-}}\frac{d\bar{F}_s}{dF_s} = \int_A \left(\frac{F_s}{F_{s-}}\phi + \frac{\bar{F}_s}{\bar{F}_{s-}}\right)\lambda(dx, s) \quad \text{a.s. } d\bar{F}_s$$
$$= \int_A \left(\left(1 + \frac{\Delta F_s}{F_{s-}}\right)\phi + \left(1 + \frac{\Delta \bar{F}_s}{\bar{F}_{s-}}\right)\right)\lambda(dx, s). \tag{17.2}$$

Now if s is a point of continuity of F then it is also a point of continuity of $\bar{F}$, and $\Delta F_s = \Delta \bar{F}_s = 0$. If $\Delta F_s \neq 0$ then the Radon–Nikodym derivative $d\bar{F}_s/dF_s = \Delta\bar{F}_s/\Delta F_s$ and the left-hand side above is:

$$\bar{\lambda}(A, s)\frac{(F_{s-} + \Delta F_s)}{\bar{F}_s}\frac{\Delta\bar{F}_s}{\Delta F_s} = \bar{\lambda}(A, s)\frac{\Delta\bar{F}_s}{\bar{F}_{s-}}\left(1 + \frac{F_{s-}}{\Delta F_s}\right).$$

Evaluating (17.2) when $A = E$, so $\bar{\lambda}(E, s) = 1 = \lambda(E, s)$

$$\frac{\Delta\bar{F}_s}{\bar{F}_{s-}} = \frac{\Delta F_s}{F_{s-}}\int_E \left(1 + \phi + \frac{\Delta F_s}{F_{s-}}\phi\right)\lambda(dx, s), \tag{17.3}$$

if $\Delta F_s \neq 0$, and we have

$$\frac{F_s}{\bar{F}_{s-}}\frac{d\bar{F}_s}{dF_s} = \int_E (1 + \phi)\lambda(dx, s)$$

if $\Delta F_s = 0$. Substituting in (17.2) we have if $(1 + (\Delta F_s/F_{s-})) \neq 0$

$$\bar{\lambda}(A, s) = \int_A \left(1 + \phi + \frac{\Delta F_s}{F_{s-}}\int_E \phi\, d\lambda\right)\lambda(dx, s) \Big/ \int_E \left(1 + \phi + \frac{\Delta F_s}{F_{s-}}\int_E \phi\right)\lambda(dx, s)$$

a.s. $d\bar{F}_s$ for $s < \bar{c}$, and for $s \leq \bar{c}$ if $\bar{F}_{\bar{c}-} \neq 0$.

Now $(1 + (\Delta F_s/F_{s-})) = 0$ only if $s = c$, $c < \infty$ and $F_{c-} \neq 0$. This situation is only of interest here if also $\bar{c} = c$ and $\bar{F}_{c-} \neq 0$. However, in this case it is easily seen that substituting

$$\phi(c, x) = (F_{c-}/\bar{F}_{c-})L(c, x) \quad \text{in (17.2)}$$

gives the correct expression for

$$\bar{\lambda}(A, c) = \lambda(A, c),$$

because

$$L(c,x)=(\Delta\bar{F}_c/\Delta F_c)(d\bar{\lambda}/d\lambda(c)).$$

Now

$$\bar{\Lambda}_t=-\int_{]0,t]}\frac{d\bar{F}_s}{\bar{F}_{s-}}=\int_{]0,t]}\frac{d\bar{F}_s}{dF_s}\frac{F_{s-}}{\bar{F}_{s-}}\,d\Lambda_s.$$

If F_t is continuous at s, again $\Delta\bar{F}_s=\Delta F_s=0$ and evaluating (17.2) for $A=E$

$$\frac{d\bar{\Lambda}_s}{d\Lambda_s}=\frac{d\bar{F}_s}{dF_s}\frac{F_s}{\bar{F}_s}=\int_E(1+\phi)\lambda(dx,s).$$

If F_t is not continuous at s,

$$\frac{d\bar{F}_s}{dF_s}=\frac{\Delta\bar{F}_s}{\Delta F_s},$$

and from (17.3) above,

$$\frac{d\bar{\Lambda}_s}{d\Lambda_s}=\frac{\Delta\bar{F}_s}{\Delta F_s}\frac{F_{s-}}{\bar{F}_{s-}}=\int_E\left(1+\phi+\frac{\Delta F_s}{F_{s-}}\phi\right)\lambda(dx,s).$$

That is

$$\bar{\Lambda}_t=\int_{]0,t]}\int_E\left(1+\phi+\frac{\Delta F_s}{F_{s-}}\phi\right)\lambda(dx,s)\,d\Lambda_s. \qquad \square$$

Notation 17.5. $\mathscr{A}$ will denote the set of right continuous, monotonic increasing (deterministic), functions Λ_t, $t\geq 0$, such that

(a) $\Lambda_0=0$,
(b) $\Delta\Lambda_u=\Lambda_u-\Lambda_{u-}\leq 1$ for all points of discontinuity u,
(c) if $\Delta\Lambda_u=1$ then $\Lambda_t=\Lambda_u$ for $t\geq u$.

If $\Lambda_t\in\mathscr{A}$ then Λ_t has a unique decomposition $\Lambda_t=\Lambda_t^c+\Lambda_t^d$, where $\Lambda_t^d=\sum_{s\leq t}\Delta\Lambda_s\in\mathscr{A}$ and $\Lambda_t^c\in\mathscr{A}$ is continuous. Note that Λ might equal $+\infty$ for finite t.

Lemma 17.6. *The formulae*

$$F_t=1-G_t,$$

$$F_t=\exp(-\Lambda_t^c)\prod_{u\leq t}(1-\Delta\Lambda_u), \tag{17.4}$$

$$\Lambda_t=-\int_{]0,t]}F_{s-}^{-1}\,dF_s, \tag{17.5}$$

define a bijection between the set $\mathscr{A}$ and the set of all probability distributions $\{G\}$ on $]0,\infty]$.

PROOF. Clearly if $\Lambda\in\mathscr{A}$ then F_t, defined by (17.4), is monotonic decreasing and right continuous. $F_0=1$ and $0\leq F_t\leq 1$. Therefore, $G_t=1-F_t$ is a probability distribution on $]0, \infty]$. Conversely, if G_t is a probability distribution, if $F_t = 1 - G_t$ and Λ_t is given by (17.5), then $\Lambda \in \mathscr{A}$.

From Theorem 13.5 (taking Ω to be a single point), F_t defined by (17.4) is the unique solution of the equation

$$dF_t = -F_{t-}\, d\Lambda_t,$$

$$F_0 = 1.$$

This shows the correspondence is a bijection. □

Remarks 17.7. If $\Lambda_t^d = 0$ and Λ_t^c is absolutely continuous with respect to Lebesgue measure, there is a measurable function α_s such that

$$\Lambda_t^c = \int_0^t \alpha_s\, ds.$$

The function α_s is often called the "rate" of the jump process. However, there are continuous increasing functions which are singular with respect to Lebesgue measure, so to discuss the optimal control of the single jump process we suppose a general "(integrated) base rate" $\Lambda_t = \Lambda_t^c + \Lambda_t^d$ is given.

Lemma 17.8. *Suppose $\bar{\Lambda}_t \in \mathscr{A}$ is a second process whose associated Stieltjes measure is absolutely continuous with respect to (the Stieltjes measure of) Λ_t. Then the associated $\bar{F}_t$ has the form*

$$\bar{F}_t = F_t \exp\Big(-\int_0^t (\alpha(s) - 1)\, d\Lambda_s^c\Big)\Pi_t.$$

Here $\alpha(s)$ is the Radon–Nikodym derivative, F_t is defined by (17.4) *and*

$$\Pi_t = \prod_{u \le t} \frac{(1 - \alpha(u)\Delta\Lambda_u^d)}{(1 - \Delta\Lambda_u^d)}.$$

Furthermore, $\alpha(u)\Delta\Lambda_u^d \le 1$, and if $\alpha(u)\Delta\Lambda_u^d = 1$ then $\alpha(t) = 0$ for $t \ge u$.

PROOF. By hypothesis

$$\bar{\Lambda}_t = \int_0^t \alpha(s)\, d\Lambda_s^c + \sum_{u \le t} \alpha(u)\Delta\Lambda_u^d,$$

so from (17.4)

$$\bar{F}_t = \exp\Big(-\int_0^t \alpha(s)\, d\Lambda_s^c\Big) \prod_{u \le t} (1 - \alpha(u)\Delta\Lambda_u^d)$$

$$= F_t \exp\Big(\Big(-\int_0^t (\alpha(s) - 1)\, d\Lambda_s^c\Big) \prod_{u \le t} \frac{(1 - \alpha(u)\Delta\Lambda_u^d)}{(1 - \Delta\Lambda_u^d)}.$$

The conditions on α follow from Lemma 17.6 and the definition of $\mathscr{A}$. □

Remarks 17.9. Because $-d\bar{F}_t = \bar{F}_{t-}\, d\bar{\Lambda}_t = (\bar{F}_t/F_{t-})\alpha(t)F_{t-}\, d\Lambda_t = -(\bar{F}_{t-}/F_{t-})\alpha(t)\, dF_t$, the probability distribution associated with $\bar{F}_t$ above is certainly absolutely continuous with respect to that associated with F_t. To ensure the converse we could require that, for some positive integer n,

$$1/n \leq \alpha(s) \leq \min(n, (n^{-1}-1)F_{s-}/\Delta F_s)$$

for all s.

The above discussion only concerns the rate Λ_t, describing when the jump occurs. Consider now the other component λ of the Lévy system, which describes where the jump goes. Because $(E, \mathscr{E})$ is a Lusin space the $\lambda(\cdot, s)$ can be chosen to be a regular family of conditional probability distributions satisfying

(a) $\lambda(A, s) \geq 0$ for $A \in \mathscr{E}$, $s > 0$,
(b) for each $A \in \mathscr{E}$ $\lambda(A, \cdot)$ is Borel measurable,
(c) for all $s \in]0, c[$ (except perhaps on a set of $d\Lambda$-measure 0), $\lambda(\cdot, s)$ is a probability measure on $(E, \mathscr{E})$, and if $c < \infty$ and $\Lambda_{c-} < \infty$ then $\lambda(\cdot, c)$ is a probability measure.

Lemma 17.10. *There is a bijection between probability measures μ on $(\Omega, \mathscr{B} \times \mathscr{E})$ and Lévy systems (λ, Λ), where λ satisfies* (a), (b), (c) *above and $\Lambda \in \mathscr{A}$.*

PROOF. Definition 15.3 indicates how a Lévy system is determined by a measure μ. Conversely, given a pair (λ, Λ), because $\Lambda \in \mathscr{A}$ we can determine a function F_t by (17.5). For $A \in \mathscr{E}$ define

$$\mu(]0, t] \times A) = -\int_{]0,t]} \lambda(A, s)\, dF_s. \qquad \square$$

Remark 17.11. We now establish the converse of theorem 17.4, that if the Lévy systems of two measures $\bar{\mu}$, μ, on $(\Omega, \mathscr{B} \times \mathscr{E})$ are absolutely continuous then the measures are absolutely continuous.

Theorem 17.12. *Suppose μ, $\bar{\mu}$ have Lévy systems (λ, Λ) and $(\bar{\lambda}, \bar{\Lambda})$. Write*

$$\bar{c} = \inf\{t : \bar{F}_t = 0\},$$

and suppose $\bar{c} \leq c$, $d\bar{\Lambda}_t \ll d\Lambda_t$ on $]0, \bar{c}]$ and $\bar{\lambda}(\cdot, t) \ll \lambda(\cdot, t)$ a.e. $d\Lambda_t$. Then $\bar{\mu} \ll \mu$ with Radon–Nikodym derivative

$$L(t, z) = \alpha(t)\beta(t, z) \exp\left(-\int_0^t (\alpha(s) - 1)\, d\Lambda_s^c\right) \Pi_{t-} I_{t \leq \bar{c}},$$

where, as in Lemma 17.8,

$$\Pi_t = \prod_{u \leq t} \frac{(1 + (\Delta F_u/F_{u-})\alpha(u))}{(1 + (\Delta F_u/F_{u-})}.$$

Here

$$\alpha(t)=\frac{d\bar{\Lambda}_t}{d\Lambda_t} \quad \text{and} \quad \beta(t,z)=\frac{d\bar{\lambda}}{d\lambda}(z,t).$$

PROOF. Define $L(t,z)$ by the above expression and write

$$\eta(t)=\exp\left(-\int_0^t (\alpha(s)-1)\,d\Lambda_s^c\right).$$

Then, because $\int_E \beta(t,z)\,d\lambda=1$ a.s.

$$E[L(T,z)]=-\int_{]0,\bar{c}]} \alpha(t)\eta(t)\Pi_{t-}\,dF_t.$$

From Lemma 17.8 and equations (17.4) and (17.5)

$$\eta(t)\Pi_{t-}=\bar{F}_{t-}/F_{t-}.$$

As measures on $[0,\infty]$

$$d\bar{\Lambda}_t=-\frac{d\bar{F}_t}{\bar{F}_{t-}}=-\alpha(t)\frac{dF_t}{F_{t-}}=\alpha(t)\,d\Lambda_t$$

so

$$E[L(T,z)]=-\int_{]0,\bar{c}]} \alpha(t)\bar{F}_{t-}\frac{dF_t}{F_{t-}}$$

$$=-\int_{]0,\bar{c}]} \bar{F}_{t-}\frac{d\bar{F}_{t-}}{\bar{F}_{t-}}=\bar{F}_0-\bar{F}_{\bar{c}}=1.$$

A probability measure $\mu^*\ll\mu$ can, therefore, be defined on $(\Omega,\mathscr{B}\times\mathscr{E})$ by putting $d\mu^*/d\mu=L$. For $t<\bar{c}$ we have

$$L_t=E[L|\mathscr{F}_t]=L(T,z)I_{t\geq T}+I_{t<T}F_t^{-1}\int_{]t,\bar{c}]\times E} L(s,x)\,d\mu.$$

By similar calculations to those above the latter term is

$$F_t^{-1}\int_{]t,\bar{c}]} \alpha(s)\eta(s)\Pi_{s-}\,dF_s=\bar{F}_t/F_t=\eta(t)\Pi_t,$$

so

$$L_t=\alpha(T)\beta(T,z)\eta(T)\Pi_{T-}I_{t\geq T}+I_{t<T}\eta(t)\Pi_t.$$

In the notation of Theorem 17.3.

(a)
$$\phi(t,x)=\frac{F_{t-}}{\bar{F}_{t-}}\left(L(t,x)-\frac{\bar{F}_t}{F_t}\right)$$

$$=\{\alpha(t)\beta(t,x)\Pi_{t-}\eta(t)I_{t\leq\bar{c}}-\eta(t)\Pi_t\}\eta(t)^{-1}\Pi_{t-}^{-1}$$

$$=\alpha(t)\beta(t,x)I_{t\leq\bar{c}}-\frac{(1+\Delta F_t/F_{t-}\alpha(t))}{(1+\Delta F_t/F_{t-})},$$

for $t<\bar{c}$.

(b) $\phi(t,x)=0$ for $t>\bar{c}$, and for $t=\bar{c}$ if $\bar{c}=\infty$ or $\bar{F}_{\bar{c}-}=0$,
(c) $\phi(\bar{c},x)=L(\bar{c},x)$ if $\bar{c}<\infty$ and $\bar{F}_{\bar{c}-}\neq 0$, that is, substituting, in this case

$$\phi(\bar{c},x)=\alpha(\bar{c})\beta(\bar{c},x).$$

The Lévy system (λ^*, Λ^*) associated with μ^* is then defined by

$$\lambda^*(A,s)=\int_A\left(1+\phi+\frac{\Delta F_s}{F_{s-}}\int_E\phi\,d\lambda\right)d\lambda\Big/\int_E\left(1+\phi+\frac{\Delta F_s}{F_{s-}}\phi\right)d\lambda$$

$$\Lambda_t^*=\int_{]0,t]}\int_E\left(1+\phi+\frac{\Delta F_s}{F_{s-}}\phi\right)d\lambda\,d\Lambda_s.$$

Substituting the above expressions for ϕ we have

$$\int_E\phi\,d\lambda=\alpha(t)I_{t\le\bar{c}}-\frac{(1+(\Delta F_t/F_{t-})\alpha(t))}{(1+\Delta F_t/F_{t-})}$$

and

$$\left(1+\frac{\Delta F_t}{F_{t-}}\right)\int_E\phi\,d\lambda=\alpha(t)-1.$$

The above expressions give

$$\frac{d\Lambda_t^*}{d\Lambda_t}=\alpha(t)\quad\text{and}\quad\frac{d\lambda^*}{d\lambda}=\beta(t,x),$$

so

$$\Lambda_t^*=\bar{\Lambda}_t\quad\text{and}\quad\lambda^*=\bar{\lambda}.$$

By Lemma 17.10, $\bar{\mu}=\mu^*\ll\mu$ and the result is proved. □

Optimal Control of a Single Jump Process 17.13

We now consider a jump process control problem by supposing the process $\{z_t\}$, $t\ge 0$, is governed by an indexed family of measures $\{\mu^u : u\in\mathscr{U}\}$ on $(\Omega,\mathscr{B}\times\mathscr{E})$. Furthermore, we assume all the measures μ^u are absolutely continuous with respect to some "base measure" μ. If (λ,Λ) and (λ^u,Λ^u) are, respectively, the Lévy systems associated with μ and μ^u then $\lambda^u\ll\lambda$ and $d\Lambda^u\ll d\Lambda$. Conversely, if λ^u and Λ^u are given and $\lambda^u\ll\lambda$, $d\Lambda^u\ll d\Lambda$ then Theorem 17.12 gives the formula for $d\mu^u/d\mu$, showing that $\mu^u\ll\mu$. The measure μ^u is, therefore, determined if the Radon–Nikodym derivatives

$$\alpha^u(s)=\frac{d\Lambda^u}{d\Lambda}(s),\qquad\beta^u(s,x)=\frac{d\lambda^u}{d\lambda}\quad\text{are given.}$$

Definition 17.14. Suppose $(U,\mathscr{G})$ is a measurable space and

$$\alpha: R^+\times U\to R^+,$$

$$\beta: R^+\times E\times U\to R^+$$

are measurable functions satisfying the following conditions for all $(s, x, u) \in R^+ \times E \times U$:

(a) $0 < c_1 \leq \alpha(s, u) \leq c(s)$,
(b) $0 < c_2 \leq \beta(s, x, u) \leq c_3$,
(c) $\int_E \beta(s, x, u)\lambda(s, dx) = 1$.

Here c_1, c_2, c_3 are constants and $c(\cdot)$ is a measurable function such that $c(s) < \min(c_4, 1/\Delta\Lambda(s))$ for each s and some bounded $c_4 > 0$. We suppose that α and β play a role analogous to that of the drift coefficient $f(t, x, u)$ in the continuous case discussed in the previous chapter.

Definition 17.15. The set of admissible controls $\{\mathscr{U}\}$ is the set of measurable functions $u: R^+ \to U$.

For $u \in \mathscr{U}$ write $\alpha^u(s) = \alpha(s, u(s))$ and $\beta^u(s, x) = \beta(s, x, u(s))$. Note that u then controls the probability Λ^u, of when the jump occurs, and λ^u, of where it goes. If $L^u(t, z)$ is defined by Theorem 17.12, a measure corresponding to $u \in \mathscr{U}$ is given by $d\mu^u = L^u(t, z)\, d\mu$. Under the above conditions μ^u and μ are mutually absolutely continuous, that is, they have the same null sets, so statements made almost surely are unambiguous.

Remarks 17.16. The predictable σ-field on $R^+ \times \Omega$ is the σ-field generated by the (real) left continuous functions. In the present situation the process stops after the single jump time $T(\omega)$, so, because they are just deterministic functions used up to time $T(\omega)$, our controls $\mathscr{U}$ are predictable functions. In the single jump control problem there is no element of "feedback". However, this appears in Section 17.27 below, where the results are extended to multi-jump processes.

Suppose a cost is associated with the jump process of the following form:

$$G(T, z) + \int_0^T \int_E c(s, x, u(s))\, d\tilde{p}^u(s).$$

Here G and c are real valued, measurable and bounded. If the control u is used the expected final cost is

$$J(u) = E_u\left[G(T, z) + \int_0^T \int_E c(s, x, u(s))\alpha^u(s)\beta^u(s, x)\lambda(dx, s)\, d\Lambda_s\right],$$

where E_u denotes the expectation with respect to the measure μ^u determined as above by the Radon–Nikodym derivatives $(\alpha^u(s), \beta^u(s, x))$. Note, however, that this apparently more general cost can be written as just a terminal cost

$$J(u) = E_u[f(T, z, u(T))], \tag{17.6}$$

where

$$f(s, x, u) = G(s, x) + c(s, x, u).$$

We suppose, therefore, that the cost is of the form (17.6), where f is real, measurable, and bounded. $J(u)$ is consequently finite for all $u \in \mathscr{U}$. The optimal control problem is to determine how $u \in \mathscr{U}$ should be chosen so that $J(u)$ is minimized. Clearly, by considering suitable μ we see that bounded terminal time problems are included in this model.

Suppose control $u \in \mathscr{U}$ is used up to time t and control $v \in \mathscr{U}$ is used from time t onwards. Then control $w \in \mathscr{U}$, if $w(s) = u(s)$, $s \leq t$ and $w(s) = v(s)$, $s > t$. The resulting expected final cost, given information $\mathscr{F}_t$, is $\psi(u, v, t) = E_w[f(T, z, w(T))|\mathscr{F}_t]$.

Lemma 17.17. *If*

$$\psi(v, t) = I_{t<T} E_w[f(T, z, v(T))|T > t],$$

then ψ is independent of u and

$$\psi(u, v, t) = f(T, z, u(T)) I_{t \geq T} + \psi(v, t).$$

PROOF.

$$\psi(u, v, t) = E_w[f(T, z, w(T))|t \geq T] I_{t \geq T} + E_w[f(T, z, w(T))|T > t] I_{T>t}$$

so the decomposition is immediate from the form of $w(s)$.

To show $\psi(v, t)$ is independent of u write

$$L^w = \frac{d\mu^w}{d\mu} = L(s, x) = L$$

and

$$L_t = E[L|\mathscr{F}_t].$$

Write $L^t = L/L_t$ (with the convention that $0/0 = 1$), so

$$L^t = I_{t \geq T} + I_{t<T} \alpha^v(T) \beta^v(T, z) \exp\left(-\int_t^T (\alpha^v(s) - 1)\, d\Lambda_s^c\right) \times \prod_{t<u<T} \frac{(1 - \alpha^v(u) \Delta\Lambda(u))}{(1 - \Delta\Lambda(u))}.$$

Clearly L^t does not depend on $u \in \mathscr{U}$ and

$$E[L^t|\mathscr{F}_t] = E[L|\mathscr{F}_t]/L_t = 1 \quad \text{a.s.}$$

By the result of Loève [60], 27.4,

$$\psi(v, t) = \frac{E[I_{t<T} L_t L^t f(T, z, v(T))|\mathscr{F}_t]}{E[L_t L^t|\mathscr{F}_t]} = E[I_{t<T} L^t(T, z) f(T, z, v(T))|\mathscr{F}_t].$$

The expectations here are with respect to the base measure μ, so the last term is independent of u. □

Definition 17.18. The *value function* for the control problem is

$$W(t) = I_{T>t}(\bigwedge E_v[f(T, z, v(T))|T>t]).$$

Here the infimum $\bigwedge$ is just the infimum of a set of real numbers. Because $T>0$ a.s. we have

$$W(0) = J^* = \inf_{u \in \mathcal{U}} J(u),$$

and certainly $W(T)=0$. Now $E_v[f(T, z, v(T))|T>t]$ is a real number so the set of random variables $\{\psi(v, t)\}$ trivially satisfies the 0-lattice property. Consequently we can, as in Corollary 16.15, apply Lemma 16.11 to the process

$$M_t^u = f(T, z, u(T))I_{t \geq T} + W(t)$$

to obtain the following "principle of optimality":

Theorem 17.19. (a) $\{M_t^u\}$ *is a submartingale on* $(\Omega, \mathcal{F}_t, \mu^u)$ *for any* $u \in \mathcal{U}$. (b) $u^* \in \mathcal{U}$ *is optimal if and only if* $\{M_t^{u^*}\}$ *is a martingale on* $(\Omega, \mathcal{F}_t, \mu^{u^*})$.

PROOF. Part (b) follows because

$$M_T^{u^*} = f(T, z, u^*(T))$$

and by the Optional Stopping Theorem 4.12

$$E_{u^*}[M_T^{u^*}] = M_0^{u^*} = W(0) = J^*.$$

M_t^u represents the minimum achievable final cost given the evolution of the process up to time t.

A Minimum Principle 17.20

Using the uniqueness of the Doob–Meyer decomposition and the above principle of optimality we now characterize an optimal control $u^* \in \mathcal{U}$. Functions, processes and measures associated with u^* will be denoted by f^*, p^*, μ^* etc.

Theorem 17.21. $u^* \in \mathcal{U}$ *is an optimal control if and only if there is a measurable function* $g: \Omega \to R$ *such that*

$$M_t^* = J^* + \int_\Omega I_{s \leq t} g(s, x)\, dq^* \quad a.s.\ \mu, \tag{17.7}$$

where the integral is a martingale under μ^* *and, at almost every point* (t, x), $u^*(\omega)$ *minimizes the Hamiltonian*

$$\alpha^u(t)\left(\int_E (g + f^u - f^*)\beta^u(t, x)\lambda(dx, t)\right). \tag{17.8}$$

Here, for $A \in \mathcal{E}$, $q^*(t, A) = p(t, A) - \tilde{p}^*(t, A)$ *are the basic martingales associated with the single jump process under measure* μ^*.

PROOF. Suppose $u^* \in \mathscr{U}$ is optimal. Then from the martingale representation result, Theorem 15.14, (17.7) is satisfied by the function g, where

$$g(s,x) = f^* - E^*[f^*] + I_{s<c}(F_s^*)^{-1} \int_{]0,s]\times E} (f^* - E^*[f^*])\, d\mu^*$$

$$= f^* - (F_s^*)^{-1} \int_{]s,\infty]\times E} f^*\, d\mu^* \quad \text{a.s. } \mu.$$

For any other control $u \in \mathscr{U}$

$$M_t^u = M_t^* + (f_t^u - f_t^*) I_{t\geq T}$$

$$= J^* + \int_{]0,t]\times E} g(dp - d\tilde{p}^u + d\tilde{p}^u - d\tilde{p}^*)$$

$$+ \int_{]0,t]\times E} (f^u - f^*)\, dq^u + \int_{]0,t]\times E} (f^u - f^*)\, d\tilde{p}^u$$

$$= J^* + \int_{]0,t]\times E} (g + f^u - f^*)\, dq^u + \int_{]0,t]\times E} (g + f^u - f^*)\, d\tilde{p}^u$$

$$- \int_{]0,t]\times E} (g + f^* - f^*)\, d\tilde{p}^*. \tag{17.9}$$

This expresses M_t^u as the sum of a local martingale and a predictable process of integrable variation. Consequently $\{M_t^u\}$ is a special semimartingale, and, from Theorem 12.38, this decomposition is unique. However, from Theorem 17.19, $\{M_t^u\}$ is a submartingale. From this representation $\{M_t^u\}$ is right continuous, and so M_t^u has a Doob–Meyer decomposition

$$M_t^u = J^* + N_t^u + A_t^u,$$

where N_t^u is a martingale on $(\Omega, \mathscr{F}_t, \mu^u)$ and A_t^u is a predictable increasing process. The final two terms in (17.9) must, therefore, be an increasing process. So, because

$$\tilde{p}^u(t,A) = \int_{]0,t\wedge T]} \alpha^u(s)\beta^u(s,x)\lambda(dx,s)\, d\Lambda_s$$

and

$$\tilde{p}^*(t,A) = \int_{]0,t\wedge T]} \alpha^*(s)\beta^*(s,x)\lambda(dx,s)\, d\Lambda_s,$$

we obtain the minimum principle (17.8).

Conversely, suppose $g: \Omega \to R$ is a measurable process such that (17.7) is satisfied and such that $u^* \in \mathscr{U}$ minimizes almost surely the Hamiltonian (17.8). Then the integral in (17.7) is a martingale under μ^* and

$$J^* = E^*[f^*] = J(u^*).$$

For any other $u \in \mathscr{U}$ write

$$A_t^u = \int_{]0,t]\times E} (g + f^u - f^*)\, d\tilde{p}^u - \int_{]0,t]\times E} (g + f^* - f^*)\, d\tilde{p}^*.$$

Because u^* minimizes the Hamiltonian, A_t^u is an increasing process, so evaluating $E_u[M_T^u]$ we have

$$J(u) = E_u[M_T^u] = J^* + E_u[A_T^u].$$

Therefore, $J^* \leq J(u)$ and $u^* \in \mathscr{U}$ is optimal. □

Corollary 17.22. *The Hamiltonian* (17.8) *can be written*

$$H(t, u) = \alpha^u(t)\Big(\int_E f^u \beta^u \, d\lambda - \eta(t)\Big),$$

where

$$\eta(t) = E^*[f^* | T > t].$$

PROOF. We have seen that

$$g = f^* - (F_t^*)^{-1} \int_{]t,\infty]\times E} f^*\, d\mu^* = f^* - E^*[f^* | T > t].$$

Substituting in (17.8) the result follows. □

Remark 17.23. Using this form of the Hamiltonian the above minimum principle appears similar to those of Pliska [70] and Rishel [71]. In fact in [71] Rishel gives a system of "adjoint equations" which are satisfied by his analog of the function $\eta(t)$. The following theorem gives the equivalent result in the present context.

Theorem 17.24. *For $t < c^*$ the function $\eta(t)$ satisfies the following equation*

$$\eta(t) - \eta(0) = \int_{]0,t]} \Big\{ \frac{\eta(s)}{1 - \alpha^*(s)\Delta\Lambda(s)} - \gamma(s) \Big\} \alpha^*(s)\, d\Lambda(s) - \sum_{s \leq t} \frac{(\alpha^*(s)\Delta\Lambda(s))^2}{1 - \alpha^*(s)\Delta\Lambda(s)} \gamma(s),$$

where

$$\gamma(s) = \int_E f^*\, d\lambda^*.$$

PROOF. For $t < c^*$ we have

$$\eta(t) = \frac{1}{F_t^*} \int_{]t,\infty]\times E} f^*\, d\mu^* = -\frac{1}{F_t^*} \int_{]t,\infty]} \int_E f^*\, d\lambda^*\, dF_s^*$$
$$= -\frac{1}{F_t^*} \int_{]t,\infty]} \gamma(s)\, dF_s^*.$$

The result is obtained by applying the product formula for Stieltjes integrals (a special case of Corollary 12.22):

$$a(t)b(t) = a(0)b(0) + \int_{]0,t]} b(s-)\,da + \int_{]0,t]} a(s-)\,db + \sum_{u \le t} \Delta a(u)\Delta b(u).$$

This gives

$$\eta(t) - \eta(0) = \int_{]0,t]} \frac{1}{F^*_{s-}} \gamma(s)\,dF^*_s - \int_{]0,t]} F^*_{s-}\eta(s-)\,d\left(\frac{1}{F^*_s}\right)$$
$$+ \sum_{s \le t} \Delta\left(\frac{1}{F^*_s}\right)\gamma(s)\Delta F^*_s. \tag{17.10}$$

Now

$$d\Lambda^*(s) = \alpha^*(s)\,d\Lambda(s) = -\frac{dF^*_s}{F^*_{s-}},$$

$$\Delta F^*_s = -F^*_{s-}\alpha^*(s)\Delta\Lambda(s),$$

and

$$F^*_s = F^*_{s-} + \Delta F^*_s = (1 - \alpha^*(s)\Delta\Lambda(s))F^*_{s-}.$$

Similarly,

$$\Delta\left(\frac{1}{F^*_s}\right) = \frac{1}{F^*_s}\alpha^*(s)\Delta\Lambda(s).$$

Substituting in (17.10) we obtain the stated formula. □

Corollary 17.25. *Suppose $\Lambda(t)$ is continuous. Then $\eta(t)$ satisfies the differential equation*

$$\frac{d\eta}{d\Lambda}(s) = \alpha^*(s)(\eta(s) - \gamma(s)) = -H(s, u^*(s)).$$

PROOF. When Λ is a continuous $\Delta\Lambda \equiv 0$ and the sum on the right disappears. $\eta(s-) = \eta(s)$ and, from Corollary 17.22, the integrand of the remaining term is just $-H(s, u^*(s))$. □

Remarks 17.26. For continuous F_t (and so continuous Λ), the minimum principle can be written in the compact form

$$H(t, u) = \alpha^u(t)\left[\left[\int_E f^u\,d\lambda^u - \eta(t)\right]\right.$$

$$H(t, u^*) = \min_{u \in U} H(t, u),$$

$$\frac{d\eta}{d\Lambda}(t) = -H(t, u^*(t)),\ \eta(0) = J^*.$$

Extension to Several Jumps 17.27

We shall show how the above results can be extended to a process which has a sequence of jumps, as in [12], [32], [33] or [49]. That is, we discuss a process $\{z_t\}$ with values in a Lusin space $(E, \mathscr{E})$ which starts at an initial position $z_0 \in E$, and at a random time $T_1(\omega)$ jumps to a random position $z_1(\omega)$. At a second random time $T_2(\omega)$ the process jumps to $z_2(\omega)$, and so on. The basic processes associated with a general right-constant jump process, where the jump times may have several accumulation points, is discussed in [30]. However, for simplicity of exposition we suppose that the jump times $\{T_k(\omega)\}$ have a single accumulation time $T_\infty(\omega)$ (which may be finite or infinite), and we suppose that there is some fixed position $z_\infty \in E$ such that $X_t(\omega) = z_\infty$ if $t > T_\infty(\omega)$. For each positive integer i write $(Y^i, \mathscr{Y}^i)$ for a copy of the measurable space $([0, \infty] \times E, \mathscr{B} \times \mathscr{E})$. Then the underlying probability space for the above process can be taken to be the product $\Omega = \prod_{i \in N} Y^i$ with the product σ-field. Write w_k for the projection of $\omega \in \Omega$ onto the first k factors $\Omega_k = \prod_{i=1}^k Y^i$. Let $\mathscr{F}_t^0 = \sigma\{z_t : s \leq t\}$, $\mathscr{F}^0 = \bigvee_t \mathscr{F}_t^0$, and suppose we have a family of conditional probability measures $\{\mu_i, i \in N\}$, which describe the jumps of the process, and which satisfy the following conditions:

(a) μ_1 is a probability measure on $\Omega_1 = Y^1$ such that $\mu_1((\{0\} \times E) \cup (R^+ \times \{z_0\})) = 0$.
(b) For $i = 2, 3, \ldots$, $\mu_i : \Omega_{i-1} \times \mathscr{Y}^i \to [0, 1]$ is a function such that $\mu_i(\cdot, \Gamma)$ is $\mathscr{F}^0_{T_{i-1}}$ measurable for each $\Gamma \in \mathscr{Y}^i$.
(c) $\mu_i(w_{i-1}(\omega), \cdot)$ is a probability measure on $(Y^i, \mathscr{Y}^i)$ for each $\omega \in \Omega$.
(d) $\mu_i(w_{i-1}(\omega),]0, t] \times E) = 0$ for $t \leq T_{i-1}(\omega)$ for each $\omega \in \Omega$.
(e) $\mu_i(w_{i-1}(\omega), R^+ \times \{z_{i-1}(\omega)\}) = 0$ for each $\omega \in \Omega$.
(f) $\mu_i(w_{i-1}(\omega), (\infty, z_\infty)) = 1$ if $T_\infty(\omega) = \infty$.

The implications of conditions (d) and (e) are that two jump times do not occur simultaneously, and the process does have a nonzero jump at each jump time. This family $\{\mu_i\}$ defines a probability measure P on $(\Omega, \mathscr{F}^0)$ as follows: for $\Gamma \in \mathscr{Y}^1$

$$P((T_1, z_1) \in \Gamma) = \mu_1(\Gamma);$$

for $\Gamma \in \mathscr{Y}^i$ and $\eta \in \Omega_{i-1}$

$$P((T_i, z_i) \in \Gamma | w_{i-1}(\omega) = \eta) = \mu_i(\eta; \Gamma).$$

Write $\mathscr{F}_t$ (resp. $\mathscr{F}$) for the completion of $\mathscr{F}_t^0$ (resp. $\mathscr{F}^0$) with respect to the measure P. For $A \in \mathscr{E}$ write

$$F_t^{kA} = F_t^{kA}(\omega) = \mu_k(w_{k-1}(\omega),]t, \infty] \times A).$$

Then F_t^{kA} is the conditional probability that $T_k > t$ and $z_k \in A$ given $\mathscr{F}_{T_{k-1}}$.

For each $\omega \in \Omega$, F_t^{kA} is right continuous and monotonic decreasing. $F_t^k = F_t^{kE}$ is the conditional probability that $T_k > t$ given $\mathscr{F}_{T_{k-1}}$, so $F_t^k = 1$ for $t \leq T_{k-1}(\omega)$. The basic martingales can now be defined as in the single jump case of Chapter 15. However, for $T_{k-1}<t\leq T_k$ the variables are now defined on $\mathcal{F}_{T_{k-1}}$, as shown by their dependence on $w_{k-1}(\omega)$.

Lemma 17.28. *For $A\in\mathcal{E}$ write*

$$p^k(t,A)=I_{t\geq T_k}I_{z_k\in A},$$

$$\tilde{p}^k(t,A)=\tilde{p}^k(w_{k-1}(\omega),t,A)=-\int_{]0,t\wedge T_k]}\frac{dF_s^{kA}}{F_{s-}^k}.$$

Then, as in theorem 15.4,

$$q^k(t,A)=p^k(t,A)-\tilde{p}^k(t,A)$$

is an $\mathcal{F}_t$ martingale.

Remarks 17.29. For each $\omega\in\Omega$ the Stieltjes measure on $([0,\infty],\mathcal{B})$ given by $F_t^{kA}(\omega)$ is absolutely continuous with respect to that given by $F_t(\omega)$, so there is a Radon–Nikodym derivative

$$\lambda_k(w_{k-1}(\omega);t,A)=\lambda_k(t,A),$$

such that

$$F_t^{kA}-F_0^{kA}=\int_{]0,t]}\lambda_k(s,A)\,dF_s^k.$$

The λ_k can be chosen to be a regular family of conditional probability measures. Write

$$\Lambda_k(t,\omega)=-\int_{]0,t]}\frac{dF_s^k}{F_{s-}^k}.$$

Definition 17.30. The family (λ_k,Λ_k), $k=1,2,\ldots$, is the Lévy system for the multi-jump process.

Suppose $(\mathcal{U},\mathcal{G})$ is a measurable space and

$$\alpha:R^+\times U\rightarrow R^+,$$

$$\beta:R^+\times E\times U\rightarrow R$$

are measurable functions satisfying the conditions of Definition 17.14.

Definition 17.31. The set of admissible controls $\{\mathcal{U}\}$ is the set of all predictable maps $u:R^+\times\Omega\rightarrow U$. The predictable σ-field on $R^+\times\Omega$ is generated by the left continuous functions. Therefore, for $T_{k-1}(\omega)<t\leq T_k(\omega)$ $u\in\mathcal{U}$ is a function only of $w_{k-1}(\omega)$. For $u\in\mathcal{U}$ write

$$\alpha^u=\alpha(s,u(s,\omega)),$$

$$\beta^u=\beta(s,x,u(s,\omega)),$$

and for $T_{k-1}(\omega)<s\leq T_k(\omega)$ we shall write α_k^u, β_k^u for the above functions.

Remarks 17.32. We have noted how a family $\{\mu_i\}$ gives rise to a probability measure P on Ω describing when and where the jumps occur. Conversely, suppose a "base measure" P is given on Ω. Then, because Ω is a Lusin space, P gives rise to a regular family of conditional probability measures $\{\mu_i, i \in N\}$ satisfying conditions (a)–(f) above. Corresponding to the family $\{\mu_i\}$ we have a Lévy system $(\lambda_k, \Lambda_k, k \in N\}$ as in Definition 17.30.

Given an admissible control $u \in \mathcal{U}$, Radon–Nikodym derivatives α^u, β^u are determined, and a new Lévy system $(\lambda_k^u, \Lambda_k^u)$ is defined by putting

$$\frac{d\Lambda_k^u}{d\Lambda_k} = \alpha_k^u, \qquad \frac{d\lambda_k^u}{d\lambda_k} = \beta_k^u.$$

These in turn give rise to a second family of conditional probability measures μ_k, $k \in N$, by putting

$$\frac{d\mu_k^u}{d\mu_k} = L_k^u.$$

Here L_k^u is given in terms of α_k^u and β_k^u by the same formula as Theorem 17.12. Because of the bounds on α and β the measures μ_k and μ_k^u are mutually absolutely continuous. The measures μ_k^u give rise to a measure P^u on Ω which is mutually absolutely continuous with respect to P. α_k^u controls the probability of when the kth jump occurs; β_k^u controls the probability of where it goes.

Definition 17.33. Motivated by the discussion after remarks 17.16 we shall suppose the cost has the form of a sum of terms, each evaluated at a jump time and position. Therefore, we suppose there is a sequence of real valued, measurable functions f_k, $k \in N$, such that f_k is defined on $\Omega_{k-1} \times R^+ \times E \times U$, and such that they are uniformly bounded $|f_k| \leq B < \infty$, for all $k \in N$. In addition, suppose there is a discount factor $0 < r < 1$.

A control $u \in \mathcal{U}$ gives rise to a measure P^u on Ω and, writing E_u for the expectation with respect to P^u, the total expected cost if control u is used is

$$J(u) = E_u\left[\sum_{k \in N} r^k f_k(w_{k-1}(\omega); T_k, z_k, u(T_k))\right].$$

The problem again is to determine how $u \in \mathcal{U}$ should be chosen to minimize $J(u)$.

Suppose control u is used up to time t and control v is used from time t onwards. Then control $w \in \mathcal{U}$, if $w(s, \omega) = u(s, \omega)$, $s \leq t$ and $w(s, \omega) = v(s, \omega)$, $s > t$. The total expected cost corresponding to w and given $\mathcal{F}_t$ is

$$\begin{aligned}\psi(u, v, t, \omega) &= E_w\left[\sum_{k \in Z^+} r^k f_k^u \,\middle|\, \mathcal{F}_t\right] \\ &= \sum_{k \in N} r^k f_k(w_{k-1}(\omega), T_k, z_k, u(T_k))I_{t \geq T_k} + \psi(v, t).\end{aligned}$$

Here $\psi(v, t, \omega) = E_v[\sum_{k \in N} r^k f_k^v I_{t<T_k} | \mathscr{F}_t]$ and, as in Lemma 17.17, $\psi(v, t)$ is independent of u. The infimum

$$W(t) = \bigwedge_{v \in \mathscr{U}} \psi(v, t, \omega)$$

exists in the lattice $L^1(\Omega, P)$ and then

$$M_t^u = \sum_{k \in N} r^k f_k^u I_{t \geq T_k} + W(t)$$

represents the minimum achievable cost, given the evolution of the process up to time t.

Consider $u_1, u_2 \in \mathscr{U}$. Then the set $A = \{\omega : \psi(u_1, t) \leq \psi(u_2, t)\}$ is in $\mathscr{F}_t$. For $s \in]t, \infty]$ we can define

$$u_3(s, \omega) = \begin{cases} u_1(s, \omega), & \omega \in A, \\ u_2(s, \omega), & \omega \in A^c. \end{cases}$$

Then $u_3 \in \mathscr{U}$ and $\psi(u_3, t) \leq \psi(u_1, t) \wedge \psi(u_2, t)$ a.s.

Consequently the family $\{\psi(u, t), u \in \mathscr{U}\}$ satisfies the 0-lattice property, and as in Corollary 16.15 and Theorem 17.19 we can deduce the following "Principle of Optimality".

Theorem 17.34. (a) *for any $u \in \mathscr{U}$, $\{M_t^u\}$ is a submartingale on $(\Omega, \mathscr{F}_t, P^u)$.* (b) *$u^* \in \mathscr{U}$ is optimal if and only if $\{M_t^{u^*}\}$ is a martingale on $(\Omega, \mathscr{F}_t, P^{u^*})$.*

Remarks 17.35. Suppose $u^* \in \mathscr{U}$ is an optimal control and write M_t^* for $M_t^{u^*}$, P^* for P^{u^*} etc. Then

$$M_t^* = M_{t \wedge T_1}^* + \sum_{k=2}^{\infty} (M_{t \wedge T_k}^* - M_{T_{k-1}}^*) I_{t \geq T_{k-1}}, \tag{17.11}$$

because this is an identity if $t < T_\infty$, and the right-hand side equals $\lim_k M_{T_k} = M_{T_\infty -}$ if $t \geq T_\infty$. However, from our assumption that $X_t(\omega) = z_\infty$ if $t > T_\infty(\omega)$ we see

$$\mathscr{F}_{T_\infty -} = \mathscr{F}, \quad \text{so } M_{T_\infty}^* = E[M_{T_\infty}^* | \mathscr{F}_{T_\infty -}] = M_{T_\infty -}^*.$$

Therefore, (17.11) holds generally.

Define

$$X_t^1 = M_{t \wedge T_1}^*,$$
$$X_t^k = M_{(t + T_{k-1}) \wedge T_k} - M_{T_{k-1}}, \qquad k = 2, 3, \ldots.$$

Then from (17.11)

$$M_t^* = \sum_{k=1}^{\infty} X_{(t - T_{k-1}) \vee 0}^k.$$

Each $\{X^k\}$ is a uniformly integrable martingale which is zero up to time T_{k-1} and which is stopped at time T_k. Working on the space $\Omega_{k-1} \times Y$, that is, given (T_i, z_i), $i < k$, we can, as in Theorem 15.14, obtain a "conditional" representation for each of the martingales X^k given the values of T_i and z_i for $1 \le i \le k-1$. In fact, write

$$S_k^* = \sum_{n=k}^{\infty} r^n f_n^*,$$

$$\gamma_k = r^k f_k^* + E^*[S_{k+1}^* | \mathcal{F}_{T_k}],$$

$$g_k = \gamma_k - (F_t^{*k})^{-1} \int_{]s,\infty]} \gamma_k \, d\mu_k^*,$$

$$\tilde{p}_k^*(t, A) = \int_{]t \wedge T_{k-1}, t \wedge T_k] \times E} \alpha_k^* \beta_k^* \, d\lambda_k \, d\Lambda_k,$$

$$q_k^*(t, A) = p_k(t, A) - \tilde{p}_k^*(t, A).$$

Then

$$X_t^k = \int_{]0,t]} g_k \, dq_k^*$$

and

$$M_t^* = J^* + \sum_{k \in N} \int_{]t \wedge T_{k-1}, t \wedge T_k]} g_k \, dq_k^*.$$

Suppose $u \in \mathcal{U}$ is another admissible control. Then

$$\begin{aligned} M_t^u &= E_u[M_\infty^* | \mathcal{F}_t] \\ &= J^* + \sum_{k \in N} r^k (f_k^u - f_k^*) I_{t \ge T_k} + \sum_{k \in N} \int_{]t \wedge T_{k-1}, t \wedge T_k]} g_k \, dq_k^u \\ &\quad + \sum_{k \in N} \int_{]t \wedge T_{k-1}, t \wedge T_k] \times E} g_k (\alpha_k^u \beta_k^u - \alpha_k^* \beta_k^*) \, d\lambda_k \, d\Lambda_k, \end{aligned} \tag{17.12}$$

where

$$q_k^u(t, A) = p_k(t, A) - \tilde{p}_k^u(t, A)$$

and

$$\tilde{p}_k^u(t, A) = \int_{]t \wedge T_{k-1}, t \wedge T_k] \times E} \alpha_k^u \beta_k^u \, d\lambda_k \, d\Lambda_k.$$

Certainly $\{M_t^u\}$ is right continuous. However, under measure P^u it is a submartingale, and so it has a unique Doob–Meyer decomposition

$$M_t^u = J^* + N_t^u + A_t^u,$$

where $\{N_t^u\}$ is a martingale on $(\Omega, \mathcal{F}_t, P^u)$ and $\{A_t^u\}$ is a predictable increasing process. We can write

$$\begin{aligned} \sum_{k \in N} r^k (f_k^u - f_k^*) I_{t \ge T_k} &= \sum_{k \in N} r^k \int_{]t \wedge T_{k-1}, t \wedge T_k] \times E} (f_k^u - f_k^*) \, dq_k^u \\ &\quad + \sum_{k \in N} r^k \int_{]t \wedge T_{k-1}, t \wedge T_k] \times E} (f_k^u - f_k^*) \, d\tilde{p}_k^u. \end{aligned} \tag{17.13}$$

$\{M_t^u\}$ is a special semimartingale, so substituting (17.13) in (17.12) the integrals with respect to the q_k^u give martingales on $(\Omega, \mathcal{F}_t, P^u)$ and so correspond to the martingale $\{N_t^u\}$ in the Doob–Meyer decomposition. By the uniqueness of the decomposition of special semimartingales the integrals with respect to the $\tilde{p}_k^u$ must equal the increasing process $\{A_t^u\}$. Noting the form of the g_k we can state the following minimum principle:

Theorem 17.36. *For $t \in]\!]T_{k-1}, T_k]\!]$ the optimal control $u^* \in \mathcal{U}$ is, almost surely, obtained by minimizing the Hamiltonian*

$$\alpha_k^u(s)\left(\int_E (r^k f_k^u + E^*[S_{k+1}^* | \mathcal{F}_{T_k}])\beta_k^u \, d\lambda_k - (F_s^{*k})^{-1} \int_{]s,\infty]\times E} \gamma_k \, d\mu_k^*\right).$$

As in the single jump case

$$(F_s^{*k})^{-1} \int_{]s,\infty]\times E} \gamma_k \, d\mu_k^*$$

is the minimum remaining cost, given that $T_{k-1} < s \leq T_k$ and given $\mathcal{F}_s$.

CHAPTER 18

Filtering

In this chapter we suppose there is a signal process $\{x_t\}$ which describes the state of a system, but which cannot be observed directly. Instead we can only observe some noisy function $\{y_t\}$ of $\{x_t\}$. Our object is to obtain an expression for the "best estimate" of x_t, (or of $\phi(x_t)$ for ϕ in a large enough class of functions), given the observations up to time t, that is, given the σ-field

$$\mathcal{Y}_t = \sigma\{y_s : s \leq t\}.$$

The most successful result of this kind was that obtained for linear systems by Kalman [52] and Kalman and Bucy [53] in 1960 and 1961, respectively. This has been applied in many fields and a proof is given below. Attempts have been made to extend this result to nonlinear systems and we shall describe the "innovations" and "reference probability" approaches to nonlinear filtering.

Equations giving the evolution of the conditional distribution of x_t were obtained in the 1960s by, for example, Bucy [9], Kushner [58], Shiryayev [73], Stratonovich [75], and Wonham [79]. Zakai [81] showed how these results could be obtained in a simpler manner in 1969 using his "reference probability" method. Kailath [54] defined the "innovations" approach to linear filtering in 1968, and it was quickly applied to the nonlinear case. It soon became clear that the filtering problem should be formulated in terms of martingales and the general theory of processes. The definitive result using the innovations approach was given by Fujisaki *et al.* [41] in 1972. Below we give new proofs of the general nonlinear filtering equation (18.6), and also the equation for the unnormalized conditional density (18.30), using the canonical decomposition of special semimartingales.

Recent work in nonlinear filtering has been concerned, *inter alia*, with the following problems:

(a) the determination of finite dimensional nonlinear filters, (Beneš [6]),
(b) obtaining "robust" or "pathwise continuous" solutions of the filtering equations (Davis [15, 16], and Elliott and Kohlmann [35]),
(c) developing a rigorous treatment of the theory of stochastic partial differential equations (Pardoux [69], Kunita [56]),
(d) using Lie algebraic methods (Brockett [8]),
(e) applying integration over function space, and group representations (Mitter [66]).

See also the nice introduction of Davis and Marcus [20]. First note the simple fact that if ϕ is some square integrable function of the history of the signal $\{x_s: s \leq t\}$, then the "best estimate" (in mean square) of $\phi(x)$ given the observations up to time t is $E[\phi(x)|\mathcal{Y}_t]$. Roughly then, the objective of our theory is to determine an expression for the conditional probability distribution Π_t of x_t given $\mathcal{Y}_t$, and to give this in a form where it is updated recursively in a memoryless manner. This will be done by expressing Π_t in terms of a differential equation. To motivate the ideas of filtering we first, in the following example, recall some classical theorems concerning the Kolomogorov forward and backward equations from the theory of Markov diffusions. These results can be thought of as filtering with no observations at all.

Example 18.1. Suppose the signal process $\{X_t\}$ is a d-dimensional homogeneous Markov process which is the unique strong solution, as in Theorem 14.27 (see Remark 14.35), of the stochastic differential equation

$$dX_t = g(X_t)\,dt + \sigma(X_t)\,dB_t, \qquad 0 \leq t \leq T, \tag{18.1}$$

with initial condition X_0 independent of $\mathcal{F}^B_{0,T}$ (see Remarks 14.17).

Associated with $\{X_t\}$ we have, as in Definitions 14.28 and 14.34, a family of operators $\{T_t\}$, $0 \leq t \leq T$, defined on $B_0(R^d)$ by

$$\begin{aligned}(T_t f)(x) &= E[f(X_t)|x_0 = x] \\ &= \int_{R^d} f(y) P(0, x, t, dy),\end{aligned}$$

where $P(0, x, t, dy)$ is the stationary transition probability associated with $\{X_t\}$. The corresponding extended generator is

$$L_t = L = \tfrac{1}{2} \sum_{i,j=1}^{d} a^{ij}(x) \frac{\partial^2}{\partial x^i\, \partial x^j} + \sum_{i=1}^{d} g^i(x) \frac{\partial}{\partial x^i},$$

where $a = (a^{ij})$ is the $d \times d$ matrix $\sigma\sigma'$.

Write $\mathcal{M}(R^d)$ for the set of probability measures on R^d and

$$\langle \phi, \mu \rangle = \int_{R^d} \phi(x)\, d\mu(x),$$

for the inner product of $\phi \in B_0(R^d)$ and $\mu \in \mathcal{M}(R^d)$. If $\Pi \in \mathcal{M}(R^d)$, is the probability distribution of the initial value x_0 of the process, we can define an adjoint semigroup on $\mathcal{M}(R^d)$ which maps Π into the probability distribution Π_t of X_t. In fact $\Pi_t = U_t\Pi$, where for any Borel set $A \subset R^d$

$$\begin{aligned}(U_t\Pi)(A) &= \Pi_t(A)\\ &= P\{X_t \in A | x_0 \text{ has distribution } \Pi\}\\ &= \langle T_t I_A, \Pi \rangle.\end{aligned}$$

Here as usual, I_A is the indicator function of A. Therefore, U_t is the adjoint map of T_t with respect to the above inner product, so for $\phi \in B_0(R^d)$ and $\Pi \in \mathcal{M}(R^d)$

$$\begin{aligned}\langle \phi, U_t\Pi \rangle &= \langle T_t\phi, \pi \rangle\\ &= E[\phi(X_t)].\end{aligned}$$

From Theorem 14.31(d) we have for $f \in \mathscr{C}_b^2(R^d)$ that

$$\begin{aligned}\frac{d}{dt}(T_tf)(x) &= \lim_{h \searrow 0} h^{-1}((T_{t+h}f)(x) - (T_tf)(x))\\ &= \lim_{h \searrow 0} h^{-1}(T_h(T_tf)(x) - (T_tf)(x))\\ &= L(T_tf)(x) = T_t(Lf)(x).\end{aligned} \tag{18.2}$$

This is a form of the so-called Kolmogorov backward equation for the process, and on integrating we have Dynkin's formula

$$(T_tf)(x) - f(x) = \int_0^t (LT_sf)(x)\, ds. \tag{18.3}$$

(Again, this is just a special case of part (c) of Theorem 14.31.) (See Definition 14.34.)

For $f \in \mathscr{C}_b^2(R^d)$ and $\Pi \in \mathcal{M}(R^d)$

$$\begin{aligned}\langle T_tf, \Pi \rangle &= \langle f, U_t\Pi \rangle\\ &= \langle f, \Pi_t \rangle,\end{aligned}$$

so in sense of Schwartz distributions, from (18.2),

$$\frac{d}{dt}\Pi_t = L^*\Pi_t, \qquad \Pi_0 = \Pi, \tag{18.4}$$

where L^* is the adjoint of L.

This is a form of the Kolmogorov forward equation for the process because it describes the evolution of the distribution Π_t of X_t. Another form for the forward equation can be obtained under the following conditions.

Suppose the homogeneous Markov process solution $\{X_t\}$ of (18.1) has (stationary) transition probabilities $P(s, x, t, A)$ each of which is absolutely continuous with respect to Lebesgue measure on R^d. That is, suppose there is a density $p(s, x, t, y)$ such that

$$P(s, x, t, A) = \int_A p(s, x, t, y)\, dy$$

for every Borel set $A \subset R^d$. Furthermore, suppose:

(a) for $t - s > \delta > 0$, $p(s, x, t, y)$ is continuous and bounded in s, t and x;
(b) the partial derivatives

$$\frac{\partial p}{\partial s}, \frac{\partial^i p}{\partial x_i}, \frac{\partial^2 p}{\partial x_i\, \partial x_j} \quad \text{exist.}$$

Then for $0 < s < t$, p satisfies the Kolmogorov forward equation

$$\begin{aligned}\frac{\partial p}{\partial t} &= L^* p \\ &= -\sum_{i=1}^{d} \frac{\partial}{\partial y_i}(g_i(y)p(s, x, t, y)) \\ &\quad + \tfrac{1}{2} \sum_{i,j=1}^{d} \frac{\partial^2}{\partial y_i\, \partial y_j}(a^{ij}(y)p(s, x, t, y),\end{aligned} \tag{18.5}$$

together with the initial condition

$$\lim_{t \searrow s} p(s, x, t, y) = \delta(y - x),$$

(the Dirac δ-"function" at $y - x \in R^d$).

To see this, suppose $\phi \in \mathscr{C}_0^2(R^d)$, (the twice continuously differentiable functions with compact support). Then

$$\begin{aligned}(T_{t-s}\phi)(x) &= \int_{R^d} \phi(y)p(s, x, t, y)\, dy \\ &= \int_{R^d} \phi(y)p(0, x, t-s, y)\, dy,\end{aligned}$$

so from the backward equation (18.2):

$$\begin{aligned}\frac{\partial}{\partial t}\int_{R^d} \phi(y)p(s, x, t, y)\, dy &= \int_{R^d} L\phi(y) \,.\, p(s, x, t, y)\, dy \\ &= \int_{R^d} \phi(y) \frac{\partial}{\partial t} \,.\, p(s, x, t, y)\, dy.\end{aligned}$$

Because ϕ has compact support we can perform various integrations by parts to obtain identities such as

$$\int_{R^d} p(s,x,t,y)g_i(y)\frac{\partial\phi}{\partial y_i}(y)\,dy = -\int_{R^d}\phi(y)\frac{\partial}{\partial y_i}(g_i(y)p(s,x,t,y))\,dy,$$

$$\int_{R^d} p(s,x,t,y)a^{ij}(y)\frac{\partial^2\phi}{\partial y_i\,\partial y_j}(y) = \int_{R^d}\phi(y)\frac{\partial^2}{\partial y_i\,\partial y_j}(a^{ij}(y)p(s,x,t,y))\,dy.$$

Therefore, for every $\phi \in \mathscr{C}_0^2(R^d)$

$$\int_{R^d}\phi(y)\left\{\frac{\partial}{\partial t}p(s,x,t,y)+\sum_{i=1}^{d}\frac{\partial}{\partial y_i}(g_i(y)p(s,x,t,y))\right.$$
$$\left.-\tfrac{1}{2}\sum_{i,j=1}^{d}\frac{\partial^2}{\partial y_i\,\partial y_j}(a^{ij}(y)p(s,x,t,y))\right\}dy=0,$$

so the expression in braces is zero, that is

$$\frac{\partial p}{\partial t}=L^*p.$$

The boundary condition is immediate because for $t>s$:

$$E[\phi(X_t)|X_s=x]=(T_{t-s}\phi)(x)$$
$$=\int_{R^d}\phi(y)p(s,x,t,y)\,dy.$$

Sufficient conditions for the existence of a smooth density p satisfying the above conditions are, for example, that the coefficient functions g and σ in (18.1) have continuous derivatives up to the third order, which satisfy a growth condition. (See Gihman and Skorohod [42], p. 99.)

Remark 18.2. The object of filtering theory can be considered to be the derivation of forward equations, generalizing (18.4) and (18.5), for the conditional distribution Π_t of X_t, given the observed σ-field $\mathscr{Y}_t=\sigma\{y_s : s\le t\}$.

The Innovations Approach 18.3

We shall assume that all processes are defined on a fixed probability space $(\Omega, \mathscr{F}, P)$ for time $t\in[0,T]$. We suppose there is a right continuous filtration $\{\mathscr{F}_t\}$, of sub-σ-fields of $\mathscr{F}$, and that the filtration is complete (each $\mathscr{F}_t$ contains all null sets of $\mathscr{F}$). The signal process $\{X_t\}$, about which more will be said later, is adapted to the filtration $\{\mathscr{F}_t\}$, but for the moment we wish to discuss the observation process $\{y_t\}$, $0\le t\le T$, which is also adapted to $\{\mathscr{F}_t\}$. However, writing $\mathscr{Y}_t=\sigma\{y_s : s\le t\}$ we have $\mathscr{Y}_t\subset\mathscr{F}_t$, the inclusion being strict in general.

Notation 18.4. If $\{\eta_t\}$ is any process write

$$\hat{\eta}_t=E[\eta_t|\mathscr{Y}_t].$$

Observation Process 18.5

We shall suppose that the *observation process* $\{y_t\}$, $0 \leq t \leq T$, is an m-dimensional semimartingale of the form:

$$y_t = \int_0^t z_u\, du + \int_0^t \alpha(y_u)\, dw_u, \qquad y_0 = 0 \in R^m.$$

Here:

(a) $\{w_t\}$ is a standard m-dimensional Brownian motion,
(b) $\alpha : \mathscr{C}([0, T]; R^m) \to L(R^m, R^m)$ is a nonsingular matrix function such that $\|\alpha(y)\| \geq \delta > 0$ for some δ, and which satisfies a Lipschitz condition of the form

$$\|\alpha(y_t) - \alpha(y'_t)\| \leq K |y - y'|^*_t,$$

(c) $\{z_t\}$ is some functional of the signal process (for example, $z_t = h(x_t)$), and for simplicity we suppose that

$$E\left[\int_0^T z_u^2\, du\right] < \infty.$$

Definition 18.6. The process $\{\nu_t\}$, $0 \leq t \leq T$, defined by

$$\nu_t = \int_0^t \frac{dy_u}{\alpha(y_u)} - \int_0^t \frac{\hat{z}_u}{\alpha(y_u)}\, du,$$

$$\nu_0 = 0 \in R^m,$$

is the *innovations process*. This terminology is motivated by the observation that, formally, $\nu_{t+h} - \nu_t$ represents the "new" information about z obtained from observations between t and $t+h$.

Lemma 18.7. *$\{\nu_t\}$ is a standard Brownian motion with respect to the filtration $\{\mathscr{Y}_t\}$.*

PROOF. We first prove that $\{\nu_t\}$ is a $\{\mathscr{Y}_t\}$ martingale. For $s \leq t$:

$$E[\nu_t - \nu_s | \mathscr{Y}_s] = E\left[\int_s^t \frac{dy_u - \hat{z}_u\, du}{\alpha(y_u)} \middle| \mathscr{Y}_s\right]$$

$$= E\left[\int_s^t \frac{z_u - \hat{z}_u}{\alpha(y_u)}\, du + w_t - w_s \middle| \mathscr{Y}_s\right] = 0 \quad \text{a.s.}$$

by Fubini's theorem. With respect to the filtration $\{\mathscr{F}_t\}$, $\nu_t = (\nu_t^1, \ldots, \nu_t^m\}$ is an m-dimensional semimartingale. By the differentiation rule

$$\nu_t^i \nu_t^j = \int_0^t \nu_u^i\, d\nu_u^j + \int_0^t \nu_u^j\, d\nu_u^i + \langle w^i, w^j \rangle_t.$$

Therefore, with respect to the $\{\mathscr{F}_t\}$ filtration $\langle \nu^i, \nu^j \rangle_t = \langle w^i, w^j \rangle_t = \delta_{ij} t$. This process is deterministic, so with respect to the $\{\mathscr{Y}_t\}$ filtration $\langle \nu^i, \nu^j \rangle_t = \delta_{ij} t$.

$\{\nu_t\}$ is continuous, so by Corollary 12.29, $\{\nu_t\}$ is an m-dimensional Brownian motion with respect to the filtration $\{\mathscr{Y}_t\}$. □

Remarks 18.8. Clearly $\sigma\{\nu_s : s \le t\} \subset \mathscr{Y}_t$. The "innovations conjecture" of Kailath was that the innovations process provides all the information about the observations, that is, that $\mathscr{Y}_t \subset \sigma\{\nu_s : s \le t\}$, so that the two σ-fields are equal. Recently the conjecture has been proved by Allinger and Mitter [1] under the hypotheses that the observation process $\{y_t\}$ is one dimensional, and that $\{w_t\}$ is independent of $\{z_t\}$.

From Remark 12.34 we know that all martingales on the filtration generated by a Brownian motion are stochastic integrals with respect to that Brownian motion. Consequently, if the innovation conjecture is true, any $\{\mathscr{Y}_t\}$ martingale is a stochastic integral with respect to $\{\nu_t\}$. However, without assuming the validity of the innovations conjecture, we can again apply the martingale representation result of Fujusaki *et al.* [41] proved in Theorem 16.22. In that theorem $\mathscr{F}_t$ is the completion of $\sigma\{B_s : s \le t\} = \sigma\{x_s : s \le t\}$ because $dB_t = \sigma(t,x)^{-1}\,dx_t$, and under measure P, for $0 \le t \le 1$,

$$dx_t = f(t,x)\,dt + \sigma(t,x)\,dw_t,$$

with initial condition x_0. Theorem 16.22 then showed that every square integrable P martingale $\{N_t\}$ with respect to the filtration $\{\mathscr{F}_t\}$ is a stochastic integral of the form

$$N_t = \int_0^t \gamma_s\,dw_s.$$

Because all processes are continuous, all local martingales (and *a fortiori*, all martingales), are locally square integrable, and so have a representation of this form where the predictable integrand $\{\gamma_s\}$ is locally square integrable. In the present situation $\mathscr{Y}_t = \sigma\{y_s : s \le t\}$, and under measure P, for $0 \le t \le T$,

$$dy_t = \hat{z}_t\,dt + \alpha(y_t)\,d\nu_t,$$

with initial condition $y_0 = 0 \in R^m$. However, because α is nonsingular and Lipschitz we could consider the process $\tilde{y}_t$ defined by

$$d\tilde{y}_t = \alpha^{-1}(y_t)\,dy_t, \qquad \tilde{y}_0 = 0,$$

and

$$\sigma\{\tilde{y}_s : s \le t\} = \mathscr{Y}_t,$$

so that

$$d\tilde{y}_t = \tilde{\alpha}^{-1}(\tilde{y}_t)\hat{z}_t\,dt + d\nu_t.$$

The situation is then analogous to that described above, so that Theorem 16.22 in this context states the following:

Theorem 18.9. *If $\{N_t\}$ is a locally square integrable martingale with respect to the filtration $\{\mathscr{Y}_t\}$ then there is a $\{\mathscr{Y}_t\}$ predictable m-dimensional process*

$\{\gamma_s\}$, $0 \le s \le T$, *and an increasing sequence of stopping times* $\{T_n\}$ *such that*:

$$\lim_n T_n = T \quad a.s.$$

$$E\left[\int_0^{T_n} |\gamma_u|^2 \, du\right] < \infty$$

and, for $0 \le t \le T$,

$$N_t = E[N_0] + \int_0^t \gamma_u \, dv_u \quad a.s.$$

Remarks 18.10. To obtain the general filtering equation we shall consider a real $\{\mathscr{F}_t\}$ semimartingale $\{\xi_t\}$, $0 \le t \le T$, and obtain a stochastic differential equation satisfied by $\hat{\xi}_t$. The kind of semimartingale we have in mind is some real valued function ϕ of the signal process $\{x_t\}$. The differential equation we obtain will provide the recursive and memoryless filter for $\hat{\xi}_t$.

Theorem 18.11. *Suppose* $\{\xi_t\}$, $0 \le t \le T$, *is a real* $\{\mathscr{F}_t\}$ *semimartingale of the form* $\xi_t = \xi_0 + \int_0^t \beta_u \, du + N_t$. *Here* $E[\xi_0^2] < \infty$, $E[\int_0^T \beta_u^2 \, du] < \infty$ *and* $\{N_t\}$ *is a square integrable* $\{\mathscr{F}_t\}$ *martingale.*

The observation process is of the form:

$$dy_t = z_t \, dt + \alpha(y_t) \, dw_t = \hat{z}_t \, dt + \alpha(y_t) \, dv_t,$$

$$y_0 = 0 \in R^m, \quad \text{Here, } \{z_t\} \text{ and } \alpha \text{ are as in } 18.5,$$

$\{w_t\}$, $0 \le t \le T$, *is a standard m-dimensional Brownian motion and* $\langle N, w^i \rangle_t = \int_0^t \lambda_u^i \, du$, $1 \le i \le m$. *Then* $\{\hat{\xi}\}$ *is given by the stochastic differential equation*

$$\hat{\xi}_t = \hat{\xi}_0 + \int_0^t \hat{\beta}_u \, du + \int_0^t (\hat{\lambda}_u + \alpha^{-1}(y_u)(\widehat{\xi_u z_u} - \hat{\xi}_u \hat{z}_u))' \, dv_u. \tag{18.6}$$

Proof. The proof below is an extension of an idea of Wong [77] and uses the unique decomposition of special semimartingales. Define

$$M_t = \hat{\xi}_t - \hat{\xi}_0 - \int_0^t \hat{\beta}_u \, du.$$

Then for $0 \le s \le T$:

$$E[M_t - M_s | \mathscr{Y}_s] = E\left[\hat{\xi}_t - \hat{\xi}_s - \int_s^t \hat{\beta}_u \, du \middle| \mathscr{Y}_s\right].$$

However

$$\begin{aligned} E[\hat{\xi}_t - \hat{\xi}_s | \mathscr{Y}_s] &= E[\xi_t - \xi_s | \mathscr{Y}_s] \\ &= E\left[\int_s^t \beta_u \, du \middle| \mathscr{Y}_s\right] + E[N_t - N_s | \mathscr{Y}_s] \end{aligned}$$

$$= E\left[\int_s^t E[\beta_u|\mathcal{Y}_u]\,du\middle|\mathcal{Y}_s\right] + E[E[N_t - N_s|\mathcal{F}_s]\mathcal{Y}_s]$$

$$= E\left[\int_s^t \hat{\beta}_u\,du\middle|\mathcal{Y}_s\right],$$

because $\{N_t\}$ is an $\{\mathcal{F}_t\}$ martingale. Therefore, $\{M_t\}$ is locally a square integrable martingale, so by Theorem 18.9 there is a $\{\mathcal{Y}_t\}$ predictable process $\{\gamma_u\}$ such that

$$M_t = \int_0^t \gamma'_u\,d\nu_u \quad \text{a.s.}$$

and we can write

$$\hat{\xi}_t = \hat{\xi}_0 + \int_0^t \hat{\beta}_u\,du + \int_0^t \gamma'_u\,d\nu_u. \tag{18.7}$$

(The prime here denotes the transpose of column vector γ_u to row vector γ'_u.) We now wish to determine γ_u.

$\xi \in R$ and $y \in R^m$, but by the differentiation rule

$$\xi_t y_t = \xi_0 y_0 + \int_0^t \xi_u(z_u\,du + \alpha(y_u)\,dw_u) + \int_0^t y_u(\beta_u\,du + dN_u) + \langle N, y\rangle_t. \tag{18.8}$$

However,

$$\langle N, y\rangle = \int_0^t \alpha(y^u)\lambda_u\,du.$$

The integrals

$$H_1(t) = \int_0^t \xi_u\alpha(y_u)\,dw_u,$$

$$H_2(t) = \int_0^t y_u\,dN_u,$$

are locally square integrable $\{\mathcal{F}_t\}$ martingales. The conditions on the components of ξ_t imply that $E[\xi_t^2] < \infty$ for all $t \in [0, T]$. Write

$$S_n = \sup\{u : \|\alpha(y_u)\| \le n\}$$

$$T_n = \sup\{u : |y_u| \le n\}$$

so that S_n and T_n are $\{\mathcal{Y}_t\}$ stopping times. Then for $t \in [0, T]$:

$$E\left[\left(\int_0^{t\wedge S_n} \xi_u\alpha(y_u)\,dw_u\right)^2\right] \le n^2 \int_0^t E[\xi_u^2]\,du < \infty$$

and

$$E\left[\left(\int_0^{t\wedge T_n} y_u\,dN_u\right)^2\right]\le n^2E\left[\int_0^T d\langle N,N\rangle_u\right]<\infty.$$

Consequently, the processes $\hat{H}_1, \hat{H}_2$ are locally square integrable martingales with respect to the filtration $\{\mathcal{Y}_t\}$.

Consider the processes

$$K_1(t)=\int_0^t \xi_u z_u\,du$$

$$K_2(t)=\int_0^t y_u\beta_u\,du$$

$$K_3(t)=\int_0^t \alpha(y_u)\lambda_u\,du.$$

Then by a calculation similar to that for M above the processes:

$$\tilde{K}_1(t)=\hat{K}_1(t)-\int_0^t \widehat{\xi_u z_u}\,du$$

$$\tilde{K}_2(t)=\hat{K}_2(t)-\int_0^t y_u\hat{\beta}_u\,du$$

$$\tilde{K}_3(t)=\hat{K}_3(t)-\int_0^t \alpha(y_u)\hat{\lambda}_u\,du$$

are local martingales with respect to the filtration $\{\mathcal{Y}_t\}$. Therefore, from (18.8)

$$\begin{aligned}\widehat{\xi_t y_t}=\hat{\xi}_t y_t &=\hat{\xi}_0 y_0+\hat{H}_1(t)+\hat{H}_2(t)\\ &+\tilde{K}_1(t)+\int_0^t \widehat{\xi_u z_u}\,du+\tilde{K}_2(t)+\int_0^t y_u\hat{\beta}_u\,du\\ &+\tilde{K}_3(t)+\int_0^t \alpha(y_u)\hat{\lambda}_u\,du.\end{aligned}\tag{18.9}$$

Because this represents $\hat{\xi}_t y_t$ as the sum of local martingales plus continuous (and so predictable), bounded variation processes, we see $\hat{\xi}_t y_t$ is a special semimartingale with respect to the filtration $\{\mathcal{Y}_t\}$. However, using (18.7) and the differential rule

$$\begin{aligned}\hat{\xi}_t y_t=\hat{\xi}_0 y_0 &+\int_0^t \hat{\xi}_u(\hat{z}_u\,du+\alpha(y_u)\,d\nu_u)\\ &+\int_0^t y_u(\hat{\beta}_u\,du+\gamma'_u\,d\nu_u)+\int_0^t \alpha(y_u)\gamma_u\,du.\end{aligned}\tag{18.10}$$

The integrals with respect to $\{\nu_t\}$ are again local martingales, and the remaining integrals give continuous, and so predictable, processes. The two canonical decompositions of the special semimartingale $\hat{\xi}_t y_t$ must be the same, so equating the integrands in the bounded variation terms:

$$\widehat{\xi_u z_u} + \alpha(y_u)\hat{\lambda}_u = \hat{\xi}_u \hat{z}_u + \alpha(y_u)\gamma_u \quad \text{a.s.}$$

Therefore

$$\gamma_u = \alpha(y_u)^{-1}(\widehat{\xi_u z_u} - \hat{\xi}_u \hat{z}_u) + \hat{\lambda}_u.$$

Substituting in (18.7) the result follows. □

Theorem 18.12. *Suppose that the signal process $\{X_t\}$ is, as in Example* 18.1, *the unique strong solution of the stochastic differential equation*

$$dX_t = g(X_t)\,dt + \sigma(X_t)\,dB_t, \qquad 0 \le t \le T,$$

with initial condition x_0 independent of $\mathscr{F}^B_{0,T}$. Suppose that ϕ is a twice continuously differentiable function on R^d so that

$$\phi(X_t) = \phi(X_0) + \int_0^t L\phi(X_u)\,du + \int_0^t \nabla\phi\,.\,\sigma(X_u)\,dB_u.$$

Furthermore, suppose the terms on the right satisfy the same integrability conditions as the components of ξ_t. The observation process $\{y_t\}$ will be as above with $z_t = h(X_t)$, a bounded function, and suppose

$$\langle B^i, w^j\rangle_t = \int_0^t \rho_u^{ij}\,du, \qquad 1 \le i \le d, \quad 1 \le j \le m.$$

For $f \in L^2_{\mathrm{loc}(R^d)}$ write

$$\Pi_t(f) = E[f(X_t)|\mathscr{Y}_t].$$

Then

$$\Pi_t(\phi) = \Pi_0(\phi) + \int_0^t \Pi_u(L\phi)\,du$$

$$+ \int_0^t \{\Pi_u(\nabla\phi\,.\,\sigma\,.\,\rho) + \alpha^{-1}(y_u)(\Pi_u(\phi h) - \Pi_u(\phi)\Pi_u(h))\}'\,d\nu_u. \qquad (18.11)$$

PROOF. $\phi(X_t)$ plays the role of the semimartingale ξ_t in Theorem 18.11 above.

$$N_t = \int_0^t \nabla\phi\,.\,\sigma(X_u)\,dB_u$$

is a martingale, and for $1 \le j \le m$

$$\langle N, w^j\rangle_t = \int_0^t \nabla\phi\,.\,\sigma(X_u)\,.\,\rho_u^j\,.\,du,$$

where

$$\rho_u^j = (\rho_u^{1j}, \ldots, \rho_u^{dj}).$$

Therefore, in the notation of Theorem 18.11

$$\lambda_u^j = \nabla\phi \,.\, \sigma(X_u) \,.\, \rho_u^j.$$

Also

$$\hat{z}_u = \Pi_u(h).$$

Substituting in the formula of Theorem 18.11 gives the stated result. □

Remarks 18.13. Because the right-hand side of equation (18.6) involves $\widehat{\xi_u z_u}$, (as well as $\hat{\xi}_u$), it is not recursive in $\hat{\xi}_u$. However, the formula of Theorem 18.12 can be considered as a recursive stochastic differential equation for Π_t, the conditional probability distribution of X_t given $\mathcal{Y}_t$, because

$$\Pi_t(\phi) = E[\phi(X_t)|\mathcal{Y}_t] = \langle \phi, \Pi_t \rangle,$$

for twice continuously differentiable ϕ with compact support. However, this is then a stochastic differential equation with a variable in the infinite dimensional space of probability measures. Only in certain special cases is it possible to obtain finite dimensional recursive filters, even for the conditional mean $\hat{X}_t$.

Suppose that the signal and observation noise are independent, so that $\langle B^i, w^j \rangle_t = 0$, and that $\alpha(y_u) = AI$, where $A > 0$ and I is the $d \times d$ identity matrix. Then as $A \to \infty$ the observations become infinitely noisy, so giving no additional information about the signal, and equation (18.11) reduces to

$$\Pi_t(\phi) = \Pi_0(\phi) + \int_0^t \Pi_u(L\phi)\, du,$$

the same equation as given by Dynkin's formula (18.3) for the unconditional expectation $E[\phi(X_t)]$.

Corollary 18.14. *Suppose that the signal and observation noise are independent and that the conditional distribution of* X_t given $\mathcal{Y}_t$, $P[X_t \le x|\mathcal{Y}_t]$ *has a density* $\hat{p}(t, x) = dP[X_t \le x|\mathcal{Y}_t]/dx$ *which satisfies suitable differentiability hypotheses, similar to those of p in Example* 18.1. (*See Section* 8.6 *of Lipster and Shiryayev* [59].) *Then with the hypotheses and notation of Theorem* 18.12

$$\Pi_t(h) = \int_{R^d} h(x)\hat{p}(t, x)\, dx.$$

Using integration by parts

$$\Pi_t(L\phi) = \int_{R^d} L\phi(x)\hat{p}(t, x)\, dx = \langle L\phi, p \rangle = \langle \phi, L^* p \rangle.$$

Equation (18.11) *holds for all twice continuously differentiable functions with compact support, so substituting we obtain the following recursive, infinite dimensional equation for* $\hat{p}$:

$$d\hat{p} = L^*\hat{p}\,dt + \alpha^{-1}\,.\,\hat{p}(h - \Pi_t(h))\,dv_t. \tag{18.12}$$

This equation is the analog of the Kolmogorov forward equation (18.5), *and if* $\alpha = AI$, *as above, it converges to it as* $A \to \infty$. *Unfortunately* (18.12) *is further complicated by the term* $\Pi_t(h)$.

Remarks 18.15. Suppose the signal and observation noise are independent, that $d = 1$ and $\phi(x) = x$. Substituting in (18.11) we obtain the following equation for the conditional mean $\hat{X}_t = \Pi_t(x)$:

$$\hat{X}_t = \hat{X}_0 + \int_0^t \Pi_u(g)\,du + \int_0^t (\Pi_u(hx) - \hat{X}_u\Pi_u(h))\,dv_u. \tag{18.13}$$

Therefore, to calculate $\hat{X}_t$ we need to know $\Pi_u(g)$, $\Pi_u(hx)$, and $\Pi_u(h)$ so the equation is not, in general, recursive. One situation where a recursive, finite dimensional filter for $\hat{X}_t$ is obtained, is when the equations for the signal and observation are both linear, with Brownian motion noise. The Kalman–Bucy filter is then obtained, and we describe its derivation in the next result.

Theorem 18.16 (Kalman–Bucy Filter). *Suppose for simplicity of exposition that both the signal process* $\{X_t\}$ *and the observation process* $\{y_t\}$ *are one-dimensional and given by the following linear equations*:

$$X_t = X_0 + \int_0^t aX_u\,du + bB_t,$$

$$y_t = \int_0^t cX_u\,du + w_t, \qquad 0 \le t \le T,$$

where $\{B_t\}$ *and* $\{w_t\}$ *are independent Brownian motions, so* $\langle B, w\rangle_t = 0$, a, b, c *are constants, and* X_0 *is a Gaussian random variable independent of* $\mathcal{F}^B_{0,T}$.

Then $\hat{X}_t$ *is given by the following finite dimensional recursive filtering equation*:

$$\hat{X}_t = \hat{X}_0 + \int_0^t a\hat{X}_u\,du + c\int_0^t P_u\,dv_u. \tag{18.14}$$

Here $\{v_u\}$ *is the innovations process given by* $dv_u = dy_u - c\hat{X}_u\,du$ *and* $P_t = E[(X_t - \hat{X}_t)^2 | \mathcal{Y}_t]$ *is the conditional convariance of the error. Furthermore,* P_t *is the solution of the deterministic equation*

$$\frac{dP_t}{dt} = 2aP_t + b^2 - c^2P_t^2.$$

PROOF. Substituting in (18.13) we have immediately

$$\hat{X}_t = \hat{X}_0 + \int_0^t a\hat{X}_u\,du + c\int_0^t [\Pi_u(X_u^2) - \hat{X}_u^2]\,d\nu_u, \tag{18.15}$$

where

$$d\nu_u = dy_u - \hat{X}_u\,du$$

as usual describes the innovations process. That

$$\begin{aligned} P_t &= \Pi_t(X_t^2) - \hat{X}_t^2 \\ &= E[(X_t - \hat{X}_t)^2 | \mathcal{Y}_t] \end{aligned}$$

is nonrandom follows because the process $\{X_t, y_t\}$ is Gaussian. In fact we can explicitly express X_t as

$$X_t = e^{at}\left(X_0 + b\int_0^t e^{-au}\,dB_u\right).$$

Then because

$$y_t = \int_0^t cX_u\,du + w_t,$$

both X_t and y_t are limits of sums of Gaussian random variables, and so are Gaussian. For $t \in [0, T]$ consider an increasing sequence $\{\Pi_n\}$ of partitions of $[0, t]$, where

$$\Pi_n = \{0 = t_0^n < t_1^n < \cdots < t_{N_n}^n = t\}$$

and $\bigcup_n \Pi_n$ is dense in $[0, t]$. Then

$$E[X_t \in A | y_\tau, \tau \in \Pi_n]$$

is Gaussian. Because $\{y_s\}$, $0 \le s \le t$, is continuous, $\mathcal{Y}_t$ is the limit of the increasing family of σ-fields $\sigma\{y_\tau : \tau \in \Pi_n\}$. Therefore, by the martingale convergence of Corollary 3.19,

$$E[X_t | \mathcal{Y}_t] = \hat{X}_t = \lim_n E[X_t | y_\tau, \tau \in \Pi_n]$$

and so $\hat{X}_t$ is Gaussian.

Write $\mathcal{L}(y, t)$ for the subspace of $L^2(\Omega, \mathcal{F}, P)$ spanned by the random variables $\{y_s : 0 \le s \le t\}$. Then $\hat{X}_t$ is the projection of X_t onto $\mathcal{L}(y, t)$ and $X_t - \hat{X}_t$ is orthogonal to $\mathcal{L}(y, t)$. Because we are dealing with Gaussian random variables $X_t - \hat{X}_t$ is, therefore, independent of the random variables $\{y_s : 0 \le s \le t\}$, that is $X_t - \hat{X}_t$ is independent of $\mathcal{Y}_t$. Consequently,

$$\begin{aligned} E[(X_t - \hat{X}_t)^2 | \mathcal{Y}_t] &= E[(X_t - \hat{X}_t)^2] \\ &= P_t, \end{aligned}$$

so P_t is nonrandom. Also

$$\begin{aligned} E[(X_t - \hat{X}_t)^3 | \mathcal{Y}_t] &= E[(X_t - \hat{X}_t)^3] \\ &= 0. \end{aligned} \tag{18.16}$$

From (18.15) we have that

$$(\hat{X}_t)^2 = (\hat{X}_0)^2 + 2\int_0^t \hat{X}_u \, d\hat{X}_u + c^2 \int_0^t P_u^2 \, du$$

$$= (\hat{X}_0)^2 + \int_0^t (2a(\hat{X}_u)^2 + c^2 P_u^2)\, du + 2c\int_0^t \hat{X}_u P_u \, dv_u. \qquad (18.17)$$

Also, substituting in the general filtering equation (18.11) for $\phi = x^2$

$$\Pi_t(X_t^2) = E[X_t^2 \,|\, \mathscr{Y}_t]$$

$$= \Pi_0(X_0^2) + \int_0^t (2a\Pi_u(X_u^2) + b^2)\, du + \int_0^t c(\Pi_u(X_u^3) - \hat{X}_u \Pi_u(X_u^2))\, dv_u. \qquad (18.18)$$

Subtracting (18.17) from (18.18)

$$P_t = \Pi_t(X_t^2) - (\hat{X}_t)^2$$

$$= P_0 + \int_0^t (2aP_u + b^2 - c^2 P_u^2)\, du + c\int_0^t (\Pi_u(X_u^3) + 2(\hat{X}_u)^3 - 3\hat{X}_u \Pi(X_u^2))\, dv_u. \qquad (18.19)$$

However, we have observed above that P_t is nonrandom, so the integrand in the stochastic integral with respect must be zero. This is so, because it is

$$E[(X_u - \hat{X}_u)^3 | \mathscr{Y}_u] = \Pi_u(X_u^3) + 2(\hat{X}_u)^3 - 3\hat{X}_u \Pi(x_u^2),$$

which is zero by (18.16). Therefore,

$$P_t = P_0 + \int_0^t (2aP_u + b^2 - c^2 P_u^2)\, du. \qquad (18.20)$$

□

Remarks 18.17. P_t represents the "tracking error" or "gain". The fact that it is deterministic relies very heavily on the Gaussian nature of the X and y processes. The nonlinear (quadratic) equation (18.20) satisfied by P_t is known as a Ricatti equation. Standard results for ordinary differential equations imply it has a unique solution. We see from equations (18.15) and (18.18), how in order to compute conditional moments of X_t a knowledge of higher conditional moments is required. (For example, in (18.18) to obtain $\Pi_t(X_t^2)$ we need $\Pi_t(X_t^3)$.) However, in the (conditionally) Gaussian case all higher moments $\Pi_t(X_t^n)$ can be expressed in terms of $\Pi_t(X_t^2)$ and $\Pi_t(X_t) = \hat{X}_t$. This situation is extensively investigated in the books of Lipster and Shiryayev [59], and Kallianpur [55], where the analogous Kalman–Bucy formulae for $\hat{X}_t$ and P_t are given, when X_t and y_t are described by linear vector equations with deterministic coefficients.

Example 18.18. As a final application of the general filtering equation (18.6) consider the filtering of a finite state Markov process $\{X_t\}$, as discussed by Wonham [79]. In this example we have a state space S with just finitely many points $s_1, \ldots, s_N$. p_t^i will denote the probability that $X_t = s_i$. We shall suppose the probability (column) vector, (i.e. probability measure on S), $p_t = (p_t^1, \ldots, p_t^N)'$ satisfies the following analog of the Kolmogorov forward equation (18.4):

$$\frac{dp_t}{dt} = Ap_t \tag{81.21}$$

where $A = (a^{ij})$ is a constant $n \times n$ matrix. Let ϕ_i be the function on S defined by

$$\phi_i(x) = \begin{cases} 1, & \text{if } x = s_i \\ 0, & \text{if } x \neq s_i \end{cases}$$

and $\phi(x)$ be the column vector $(\phi_1(x), \ldots, \phi_N(x))$. Then $p_t = E[\phi(X_t)]$ and from (18.21)

$$\phi(X_t) - \phi(X_0) - \int_0^t A\phi(X_u)\, du$$

is a martingale N_t. The process whose conditional expectation is required, therefore, is represented by:

$$\phi(X_t) = \phi(X_0) + \int_0^t A\phi(X_u)\, du + N_t.$$

We suppose the observation process $\{y_t\}$ is scalar and given by an equation

$$y_t = \int_0^t h(X_u)\, du + w_t, \qquad 0 \leq t \leq T,$$

where $\{w_t\}$ is a standard scalar Brownian motion independent of $\{X_t\}$, and so of $\{N_t\}$. Here $h: S \to R$; write H for the column vector $(h(s_1), \ldots, h(s_N))$ and B for the diagonal matrix whose diagonal is H.

We wish to determine the conditional probabilities

$$\hat{p}_t^i = P[X_t = s_i | \mathcal{Y}_t] = \hat{\phi}(X_t).$$

Now

$$\begin{aligned} \hat{h}(X_u) &= E[h(X_u) | \mathcal{Y}_u] \\ &= \sum_{i=1}^N h(s_i) p_u^i, \end{aligned}$$

so the innovation process $\{\nu_u\}$ is given by

$$\begin{aligned} d\nu_u &= dy_u - \hat{h}(X_u)\, du \\ &= dy_u - \langle H', \hat{p}_u \rangle\, du, \end{aligned}$$

where

$$\hat{p}_u = (\hat{p}_u^1, \ldots, \hat{p}_u^N)'.$$

Substituting in (18.6) we have the following system of stochastic equations:

$$\hat{p}_t = \hat{p}_0 + \int_0^t A\hat{p}_u \, du + \int_0^t (B - \langle H', \hat{p}_u \rangle I) p_u \, d\nu_u.$$

Here I is the $N \times N$ identity matrix. These equations provide a recursive description for the conditional probability distributions.

The Reference Probability Method 18.19

This approach was first considered by Zakai [81]. Again we assume all processes are defined on a fixed probability space $(\Omega, \mathcal{F}, P)$ for time $t \in [0, T]$. We suppose there is a right continuous filtration $\{\mathcal{F}_t\}$, of sub-σ-fields of $\mathcal{F}$, and that each $\mathcal{F}_t$ contains all null sets of $\mathcal{F}$.

The signal process $\{X_t\}$ will be, as in Example 18.1, a d-dimensional homogeneous Markov process, which is the unique strong solution of the stochastic differential equation

$$dX_t = g(X_t)\, dt + \sigma(X_t)\, dB_t, \qquad 0 \le t \le T,$$

with initial condition X_0 independent of $\mathcal{F}_{0.T}^B = \sigma\{B_s : 0 \le s \le T\}$. The infinitesimal generator of $\{X_t\}$ is, as in Example 18.1,

$$L = \tfrac{1}{2} \sum_{i,j=1}^d a^{ij}(x) \frac{\partial^2}{\partial x^i \, \partial x^j} + \sum_{i=1}^d g^i(x) \frac{\partial}{\partial x^i}.$$

We shall suppose the observation process $\{y_t\}$ is defined by the m-dimensional system of equations

$$\begin{aligned} dy_t &= h(X_t)\, dt + \alpha(y_t)\, dw_t, \qquad 0 \le t \le T, \\ y_0 &= 0 \in R^m. \end{aligned} \tag{18.22}$$

Here $\alpha : \mathscr{C}[0, T]; R^m) \to L(R^m, R^m)$ is a nonsingular matrix, $h : R \to R$ is bounded and measurable, $\{w_t\}$ is a standard m-dimensional Brownian motion and as in Theorem 18.12,

$$\langle B^i, w^j \rangle_t = \int_0^t \rho_u^{ij} \, du, \qquad 1 \le i \le d, \quad 1 \le j \le m.$$

As before write $\mathscr{Y}_t = \sigma\{y_s : s \le t\}$.

For $\phi \in \mathscr{C}_b^2(R^d)$ we have from Theorem 18.12

$$\begin{aligned} E[\phi(X_t) | \mathscr{Y}_t] = \Pi_t(\phi) = \Pi_0(\phi) + \int_0^t \Pi_u(L\phi)\, du + \int_0^t \{\Pi_u(\nabla\phi \,.\, \sigma \,.\, \rho) \\ + \alpha^{-1}(y_u)(\Pi_u(\phi h) - \Pi_u(\phi)\Pi_u(h))\}'\, d\nu_u. \end{aligned} \tag{18.23}$$

Here, as above,

$$dv_u = \alpha(y_u)^{-1}(dy_u - \Pi_u(h)\,du).$$

In this section, instead of $\Pi_t(\phi)$, we shall consider an "unnormalized" conditional expectation and show it satisfies a less complicated equation than (18.11).

First we introduce a new probability measure P_0 on $(\Omega, \mathcal{F})$ defined by $dP_0/dP = \Lambda_T^{-1}$ where

$$\begin{aligned}\Lambda_t^{-1} &= \mathscr{E}\left(\int_0^t -\{\alpha^{-1}(y_u)h(X_u)\}'\,dw_u\right\}\\ &= \exp\left\{-\int_0^t (\alpha^{-1}(y_u)h(X_u))'\,dw_u - \tfrac{1}{2}\int_0^t |\alpha^{-1}(y_u)h(X_u)|^2\,du\right\}\\ &= \exp\left\{-\int_0^t (\alpha^{-1}(y_u)h(X_u))'\alpha^{-1}(y_u)\,dy_u + \tfrac{1}{2}\int_0^t |\alpha^{-1}(y_u)h(X_u)|^2\,du\right\}.\end{aligned}$$

Consequently,

$$\Lambda_t = \exp\left\{\int_0^t (\alpha^{-1}(y_u)h(X_u))'\alpha^{-1}(y_u)\,dy_u - \tfrac{1}{2}\int_0^t |\alpha^{-1}(y_u)h(X_u)|^2\,du\right\}.$$

By Girsanov's theorem, (Corollary 13.25), under P_0 the processes $v_t = (v^1, \ldots, v^d)$ and $\tilde{y}_t$ are standard d- and m-dimensional Brownian motions, respectively, where

$$dv_t^i = dB_t^i + \langle \rho^i, \alpha^{-1}h\rangle\,dt.$$

and

$$\begin{aligned}d\tilde{y}_t &= \alpha^{-1}\,dy = dw_t + \alpha^{-1}h\,dt\\ &= dv_t + \alpha^{-1}(y_t)\Pi_t(h)\,dt.\end{aligned}$$

Furthermore, under P_0

$$\langle v^i, \tilde{y}^j\rangle_t = \int_0^t \rho_u^{ij}\,du.$$

Now Λ_t is an $\{\mathcal{F}_t\}$ martingale under P_0 and

$$\Lambda_t = E_0\left[\frac{dP}{dP_0}\middle|\mathcal{F}_t\right]$$

where E_0 denotes expectation with respect to P_0. From Loève [60], § 27.4 for any integrable function ϕ

$$\begin{aligned}\Pi_t(\phi) &= E[\phi(X_t)|\mathcal{Y}_t]\\ &= \frac{E_0[\Lambda_t\phi(X_t)|\mathcal{Y}_t]}{E_0[\Lambda_t|\mathcal{Y}_t]}\\ &= \frac{\sigma_t(\phi)}{\sigma_t(1)},\end{aligned} \tag{18.24}$$

where

$$\sigma_t(\phi) = E_0[\Lambda_t \phi(X_t)|\mathcal{Y}_t]$$

is the unnormalized conditional expectation we shall now investigate. By definition

$$\sigma_t(\phi) = \sigma_t(1) \,.\, \Pi_t(\phi),$$

so, as we already have an expression (18.11) for $\Pi_t(\phi)$, we shall derive an equation for $\sigma_t(\phi)$ by obtaining an equation for $\sigma_t(1) = E_0[\Lambda_t|\mathcal{Y}_t]$ and using the differentiation rule for the product.

Notation 18.20. Write

$$\hat{\Lambda}_t = E_0[\Lambda_t|\mathcal{Y}_t].$$

(Note the expectation here is with respect to measure P_0.)

Theorem 18.21.

$$\hat{\Lambda}_t = \exp\left\{\int_0^t (\alpha^{-1}(y_u)\Pi_u(h))'\alpha^{-1}(y_u)\,dy_u - \tfrac{1}{2}\int_0^t |\alpha^{-1}(y_u)\Pi_u(h)|^2\,du\right\}.$$

Here

$$\Pi_u(h) = E[h(X_u)|\mathcal{Y}_u],$$

and is an expectation with respect to measure P.

PROOF. We have already noted that, under P_0, Λ_t is an $\{\mathcal{F}_t\}$ martingale and

$$\Lambda_t = \mathscr{E}\left(\int_0^t (\alpha^{-1}(y_u)h(X_u))'\alpha^{-1}(y_u)\,dy_u\right)$$

so

$$\Lambda_t = 1 + \int_0^t \Lambda_u(\alpha^{-1}(y_u)h(X_u))'\alpha^{-1}(y_u)\,dy_u. \tag{18.25}$$

Now $\hat{\Lambda}_t$ is a $\{\mathcal{Y}_t\}$ martingale under P_0; in fact, as in the remarks preceding Theorem 18.9, $\hat{\Lambda}_t$ is locally square integrable. Therefore, from Theorem 18.9 there is a locally square integrable, $\{\mathcal{Y}_t\}$-predictable, m-dimensional, process $\{\eta_s\}$, such that for all $t \in [0, T]$:

$$\hat{\Lambda}_t = 1 + \int_0^t \eta'_u \,.\, \alpha^{-1}(y_u)\,dy_u. \tag{18.26}$$

As in Theorem 18.11 we shall identify $\{\eta_u\}$ by using the unique decomposition of special semimartingales. Using the differentiation rule for Λ_t and $y_t \in R^m$ we have from (18.22) and (18.25):

$$\Lambda_t y_t = \int_0^t \Lambda_u\,dy_u + \int_0^t y_u\Lambda_u(\alpha^{-1}(y_u)h(X_u))'\alpha^{-1}(y_u)\,dy_u$$

$$+ \int_0^t \Lambda_u(\alpha^{-1}(y_u)h(X_u))'\alpha(y_u)\,du. \tag{18.27}$$

Because h and α^{-1} are bounded we have for some constant K

$$E_0[\Lambda_t^2] \leq 2 + 2K \int_0^t E_0[\Lambda_u^2]\, du$$

for $t \in [0, T]$. Therefore, by Gronwall's inequality, Lemma 14.20, Λ_t is a square integrable $\{\mathcal{F}_t\}$ martingale under P_0. Consider the $\{\mathcal{Y}_t\}$ stopping times $\{S_n\}$ and $\{T_n\}$ of Theorem 18.11 and the processes

$$H_1(t) = \int_0^t \Lambda_u\, dy_u$$

$$H_2(t) = \int_0^t y_u \Lambda_u (\alpha^{-1}(y_u) h(X_u))' \alpha^{-1}(y_u)\, dy_u.$$

Then, using Doob's inequality, Theorem 4.2 (5), we have, as in Theorem 18.11, $E[H_1(t \wedge S_n)^2] < \infty$ and $E[H_2(t \wedge T_n)^2] < \infty$, so H_1 and H_2 are locally square integrable martingales under measure P_0 with respect to the filtration $\{\mathcal{F}_t\}$. Because the stopping times are $\{\mathcal{Y}_t\}$ measurable the processes

$$\hat{H}_1(t) = E_0[H_1(t) | \mathcal{Y}_t]$$

$$\hat{H}_2(t) = E_0[H_2(t) | \mathcal{Y}_t]$$

are locally square integrable martingales under measure P_0 with respect to the filtration $\{\mathcal{Y}_t\}$. Write

$$K(t) = \int_0^t \Lambda_u (\alpha^{-1}(y_u) h(X_u))' \alpha(y_u)\, du$$

and

$$\hat{K}(t) = E_0[K(t) | \mathcal{Y}_t].$$

Then

$$\tilde{K}(t) = \hat{K}(t) - \int_0^t (\alpha^{-1}(y_u) \widehat{\Lambda_u h}(X_u))' \alpha(y_u)\, du$$

is a local martingale under measure P_0 with respect to the filtration $\{\mathcal{Y}_t\}$. From (18.27), therefore,

$$E_0[\Lambda_t y_t | \mathcal{Y}_t] = \hat{\Lambda}_t y_t = \hat{H}_1(t) + \hat{H}_2(t) + \tilde{K}(t)$$
$$+ \int_0^t (\alpha^{-1}(y_u) \widehat{\Lambda_u h}(X_u))' \alpha(y_u)\, du. \qquad (18.28)$$

This represents $\hat{\Lambda}_t y_t$ as the sum of a local martingale and a continuous (and so predictable) process of bounded variation. Consequently, $\hat{\Lambda}_t y_t$ is a special semimartingale and this representation is unique. However, from (18.26) and Definition 18.6:

$$\hat{\Lambda}_t y_t = \int_0^t \hat{\Lambda}_u\, dy_u + \int_0^t y_u (\eta'_u \alpha^{-1}(y_u))\, dy_u + \int_0^t \eta'_u \alpha(y_u)\, du.$$

Again the first two integrals are local $\{\mathcal{Y}_t\}$ martingales under measure P_0. By the uniqueness of the decomposition of special semimartingales

$$\eta_u = \alpha^{-1}(y_u)\widehat{\Lambda_u h}(X_u)$$
$$= \alpha^{-1}(y_u)E_0[\Lambda_u h(X_u)|\mathcal{Y}_t].$$

However, from (18.24) this is

$$= \alpha^{-1}(y_u)\hat{\Lambda}_u\Pi_u(h).$$

Substituting in (18.26)

$$\hat{\Lambda}_t = 1 + \int_0^t \hat{\Lambda}_u(\alpha^{-1}(y_u)\Pi_u(h))'\alpha^{-1}(y_u)\,dy_u. \tag{18.29}$$

From Theorem 13.5 this equation has the unique solution

$$\hat{\Lambda}_t = \exp\left\{\int_0^t (\alpha^{-1}(y_u)\Pi_u(h))'\alpha^{-1}(y_u)\,dy_u - \tfrac{1}{2}\int_0^t |\alpha^{-1}(y_u)\Pi_u(h)|^2\,du\right\},$$

and the theorem is proved. □

Theorem 18.22. *For any $\phi \in \mathscr{C}_b^2(R^d)$, $\sigma_t(\phi)$ satisfies the equation:*

$$\sigma_t(\phi) = \sigma_0(\phi) + \int_0^t \sigma_u(L\phi)\,du$$
$$+ \int_0^t \{\sigma_u(\nabla\phi\,.\,\sigma\,.\,\rho) + \alpha^{-1}(y_u)\sigma_u(\phi h)\}'\alpha^{-1}(y_u)\,dy_u. \tag{18.30}$$

PROOF. We have noted that

$$\sigma_t(\phi) = \hat{\Lambda}_t\Pi_t(\phi)$$

so from (18.11) and (18.29)

$$\hat{\Lambda}_t\Pi_t(\phi) = \sigma_0(\phi) + \int_0^t \hat{\Lambda}_u\Pi_u(L\phi)\,du$$
$$+ \int_0^t \hat{\Lambda}_u\{\Pi_u(\nabla\phi\,.\,\sigma\,.\,\rho) + \alpha^{-1}(y_u)(\Pi_u(\phi h) - \Pi_u(\phi)\Pi_u(h))\}\,d\nu_u$$
$$+ \int_0^t \Pi_u(\phi)\hat{\Lambda}_u(\alpha^{-1}(y_u)\Pi_u(h))'\alpha^{-1}(y_u)\,dy_u$$
$$+ \int_0^t \hat{\Lambda}_u\{\Pi_u(\nabla\phi\,.\,\sigma\,.\,\rho) + \alpha^{-1}(y_u)$$
$$\times(\Pi_u(\phi h) - \Pi_u(\phi)\Pi_u(h))\}'(\alpha^{-1}(y_u)\Pi_u(h))\,du$$
$$= \sigma_0(\phi) + \int_0^t \sigma_u(L\phi)\,du$$
$$+ \int_0^t \{\sigma_u(\nabla\phi\,.\,\sigma\,.\,\rho) + \alpha^{-1}(y_u)\sigma_u(\phi h)\}'\alpha^{-1}(y_u)\,dy_u,$$

by (18.24). □

Remarks 18.23. Note the much simpler form of the equation (18.30) for $\sigma_t(\phi)$ compared with (18.11) for $\Pi_t(\phi)$: (18.30) is linear in σ_t, whereas (18.11) is quadratic in Π_t. In particular, when the signal noise $\{B_t\}$ is independent of the noise $\{w_t\}$ in the observation, so that the predictable quadratic covariation matrix $\rho = (\rho^{ij}) = (\langle B^i, w^j \rangle)$ is zero, the unnormalized density $\sigma_t(\phi)$ satisfies the equation

$$\sigma_t(\phi) = \sigma_0(\phi) + \int_0^t \sigma_u(L\phi)\, du + \int_0^t \{\alpha^{-1}(y_u)\sigma_u(\phi h)\}'\alpha^{-1}(y_u)\, dy_u. \quad (18.31)$$

Example 18.24. Suppose the homogeneous Markov process $\{X_t\}$ is such that, as in Corollary 18.14, the conditional distribution of X_t given $\mathcal{Y}_t$ has a smooth density $\hat{p}(t, x)$. Then we can define the *unnormalized conditional density* as

$$q(t, x) = \hat{\Lambda}_t \hat{p}(t, x),$$

so that

$$\hat{p}(t, x) = \frac{q(t, x)}{\int_{R^d} q(t, y)\, dy}.$$

Similarly to Corollary 18.14, substituting in (18.31) and integrating by parts we obtain the following stochastic partial differential equation for q:

$$dq(t, x) = L^* q(t, x)\, dt + h(x) q(t, x)\, dy_t, \quad (18.32)$$

where again L^* is the adjoint of the infinitesimal generator L. Equation (18.32) is much simpler than (18.12) obtained for $\hat{p}$. It is linear in q, it does not involve terms such as $\Pi_t(h)$, and it has the observation process $\{y_t\}$ as input.

References

[1] D. ALLINGER and S. K. MITTER. New results on the innovations problem for nonlinear filtering, *Stochastics*, **4** (1981), 339–348.

[2] J. J. BENEDETTO. *Real Variable and Integration*, Teubner: Stuttgart, 1976.

[3] V. E. BENEŠ. Existence of optimal strategies based on specified information, for a class of stochastic decision problems, *S.I.A.M. J. Contr.*, **8** (1970), 179–188.

[4] V. E. BENEŠ. Existence of optimal stochastic control laws, *S.I.A.M. J. Contr.*, **9** (1971), 446–475.

[5] V. E. BENEŠ. Full "bang" to reduce predicted miss is optimal, *S.I.A.M. J. Contr.*, **15** (1976), 52–83.

[6] V. E. BENEŠ. Exact finite dimensional filters for certain diffusions with nonlinear drift, *Stochastics*, **5** (1981), 65–92.

[7] R. BOEL and P. VARAIYA. Optimal control of jump processes, *S.I.A.M. J. Contr.*, **15** (1977), 92–119.

[8] R. W. BROCKETT. Nonlinear systems and nonlinear estimation theory. In *Stochastic Systems*, eds. M. HAZEWINTEL and J. C. WILLEMS. NATO Advanced Study Institute, Les Arcs, France, June 1980. Reidel: Dordrecht.

[9] R. S. BUCY. Nonlinear filtering, *IEEE Trans. Automat. Contr.*, **10** (1965), 198.

[10] C. S. CHOU and P. A. MEYER. Sur la représentation des martingales comme intégrales stochastiques dans les processus ponctuels. Séminaire de Probabilités IX, Lecture Notes in Math., 465. Springer-Verlag: Berlin, Heidelberg, New York, 1975.

[11] M. H. A DAVIS. On the existence of optimal policies in stochastic control, *S.I.A.M. J. Contr.*, **11** (1973), 507–594.

[12] M. H. A. DAVIS. The representation of martingales of jump processes, *S.I.A.M. J. Contr.*, **14** (1976), 623–638.

[13] M. H. A. DAVIS. Martingales of Wiener and Poisson processes, *J. London Math. Soc.* (2), **13** (1976), 336–338.

[14] M. H. A. DAVIS. Martingale methods in stochastic control. In *Stochastic Control and Stochastic Differential Systems*. Lecture Notes in Control and Information Sciences, 16. Springer-Verlag: Berlin, Heidelberg, New York, 1979.

[15] M. H. A. DAVIS. On a multiplicative functional transformation arising in nonlinear filtering theory, *Z. Wahrsch. verw. Gebiete*, **54** (1980), 125–139.
[16] M. H. A. DAVIS. Pathwise nonlinear filtering. In *Stochastic Systems*, eds. M. HAZEWINKEL and J. C. WILLEMS. NATO Advanced Study Institute, Les Arcs, France, June 1980. Reidel: Dordrecht.
[17] M. H. A. DAVIS. Some current issues in stochastic control theory, *Stochastics*, to appear.
[18] M. H. A. DAVIS and J. M. C. CLARK. On "Predicted Miss" stochastic control problems, *Stochastics*, **2** (1979), 197–209.
[19] M. H. A. DAVIS and R. J. ELLIOTT. Optimal control of a jump process, *Z. Wahrsch. verw. Gebiete*, **40** (1977), 183–202.
[20] M. H. A. DAVIS and S. I. MARCUS. An introduction to nonlinear filtering. In *Stochastic Systems*, eds. M. HAZEWINKEL and J. C. WILLEMS. NATO Advanced Study Institute, Les Arcs, France, June 1980. Reidel: Dordrecht.
[21] M. H. A. DAVIS and P. VARAIYA. Dynamic programming conditions for partially observed stochastic systems, *S.I.A.M. J. Contr.*, **11** (1973), 226–261.
[22] C. DELLACHERIE. *Capacités et Processus Stochastiques*. Springer-Verlag: Berlin, Heidelberg, New York, 1972.
[23] C. DELLACHERIE and P. A. MEYER. *Probabilités et Potentiel*, Chaps. I–IV, 2ème ed. Hermann: Paris, 1975.
[24] C. DELLACHERIE and P. A. MEYER. *Probabilités et Potentiel, Théorie des Martingales*, Chaps. V–VIII, 2ème ed. Hermann: Paris, 1980.
[25] C. DOLÉANS-DADE. Quelques applications de la formule de changement de variables pour les semimartingales. *Z. Wahrsch. verw. Gebiete*, **16** (1970), 181–194.
[26] C. DOLÉANS-DADE and P. A. MEYER. Equations differentialles stochastiques. Séminaire de Probabilités XI, Lecture Notes in Math., 581. Springer-Verlag: Berlin, Heidelberg, New York, 1977.
[27] J. L. DOOB. *Stochastic Processes*. Wiley: New York, 1953.
[28] T. E. DUNCAN and P. VARAIYA. On the solutions of a stochastic control system, *S.I.A.M. J. Contr.*, **9** (1971), 354–371.
[29] N. DUNFORD and J. T. SCHWARTZ. *Linear Operators*, Part I. Wiley-Interscience: New York, 1958.
[30] R. J. ELLIOTT. Stochastic integrals for martingales of a jump process with partially accessible jump times, *Z. Wahrsch. verw. Gebiete*, **36** (1976), 213–226.
[31] R. J. ELLIOTT. A stochastic minimum principle, *Bull. Amer. Math. Soc.*, **82** (1976), 944–946.
[32] R. J. ELLIOTT. Lévy functionals and jump process martingales, *J. Math. Anal. App.*, **57** (1977), 638–652.
[33] R. J. ELLIOTT. Innovation projections of a jump process and local martingales, *Proc. Camb. Phil. Soc.*, **81** (1977), 77–90.
[34] R. J. ELLIOTT. The optimal control of a stochastic system, *S.I.A.M. J. Contr.*, **15** (1977), 756–778.
[35] R. J. ELLIOTT and M. KOHLMANN. The variational principle and stochastic optimal control, *Stochastics*, **3** (1980), 229–241.
[36] R. J. ELLIOTT and P. VARAIYA. A sufficient condition for the optimal control of a partially observed stochastic system. In *Analysis and Optimization of Stochastic Systems*, ed. O. L. R. JACOBS. Academic Press: New York, London, 1979.
[37] W. H. FLEMING, Optimal continuous parameter stochastic control, *S.I.A.M. Review*, **11** (1969), 470–509.
[38] W. H. FLEMING. Nonlinear semigroup for controlled partially observed diffusions, *S.I.A.M. J. Contr.*, **20** (1982), 286–301.

[39] W. H. FLEMING and E. PARDOUX, Existence of optimal controls for partially observed diffusions, *S.I.A.M. J. Contr.*, **20** (1982), 261–285.
[40] W. H. FLEMING and R. W. RISHEL, *Deterministic and Stochastic Optimal Control.* Springer-Verlag: Berlin, Heidelberg, New York, 1976.
[41] M. FUJISAKI, G. KALLIANPUR and H. KUNITA. Stochastic differential equations for the nonlinear filtering problem, *Osaka J. Math.*, **1** (1972), 19–40.
[42] I. I. GIHMAN and A. V. SKOROHOD. *Stochastic Differential Equations.* Springer-Verlag: Berlin, Heidelberg, New York, 1972.
[43] I. V. GIRSANOV. On transforming a certain class of stochastic processes by absolutely continuous substitution of measures, *Theory Prob. App.*, **5** (1960), 285–301.
[44] U. G. HAUSSMANN. On the stochastic minimum principle, *S.I.A.M. J. Contr.*, **16** (1978), 236–251.
[45] U. G. HAUSSMANN. Functionals of Ito processes as stochastic integrals, *S.I.A.M. J. Contr.*, **16** (1978), 252–269.
[46] U. G. HAUSSMANN. On the integral representation of functionals of Ito processes, *Stochastics*, **3** (1979), 17–27.
[47] U. G. HAUSSMANN. Existence of partially observable optimal stochastic controls. In *Stochastic Differential Systems*, Lecture Notes in Control and Information Sciences, 36. Springer-Verlag: Berlin, Heidelberg, New York, 1980.
[48] K. ITO. Stochastic integrals. *Proc. Imp. Acad. Tokyo*, **20** (1944), 519–524.
[49] J. JACOD. Multivariate point processes; predictable projection, Radon–Nikodym derivatives, representation of martingales. *Z. Wahrsch. verw. Gebiete*, **31** (1975), 235–253.
[50] J. JACOD. Calcul stochastique et problemes de martingales, Lecture Notes in Math., 794. Springer-Verlag: Berlin, Heidelberg, New York, 1979.
[51] G. JOHNSON and L. L. HELMS. Class (D) supermartingales, *Bull. Amer. Math. Soc.*, **69** (1963), 59–62.
[52] R. E. KALMAN. A new approach to linear filtering and prediction problems, *J. Basic Eng. ASME*, **82** (1960), 33–45.
[53] R. E. KALMAN and R. S. BUCY. New results in linear filtering and prediction theory, *J. Basic. Eng. ASME*, **83** (1961), 95–108.
[54] T. KAILATH. An innovations approach to least squares estimation. Part I: Linear filtering in additive white noise. *IEEE Trans. Automat. Contr.*, **13** (1968), 646–655.
[55] G. KALLIANPUR. *Stochastic Filtering Theory*, Springer-Verlag: Berlin, Heidelberg, New York, 1980.
[56] H. KUNITA. Cauchy problems for stochastic partial differential equations arising in nonlinear filtering theory, *Syst. Contr. Lett.*, **1** (1981), 37–41.
[57] H. KUNITA and S. WATANABE. On square integrable martingales, *Nagoya Math. J.*, **30** (1967), 209–245.
[58] H. J. KUSHNER. On the differential equations satisfied by conditional probability densities of Markov processes, *S.I.A.M. J. Contr.*, **2** (1964), 106–119.
[59] R. S. LIPSTER and A. N. SHIRYAYEV. *Statistics of Random Processes*, Vols. I and II. Springer-Verlag: Berlin, Heidelberg, New York, 1977.
[60] M. LOÈVE. *Probability theory*, Vols. I and II, 4th ed. Springer-Verlag: Berlin, Heidelberg, New York.
[61] J. MEMIN. Conditions d'optimalité pour un problème de control portant sur une famille dominée de probabilités. Journée de control. Metz: Mai 1976.
[62] P. A. MEYER. *Probability and Potentials.* Blaisdell: Waltham, Mass., 1966.
[63] P. A. MEYER. Processus de Markov, Lecture Notes in Math., 26. Springer-Verlag: Berlin, Heidelberg, New York, 1967.

[64] P. A. MEYER. Un cours sur les intégrales stochastiques. Seminaire de Probabilités X, Lecture Notes in Math., 511. Springer-Verlag: Berlin, Heidelberg, New York, 1976.
[65] P. A. MEYER. A differential geometric formalism for the Ito calculus. In *Stochastic Integrals*, ed. D. WILLIAMS. Lecture Notes in Math., 851, Springer-Verlag: Berlin, Heidelberg, New York, 1981.
[66] S. K. MITTER. Lectures on nonlinear filtering and stochastic mechanics. In *Stochastic Systems*, eds. M. HAZEWINTEL and J. C. WILLEMS. NATO Advanced Study Institute, Les Arcs, France, June 1980. Reidel: Dordrecht.
[67] J. NEVEU. *Martingales*. Notes partielles d'un cours de 3ème cycle. Paris, 1970–1971.
[68] A. A. NOVIKOV. On an identity for stochastic integrals, *Theory Prob. App.*, **17** (1972), 717–720.
[69] E. PARDOUX. Stochastic partial differential equations and filtering of diffusion processes. *Stochastics*, **3** (1979), 127–167.
[70] S. R. PLISKA. Controlled jump processes, *Stochastic Processes App.*, **3** (1975), 259–282.
[71] R. RISHEL. A minimum principle for controlled jump processes. Lecture Notes in Economics and Math. Systems, 107. Springer-Verlag: Berlin, Heidelberg, New York, 1975.
[72] J. VAN SCHUPPEN and E. WONG. Transformations of local martingales under a change of law, *Ann. Prob.*, **2** (1974), 879–888.
[73] A. N. SHIRYAYEV. Some new results in the theory of controlled stochastic processes. *Trans. 4th Prague Conference on Information Theory*. Czech Acad. of Sciences: Prague, 1967.
[74] R. L. STRATONOVICH. A new representation for stochastic integrals and equations, *S.I.A.M. J. Contr.*, **4** (1966), 362–371.
[75] R. L. STRATONOVICH. *Conditional Markov Processes and Their Application to the Theory of Optimal Control*. Elsevier: New York, 1968.
[76] C. STRIEBEL. Optimal control of discrete time stochastic systems. Lecture Notes in Economics and Math. Systems, 110. Springer-Verlag: Berlin, Heidelberg, New York, 1975.
[77] E. WONG. Recent progress in stochastic processes—a survey, *IEEE. Trans. Info. Theory*, **19** (1973), 262–275.
[78] W. N. WONHAM. On the separation theorem of stochastic control, *S.I.A.M. J. Contr.*, **6** (1968), 312–326.
[79] W. N. WONHAM. Some applications of stochastic differential equations to optimal nonlinear filtering, *S.I.A.M. J. Contr.*, **2** (1965), 347–369.
[80] C. YOEURP. Decompositions des martingales locales et formules exponentielles. Séminaire de Probabilités X, Lecture Notes in Math., 511. Springer-Verlag: Berlin, Heidelberg, New York, 1976.
[81] M. ZAKAI. On the optimal filtering of diffusion processes. *Z. Wahrsch. verw. Gebiete*, **11** (1969), 230–243.
[82] S. I. MARCUS. Modeling and approximation of stochastic differential equations driven by semimartingales, *Stochastics*, **4** (1981), 223–245.
[83] J. MEMIN and A. N. SHIRYAYEV. Un critère prévisible pour l'uniforme integrabilité des semimartingales exponentielles. Séminaire de Probabilités XIII, Lecture Notes in Math., 721. Springer-Verlag: Berlin, Heidelberg, New York, 1979.

Notation

Symbol	Reference
a	accessible
$\mathscr{A}$	7.6, 17.5
$\mathscr{A}^+$	7.6
$\tilde{\mathscr{A}}$	15.29
$\mathscr{B}$	2.27
$B(\mathscr{B} \otimes \mathscr{F})$	6.38
$B(\Sigma_x)$	6.38
$\mathscr{B}_x$	6.19
$\mathscr{C}_p$	8.9
$\mathscr{D}$	12.45
$\mathscr{E}(X)$	13.6
$\mathscr{F}_T$	2.8
$\mathscr{F}_{T-}$	5.12
$\tilde{\mathscr{F}}$	15.19
$H \cdot M$	11.1
$\mathscr{H}^p$	9.5
$\mathscr{H}^{2,c}$	9.19
$\mathscr{H}^{2,d}$	9.20
$K[\{T_n\}]$	5.15
loc	7.36
$L^2(M)$	11.14
$L^2(\langle M \rangle)$	11.7
$\mathscr{M}, \mathscr{M}_{\text{loc}}$	9.2
$\mathscr{M}(T)$	9.21
$\langle M, M \rangle$	10.1, 10.14
$[M, M]$	10.1, 10.14, 11.35
$\langle M, N \rangle$	10.6, 11.35, 11.46
$[M, N]$	10.7, 11.35, 11.46
N	natural numbers
o	optional
p	predictable
Q	4.1
$\mathscr{S}(T)$	5.15
T_s^t	14.28
$\mathscr{T}_x$ $(x = o, a, p)$	5.8
$\mathscr{U}_r^t$	16.13
$\mathscr{V}$	7.3
$\mathscr{V}^+$	7.2
$\tilde{\mathscr{V}}^1$	15.43
$\mathscr{W}, \mathscr{W}^+$	7.1
X^c	1.16
$X \cdot A$	7.10
$[X, X]$	12.6
$\langle X, Y \rangle$	12.6
Λ	16.3, 16.31
$\hat{\eta}$	18.4
$\bar{\mu}$	17.2
μ_A	7.12
$\Pi_t(f)$	18.12
Π_x $(x = o, a, p)$	6.39
Π_x^* $(x = o, a, p)$	7.26, 15.48
$\sigma_t(\phi)$	equation (18.24)
Σ_x $(x = o, a, p)$	6.2
Σ_π	6.4
$\tilde{\Sigma}_x$ $(x = o, a, p)$	15.19
$\tilde{\Omega}$	15.19
$\Vert \; \Vert_p$	9.5

Index